FOOD SAFETY AND TOXICITY

EDITED BY
John de Vries, Ph.D.

Professor of Toxicology
Department of Natural Sciences
Open University of the Netherlands
Heerlen, The Netherlands

CRC Press
Boca Raton   New York   London   Tokyo

Library of Congress Cataloging-in-Publication Data

Food safety and toxicity / edited by John De Vries.
    p.   cm.
   Includes bibliographical references and index.
   ISBN 0-8493-9488-0 (alk. paper)
   1. Food adulteration and inspection.   2. Food industry and trade--Safety measures.
   3. Food--Toxicology.   I. De Vries, John, 1936-   .
   TX531.F568   1996
   363.19′26--dc20
                       95-50844
                           CIP

No claim to original U.S. Government works
International Standard Book Number 0-8493-9488-0
Library of Congress Card Number 95-50844
Printed in the United States of America   2 3 4 5 6 7 8 9 0
Printed on acid-free paper

Foreword

The rationale for this book has been the many changes in food science forced upon us by the revolution in microbial pathogenesis, changes in food processing and preservation techniques, changes in food regulations and laws particularly in food labelling, the incorporation of HACCP programs, the need for risk assessment, risk evaluation, and risk management, and finally societal changes in health consciousness. Although some of these areas have been addressed by occasional reviews or specialized texts, these are often hard to find or are too detailed for the non-specialist. Thus, in writing this text, the intention was to compile in a readable form, recent information and advances which would to serve as a reference for graduate students, teachers, and professionals in food-safety, -toxicology, and -chemistry, and food scientists both in government and industry.

The authors believe that the responsibility for food safety is shared by government, producers, and consumers. Problems should be handled with an integrated approach, and scientists, public information services, health care workers, producers, policy makers, and consumers should be willing to cooperate from their own responsibility. Cooperation requires communication, since public perception of food-related hazards usually do not agree with the acknowledged health risks assessed on the basis of accepted scientific criteria. For example, there is a great deal of public concern about the effect of pesticide residues or additives. In reality however, the risks posed by either in foods is minimal, certainly when compared to microbial pathogens or plant toxins.

The book is divided into three main sections, with Part 1 subdivided into two sections. Part 1A primarily describes the chain of steps and processes (the pathway) from raw material to the consumer by integrating knowledge of food chemistry, food microbiology, and food technology. There has been no attempt to cover every area in detail but to give an in-depth treatment to areas of known focus. Part 1B incorporates a unique section on behavioral and sociological dimensions, and the effects of dietary behavior on food choice. Part 2 follows the pathway of food in the body, that is, the sequence of steps or processes food components undergo within the body. This includes membrane transport, biotransformation and interaction with targets, resulting in the induction of effects. Part 3 describes the process of risk management, including risk assessment and evaluation, standard setting, and food safety policy.

Each section is composed of series of chapters arranged in similar formats for easier and more consistent reading. Some chapters contain boxed items of highly specific interest (intermezzo). The tables within each chapter contain data specific for the United States, although where appropriate some European data are presented. At the end of each chapter is a brief summary of the main points, and a minimal number of references. The authors chose not to burden the text with a litany of articles, but instead chose the most current works that could provide a greater overview, and where necessary access to the primary literature.

I believe this is a valuable text designed to examine the many current problems and changes in food safety and toxicity. I am particularly happy to see the topic of risk management extensively discussed, since this is an area of particular recent concern. Finally, the Food Science and Human Nutrition Department at the University of Florida teaches a graduate course in Food Toxicology and Foodborne Infections. I believe that this book is a most appropriate text for this course.

James A. Lindsay
U.S. Editorial Advisor

Preface

Prevention of health risks — including toxicological risks — due to food intake is central in *food safety* policy. The responsibility for food safety is shared by governments, producers, and consumers.

Food safety problems should be handled using an integrated approach. Scientists, public information service people, workers in the health care sector, producers, policymakers, and consumers should be willing to cooperate from their own responsibilities. Cooperation requires communication between the responsible parties about food safety. In today's society, it is increasingly important how the public perceives risk. Public perceptions of food-related hazards usually do not agree with the acknowledged health risks assessed on the basis of accepted scientific criteria. For example, there is much public concern about the effects of pesticide residues. Based on scientific criteria, however, the risks posed by pesticide residues in food are minimal: they are more than a hundred times smaller than those posed by toxins of plant and vegetable origin. With regard to food safety, educated consumers consider primarily the activities and processes that determine the exposure to food components.

Therefore, this textbook first sets out the *Pathway from raw material to consumer* (Part 1A), which is diagrammatically summarized in Figure 1. This pathway includes the factors determining exposure to food components.

Part 1A primarily describes the *chain* of steps and processes on the way from raw material to consumer by integrating knowledge of food chemistry, food microbiology, and food technology. Part 2 follows the *pathway of food in the body*, i.e., the sequence of steps or processes food components undergo in the body. This pathway includes membrane transport, biotransformation, and interaction with targets, resulting in the induction of effects. Part 3 concludes the book with a group of chapters treating the process of *risk management*.

The theme of food safety is usually approached from the viewpoint of the *natural sciences*. In this book, *behavioral and sociological dimensions* are also incorporated. When studying the factors determining exposure, attention is therefore also paid to the effects of dietary behavior on food choice (Part 1B). Part 3 assesses how far changes in dietary behavior are relevant to risk management. The concepts of risk perception and risk management are included.

Throughout this book, health risks associated with food intake are distinguished into two types: microbiological risks and toxicological risks. This distinction is not rigidly defined. Microbiological risks can be subdivided in risks of infection and risks of intoxication. In this book attention is focused on the toxicological aspects of food safety. It treats toxicological risks associated with food intake, including microbiological risks of intoxication.

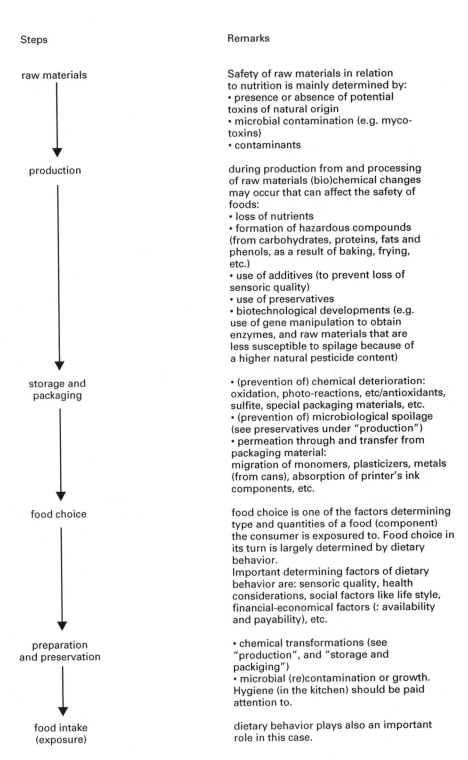

Steps	Remarks
raw materials	Safety of raw materials in relation to nutrition is mainly determined by: • presence or absence of potential toxins of natural origin • microbial contamination (e.g. myco-toxins) • contaminants
production	during production from and processing of raw materials (bio)chemical changes may occur that can affect the safety of foods: • loss of nutrients • formation of hazardous compounds (from carbohydrates, proteins, fats and phenols, as a result of baking, frying, etc.) • use of additives (to prevent loss of sensoric quality) • use of preservatives • biotechnological developments (e.g. use of gene manipulation to obtain enzymes, and raw materials that are less susceptible to spilage because of a higher natural pesticide content)
storage and packaging	• (prevention of) chemical deterioration: oxidation, photo-reactions, etc/antioxidants, sulfite, special packaging materials, etc. • (prevention of) microbiological spoilage (see preservatives under "production") • permeation through and transfer from packaging material: migration of monomers, plasticizers, metals (from cans), absorption of printer's ink components, etc.
food choice	food choice is one of the factors determining type and quantities of a food (component) the consumer is exposed to. Food choice in its turn is largely determined by dietary behavior. Important determining factors of dietary behavior are: sensoric quality, health considerations, social factors like life style, financial-economical factors (: availability and payability), etc.
preparation and preservation	• chemical transformations (see "production", and "storage and packiging") • microbial (re)contamination or growth. Hygiene (in the kitchen) should be paid attention to.
food intake (exposure)	dietary behavior plays also an important role in this case.

Figure 1 Pathway from raw material to consumer.

Contributors

Drs. T. Bruggink
Department of Allergy
Elisabeth Hospital
Haarlem, The Netherlands

Drs. A.E.M. de Hollander
Centre of Epidemiology
National Institute of Public Health and
  Environmental Protection
Bilthoven, The Netherlands

Dr. H.J.G.M. Derks
Unit Biotransformation, Pharmaco- and
  Toxicokinetics
National Institute of Public Health and
  Environmental Protection
Bilthoven, The Netherlands

Prof. John de Vries
Department of Natural Sciences
Open University of the Netherlands
Heerlen, The Netherlands

Prof. V.J. Feron
TNO Nutrition and Food Research
  Institute
Zeist, The Netherlands

Dr. E.J.M. Feskens
Centre of Epidemiology
National Institute of Public Health and
  Environmental Protection
Bilthoven, The Netherlands

Dr. C. Groen
National Institute of Public Health and
  Environmental Protection
Bilthoven, The Netherlands

Dr. J.P. Groten
TNO Nutrition and Food Research
  Institute
Zeist, The Netherlands

Dr. C.J. Henry
Office of Science and Technology
Office of Environmental Management
Department of Energy
Washington, D.C.

Ir. M.M.T. Janssen
Department of Food Science
Wageningen Agricultural University
Wageningen, The Netherlands

Prof. G.J. Kok
Department of Health Education and
  Promotion
University of Limburg
Maastricht, The Netherlands

Prof. R. Kroes
National Institute of Public Health and
  Environmental Medicine
Bilthoven, The Netherlands

Dr. James Lindsay
Food Science and Human Nutrition
  Department
University of Florida
Gainesville, Florida

Dr. R.M. Meertens
Department of Health Education and
  Promotion
University of Limburg
Maastricht, The Netherlands

Dr. M.J.R. Nout
Department of Food Science
Wageningen Agricultural University
Wageningen, The Netherlands

Drs. M. Olling
National Institute of Public Health and
  Environmental Protection
Bilthoven, The Netherlands

Dr. H.M.C. Put
Department of Food Science
Wageningen Agricultural University
Wageningen, The Netherlands

Dr. A.A.J.J.L. Rutten
TNO Nutrition and Food Research
  Institute
Zeist, The Netherlands

Dr. M. Smith
Environmental Safety Laboratory
Unilever plc, Colworth House
Sharnbrook, Bedford, UK

Dr. P. van Assema
Department of Health Education and
  Promotion
University of Limburg
Maastricht, The Netherlands

Prof. H. van Genderen
Research Institute of Toxicology
State University of Utrecht
Utrecht, The Netherlands

Dr. F.X.R. van Leeuwen
Department of Toxicology
National Institute of Public Health and
  Environmental Protection
Bilthoven, The Netherlands

Dr. M.J. van Stigt Thans
Netherlands Bureau for Food and
  Nutrition Education
s'Gravenhage, The Netherlands

Dr. H. Verhagen
TNO Nutrition and Food Research
  Institute
Zeist, The Netherlands

Dr. W.M.M. Verschuren
Department of Chronic Disease and
  Environmental Epidemiology
National Institute of Public Health and
  the Environment
Bilthoven, The Netherlands

Prof. A.G.J. Voragen
Department of Food Science
Wageningen Agricultural University
Wageningen, The Netherlands

Dr. M. J. Zeilmaker
National Institute of Public Health and
  Environmental Protection
Bilthoven, The Netherlands

Contents

Part 3. Risk management in relation to food and its components (scientific coordinator R. Kroes)

Part 1A

From raw materials to consumer: chemical, microbiological and technological aspects of food

chapter one

Introduction to the raw materials of food

M.M.T. Janssen and A.G.J. Voragen

1.1   Introduction

Food is of fundamental importance to life. It is necessary for development and functioning, including maintenance and reproduction. On average, man consumes 30 tons of food during his lifetime; this is consumed in many basic dietary versions, varying at local, national, and international levels. Also, diet is related to social class. It is easy to see the difference in character between french fries and stew on one end of the scale and delicacies such as pâté de foie gras, filet steak, and quail eggs at the other. However, digestion splits all these foods into the same basic nutrients. The differences lie in quality, shape, and flavor only.

Basically, food is a mixture of chemicals. Usually, food components are distinguished in four categories: nutrients, toxins of natural origin, contaminants, and additives. The nutrients account for more than 99.9% of the food. The main classes of nutrients are carbohydrates, proteins, fats, vitamins, and minerals, and all of them may pose toxicological risks to the consumer.

In the course of evolution, through trial and error, man has learned to handle those foods that would cause acute adverse effects. Further, he has developed processing methods to eliminate or reduce toxicity in a number of cases. Cooking and other common means of food preparation effectively destroy many of the major toxic components, particularly those found in important plant foods.

Most of the food is treated in some way to improve its shelf life, texture, palatability or appearance. It would be difficult to change this situation. So, it is important to know what happens to the various food components on the way from raw material to consumer. The food industry is a large, continuously expanding industry. Although there are people who would like to do without industrially processed food and go back to nature, this is not possible on a large scale from a socio-economical point of view. The majority of the population depend on the food manufacturing industry for their daily food supply.

1.1.1   History of food manufacturing

Since the time when man settled in one place and became dependent on cultivated crops and animal husbandry for their food, the need for storage and preservation was evident. Grain and root crops could be kept reasonably well during winter, but products of vegetable, fruit, or animal origin could not be stored for long. Through experience, man learned to preserve perishables by drying, smoking, pickling, candying, and fermenting. Gradually, food manufacturing became a craft with the emphasis still on preservation. People began to specialize in food manufacturing for other people, without understanding the (bio)chemical and microbiological mechanisms underlying the processes involved. How to bake bread, to cure ham, or to make cheese has been known since ancient times, but it was not until 1857 that Pasteur could clarify the metabolism of the microorganisms involved. In 1912, Maillard first published on the browning reaction between sugars and amino acids, now known as the Maillard reaction.

In the 19th century, industrialization set in, and society changed with it. The population began to grow and large industrial areas developed. People became separated from the sites of growth, manufacturing and preservation of their food. This development was possible because new food preservation and production methods were developed, and old and new methods were made suitable for application on a large industrial scale. For instance, in 1809 Nicolas Appert discovered that food can be preserved by heat treatment. At first, the food was heated in glass vessels. About 50 years later tins were introduced in the U.S., while in that same period, Nestlé started the production of condensed milk and powdered milk by concentrating milk through evaporation. The development of methods for the extraction of sugar from sugar beets and the production of a butter substitute from vegetable oils and cheap animal fats, i.e., margarine, also took place in the 19th century.

Initially, these industrial processes were rather unsophisticated and poorly manageable. The sensoric quality of the products was often unsatisfactory, as they lost some of their color, flavor, and texture. New insights in organic and analytical chemistry, as well as in biochemistry, the nutritional sciences, microbiology, toxicology, and technology have been applied to industrial food processing since its first steps. The modern food industry is capable of manufacturing a wide variety of safe food products of high nutritional value and good quality, with great efficiency.

1.1.2   From raw materials to consumer

Food processing can be regarded as the conversion of raw agricultural material into a form suitable for eating. The first step is collecting or harvesting the raw material. This is primarily carried out by the producer. The time of harvesting is influenced by the ultimate purpose of the raw material; quality and ripeness are important for the efficiency and result of food processing. The raw material is transported as rapidly as possible to the site of manufacturing or to the shop so that there is as little deterioration as possible.

The next step is often separating the actual foodstuff from the bulky and indigestible material. The extraction of fats and oils, sugar, flour and starch, vitamins or natural colors and flavors are examples of this step. These refining processes are carried out almost exclusively on an industrial scale.

In the next step from raw materials to consumer, the raw materials and purified components are converted to palatable food products. A diet of fruit, milk, eggs, vegetables, grains, and meat in their raw state can meet all our physiological needs, but they

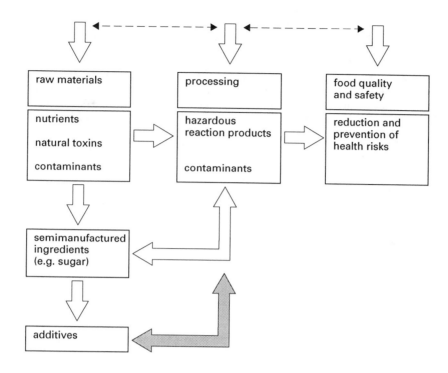

Figure 1.1 Food: from raw material to consumer.

can also be made into a wide range of tasteful and appetizing food products, which make eating them a much greater pleasure.

Traditionally, most of the food is prepared at home. The majority of modern consumers prefer to buy pretreated food which is easily stored and prepared at home. Changes in society, such as more women working away from home, falling birth rate, aging of the population, in combination with familiarity with foreign cultures, influence the modern food supply. Recently, food products with special characteristics, desirable from a nutritional or social point of view, have been developed. These include products containing less saturated and more unsaturated fat, fewer calories, less cholesterol, and more dietary fiber. In general, the food processing industry appears to be willing to gratify the consumer's wishes.

The pathway from raw material to consumer is summarized in Figure 1.1. It shows the various processing techniques that are applied, the reactions that take place, and the quality characteristics that are important.

Part 1A deals with the effects of origin, processing, manufacturing, storage, transport, and preparation of food on the toxicological risks associated with the intake of food, including the formation of hazardous products and the adverse interactions with nutrients.

Since substances of natural origin are not always harmless and the toxicological risks associated with the use of man-made products such as additives have been estimated to be minimal, the four categories of food components are discussed in the following order: natural toxins (including microbial toxins, Chapter 2) and naturally occurring antinutritives (Chapter 3), contaminants (Chapter 4), food additives and the rationale for their use (Chapter 5), and nutrients (Chapter 6). The latter are discussed with the emphasis on the effects of processing on their nature and contents.

Reference and reading list

Belitz, H.-D. and W. Grosch, (Eds.), *Food Chemistry*. Berlin, Springer Verlag, 1987.

Birch, G.G., A.G. Cameron, and M. Spencer, (Eds.), *Food Science*. Oxford, Pergamon Press, 1977.

Birch, G.G., (Ed.), *Food for the 90s*. Amsterdam, Elsevier Applied Sciences, 1990.

Friedman, M., Dietary impact of food processing, *Annu. Rev. Nutr.* 12, 119–137, 1992.

Shapiro, A., C. Mercier, Safe food manufacturing, *Sci. Total Environ.* 143, 75–92, 1994.

Troller, J. A., (Eds.), *Sanitation in food processing*. London, Academic Press, 1993.

chapter two

Natural toxins

M.M.T. Janssen, H.M.C. Put, and M.J.R. Nout

2.1   Introduction

Man's diet contains many thousands of substances, of which many are unknown. Relatively few are of nutritional significance. The majority contribute to the sensoric quality of food.

A number of toxic components of natural origin have been identified, and the mechanisms underlying their toxicities elucidated. For example, the potato contains more than 250 substances including solanine, which is known to cause neurotoxic effects in animals and man.

Usually, natural toxins are not acutely toxic, except in a few cases in animals. An example is tetrodotoxin, a neurotoxin first identified in puffer fish, a Japanese delicacy. Expert cleaning of the fish prevents transmission of the toxin to the edible parts of the fish. Yet, accidents happen each year. Most of the natural toxins, particularly those occurring in plant-derived foods, induce adverse effects only after chronic ingestion or by allergic reactions.

So far, many minor plant food components have not been chemically identified yet and, consequently, have not been evaluated for any toxic properties. Indications of their presence have been obtained from chromatographic and spectroscopic studies. It may even be expected that with the further development of analytical techniques still more components will be found, and of these more may appear to be toxic. This chapter deals with endogenous toxins of plant origin (Section 2.2) and contaminants of natural origin (Section 2.3), including toxins of microbial origin (Section 2.3.3).

2.2   Endogenous toxins of plant origin

There is no simple way of classifying toxic food components of plant origin, since this category comprises many different types of substances. These are classified on the basis of common functional groups (Sections 2.2.1 to 2.2.3), physiological action (Section 2.2.4) and type of effect (Sections 2.2.5 and 2.2.6). Several important representatives of the various types will be highlighted in this section.

2.2.1   Toxic phenolic substances

More than 800 phenolic substances have been detected in plants. Many of them contribute to the (bitter) taste and flavor of foods, and some also contribute to color.

These substances can be divided into two major groups on the basis of frequency of occurrence, structural relationship, and relative toxicity:

1. widespread phenolic plant substances, often used in the production of foods and beverages. About 25 have been identified and only a few are present in relatively high concentrations in certain plant foodstuffs; the majority are only present in trace amounts. This group includes phenolic acids such as caffeic acid, ferulic acid, gallic acid, flavonoids, lignin, hydrolyzable and condensed tannins, and derivatives. At the levels present in food, these substances are devoid of acute toxicity. Presumably, evolutionary adaptation gave animals and man the ability to detoxicate them;

2. a more heterogeneous group of highly toxic phenolic substances. Examples are coumarin, safrole, myristicin, and phenolic amines also known as catecholamines, and gossypol (see Chapter 3).

A number of phenolic substances will be discussed in more detail in the following subsections.

2.2.1.1   Flavonoids

A class of plant pigments that are widely present in human food, are the flavonoids. These pigments are polyhydroxy-2-phenylbenzo-γ-pyrone derivatives, occurring as aglycones, glycosides and methyl ethers. They are divided into six main subgroups (see Figure 2.1). Most flavonoids are present as β-glucosides. The enzymes catalyzing the formation of the active agents have not yet been found.

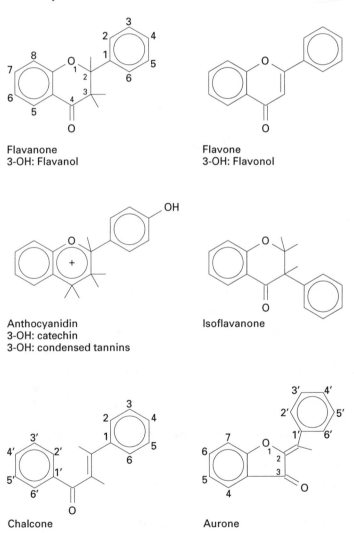

Flavanone
3-OH: Flavanol

Flavone
3-OH: Flavonol

Anthocyanidin
3-OH: catechin
3-OH: condensed tannins

Isoflavanone

Chalcone

Aurone

Figure 2.1   Flavonoid classification.

Because of their wide distribution, the effects of the flavonoids on human well-being are of considerable interest. More than 1 g of various flavonoids are ingested daily in the diet of the Western world.

A group of yellow pigments that occurs abundantly is the flavones. Examples are nobiletin, tangeretin and 3, 3', 4', 5 ,6 ,7, 8-heptamethoxyflavone. The former two are found in citrus fruits such as tangerines, mandarines and oranges, the latter in grapefruit.

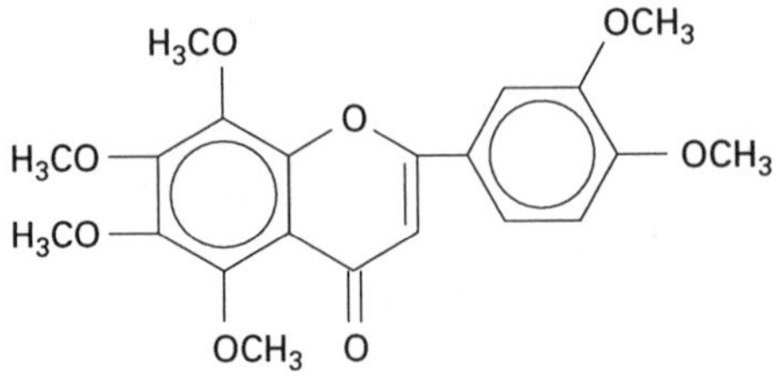

Nobiletin

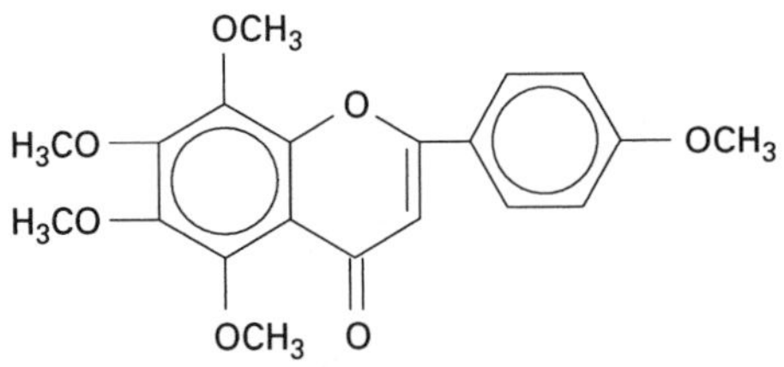

Tangeretin

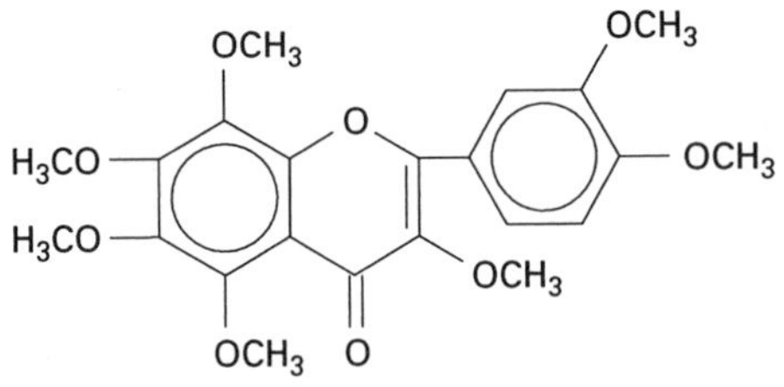

3,3',4',5,6,7,8 - Heptamethoxyflavone

The flavones are generally located in the oil vesicles of the fruit peel. Flavones are apolar, and therefore readily soluble in the oil. They can be found in the juice after pressing the peel. The oily material from orange peel can contain about 2 mg nobiletin per 100 ml oil and 0.3 mg tangeretin per 100 ml oil.

The flavones group has been extensively investigated for mutagenicity. A well-known mutagenic representative is quercetin, occurring, for example, in cereal crops. Quercetin is the only flavonoid shown to be carcinogenic in mammals after oral administration. Its structure can be found in Chapter 3.

2.2.1.2   Tannins

Tannins are a heterogeneous group of broadly distributed substances of plant origin. They are considered to include all polyhydric phenols of plant origin with a molecular mass higher than 500. Two types of tannins can be distinguished on the basis of degradation behavior and botanical distribution, namely hydrolyzable tannins and condensed tannins.

The *hydrolyzable* tannins are gallic, digallic, and ellagic acid esters of glucose or quinic acid. An example of this group is tannic acid, also known as gallotannic acid, gallotannin, or simply tannin. Tannic acid has been reported to cause acute liver injury, i.e., liver necrosis and fatty liver.

With DG =

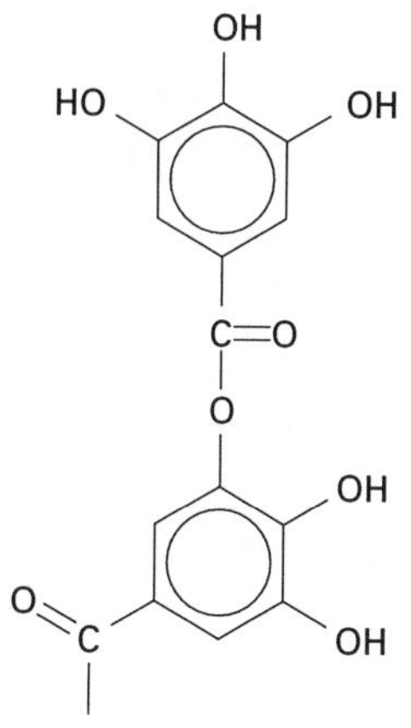

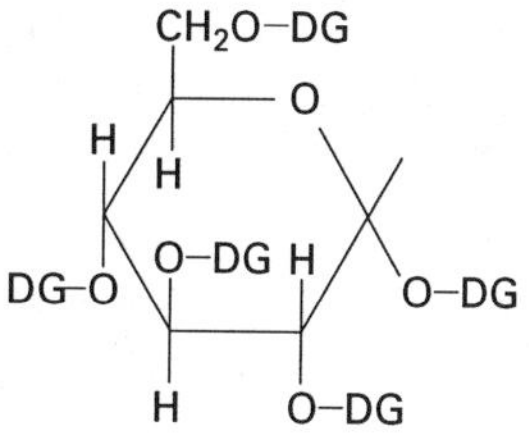

Tannic Acid

digallic

The *condensed* tannins are flavonoids. They are polymers of leukoanthocyanidins (Figure 2.1). The monomers are linked by C–C bonds between positions 4 and 6, or positions 4 and 8.

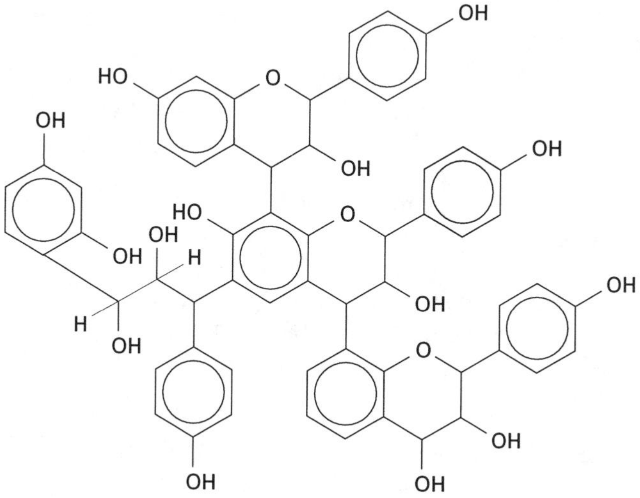

Proposed structure of
a condensed tannin

Tannins occur in many tropical fruits, including mango, dates, and persimmons. The contribution of the tannins in tea, coffee, and cocoa to the total tannin intake by humans is of particular importance. A cup of regular ground coffee was found to contain 72 to 104 mg of tannins, a cup of instant coffee 111 to 128 mg, and a cup of decaffeinated instant coffee 134 to 187 mg. One particular brand of cocoa was found to contain 215 mg tannins per cup. Tea has the highest tannin content. It has been shown that on maximum extraction, black tea leads to 431 to 450 mg tannin per cup. Green tea may yield more soluble tannins, while black tea contains tannins with a higher molecular mass, as a result of oxidation of phenolic precursors during fermentation. From these data it can be estimated that a person may easily ingest 1 g or more tannins per day. Other important sources of tannins are grapes, grape juice, and wines. The tannins in grapes are mostly of the condensed type. The highest levels are found in the skin of the fruit. On average, grapes contain 500 mg per kg, and red wines 1 to 4 g per l of wine. Tannins are also found in large amounts in ferns.

2.2.1.3 Coumarin, safrole, and myristicin

Natural toxins can also be found among the flavorings. Three examples will be discussed in this section: coumarin, a chroman derivative, and safrole and myristicin, both methylenedioxyphenyl substances.

Only a few flavor components of natural origin have been toxicologically evaluated. One reason for this may be that isolation of sufficient quantities for testing is often difficult. More is known about those flavor components that are also synthetically made. This concerns coumarin and safrole.

Coumarin   Safrole   Myristicin   OCH$_3$

Coumarin widely occurs in a number of natural flavorings, including cassis, lavender, and lovage. These flavorings are extensively used in sweets and liquors. Traces of coumarin occur in citrus oils and some edible fruits.

Safrole has been shown to cause liver tumors in rats. It is found in the oil of sassafras and in black peppers. Both coumarin and safrole are still allowed for food use in the European Community. They are prohibited in the US though, as they have been found to cause liver damage in rats.

One of the most common other methylenedioxyphenyl substances is *myristicin*. It is found in spices and herbs such as nutmeg, mace, black pepper, carrot, parsley, celery, and dill. It has been suggested that myristicin contributes to the toxicity of nutmeg. After nutmeg abuse, tachycardia, failing salivation, and excitation of the central nervous system have been reported. Nutmeg has been abused as a narcotic.

2.2.2 Cyanogenic glycosides

Cyanogenic glycosides are glycosides from which cyanide is formed by the activity of hydrolytic enzymes. They are widely spread in higher plants. More than 1000 plant species have been reported to be cyanophoric, mostly in edible plants (see Table 2.1).

Cyanide doses that are lethal to humans can easily be reached or even exceeded after the intake of a variety of cyanogenic foodstuffs. Lethal intakes by humans range from 0.5 to 3.5 mg per kg body weight. The quantities of cyanide produced by Asiatic varieties of lima beans range from 200 to 300 mg per 100 g (see Table 2.2). American varieties of lima beans produce less than 20 mg HCN per 100 g. Selected breeding of low-cyanide varieties has been started.

Fresh cassava cortex produces cyanide in quantities ranging from 1.0 to more than 60.0 mg per 100 g, depending on several conditions, including variety, source, time of harvest and field conditions. Damaged roots can contain even more cyanide, i.e., 245 g per 100 g.

Table 2.1 Cyanogenic glycosides in edible plants

Glycosides	Aglycone	Sugar	Food found
Amygdalin	D-mandelonitrile	Gentiobiose	Almonds, apple, apricot, cherry, peach, pear, plum, quince
Dhurrin	L-*p*-hydroxymandelonitrile	D-glucose	Sorghums, kaffir corns
Linamarin	α-hydroxyisobutyronitrile	D-glucose	Lima beans, flax seed, cassava or manioc
Lotoaustralin	α-hydroxy-α-methylbutyronitrile	D-glucose	Same as linamarin: cassava
Prunasin	D-mandelonitrile	D-glucose	Same as amygdalin
Sambunigrin	L-mandelonitrile	D-glucose	Legumes, elderberry
Vicianin	D-mandelonitrile	Vicianose	Common vetch, and other vicias

Table 2.2 Hydrogen cyanide contents of some foodstuffs

Food	HCN (mg/100 g)
Lima beans	210–310
Almonds	250
Sorghum sp.	250
Cassava	110
Peas	2.3
Beans	2.0
Chick peas	0.8

Cyanogenic glycosides consist of a saccharide moiety and an aglycone, a β-hydroxynitrile. The saccharide group can be a monosaccharide, e.g., glucose, or a disaccharide, e.g., gentiobiose and vicianose. Glycoside linkages can be hydrolyzed by glycosidases. The nitrile can undergo further degradation by a lyase to hydrogen cyanide and an aldehyde, a ketone or in some cases an acid.

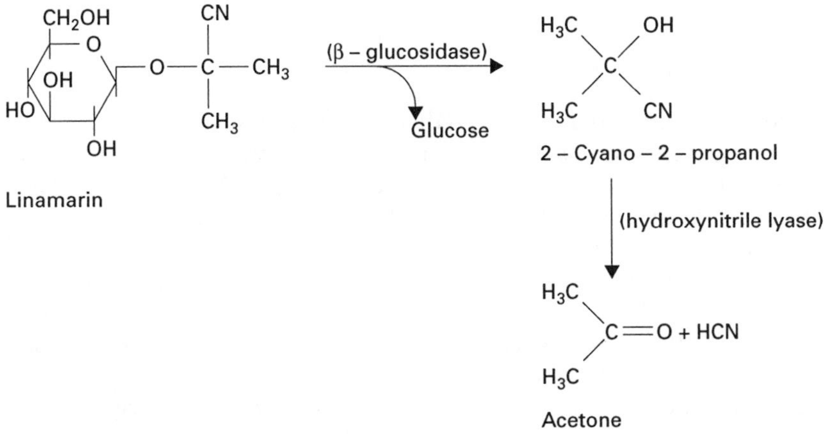

Figure 2.2 Degradation of the cyanogenic glucoside linamarin.

Glycosidases and hydroxynitrile lyase are present in plant cells. They become available when plant tissue is damaged. This inevitably occurs when food is prepared for consumption. As mentioned above, the damaged parts of cassava roots contain high concentrations of cyanide. Nevertheless, cassava, being rich in starch, remains an important food source in Africa, parts of Asia and Latin America, because preparation methods have been developed by which the cyanogenic glycosides are removed or hydrolyzed, and β-glucosidase is destroyed. The cassava is grated, soaked in water, and fermented for several days. The soaked plant tissue is then dried and pounded to flour. Such processes greatly reduce the cyanogen content of food to safe levels. For example, "gari," a fermented cassava preparation, contains an average of 1.0 mg HCN per 100 g. Consumption of cassava may lead to goiter, as the cyanide formed can be metabolized to thiocyanate by the enzyme rhodanase. High consumption of dry, unfermented cassava, containing high levels of cyanogen, accounts for the widespread incidence of goiter in parts of Africa.

Sorghum can be consumed safely, as it is free from or very poor in cyanogen. On germination the sorghum seedling may reach a concentration of 0.3 to 0.5% HCN (dry weight). The young green leaves, however, are rich in cyanogen. This is why cattle are not allowed to graze on young sorghum plants. If sorghum is packed in a silo, cellular degradation and fermentation may lead to the release and elimination of cyanide.

2.2.3 *Glucosinolates*

Glucosinolates are a particular group of substances, occurring in cruciferous plants, such as cabbage and turnips. They can be considered as natural toxins, but also as antinutritives. This Section is limited to the glucosinolates' pathway from raw material to consumer only. Glucosinolates as antinutritives are dealt with in Chapter 3. Concerning toxicity and antinutritive activity, the hydrolysis products are the active agents, not the glucosinolates themselves. Hydrolysis of glucosinolates results in the formation of isothiocyanates and nitriles. The enzyme becomes available for catalysis when cells are damaged on cutting or chewing.

Several isothiocyanates have been shown to be embryotoxic in rats, while *in vitro* studies have proved a number of them to be cytotoxic and mutagenic. Further, several nitriles have been identified as precursors of N-nitroso compounds. These will be discussed in Chapter 5.

2.2.4 *Acetylcholinesterase inhibitors*

Acetylcholinesterase inhibitors have been detected in several edible fruits and vegetables. Their active components are alkaloids. In many foodstuffs, however, they have not yet been identified. These include broccoli, Valencia oranges, sugar beet, cabbage, pepper, carrot, strawberry, apple, lima bean and radish. In potato, eggplant and tomato — members of the Solanaceae family — the principal alkaloids have been identified. The most potent inhibitors are found in potatoes, and of these the most active component is the glycoalkaloid solanine.

The toxicity of solanine has been the subject of extensive study. Oral administration results primarily in gastrointestinal and neurological symptoms.

The solanine concentration of potato tubers varies with the degree of maturity at harvest, rate of nitrogen fertilization, storage conditions, variety, and greening by exposure to light. Commercial potatoes contain 2 to 15 mg of solanine per 100 g fresh weight. Greening of potatoes may increase the solanine content to 80 to 100 mg per 100 g. Most of the alkaloid is concentrated in the skin. Sprouts may contain lethal amounts of solanine.

The identities of R in the predominant glucosinolates of a number of vegetables:

Cabbage	CH$_2$—	Indolylmethyl
Other brassicas, horseradish, black mustard	CH$_2$=CH—CH$_2$—	Allyl
White mustard	HO—⬡—CH$_2$—	p - Hydroxybenzyl
Rape	CH$_2$=CH—CHOH—CH$_2$—	2 - Hydroxybut - 3 - enyl

General structures of glucosinolates and their hydrolysis products.

Since potatoes also contain other glycoalkaloids, namely chaconine and tomatine, with biological properties similar to solanine, the symptoms seen in potato poisoning may be due to combined actions of the alkaloids. All existing and newly developed varieties of potatoes are now monitored for alkaloid content. Solanine is heat stable and insoluble in water. Hence, toxic potatoes cannot be rendered harmless by cooking. It is generally accepted that 20 mg solanine per 100 g fresh weight is the upper safety limit.

2.2.5 Biogenic amines

Natural toxins also include certain amines which can be of plant as well as of microbial origin. The latter source is dealt with in Section 2.3.3. The most important biogenic amines found in plants are listed in Table 2.3.

The dietary intake of biogenic amines may pose risks. A well-known harmful effect of all three of the phenethylamines, dopamine, norepinephrine, and tyramine is hypertension. The risk is greater when combinations of biogenic amines and monoamine oxidase (MAO) inhibitors are ingested. Monoamine oxidases mediate the oxidative deamination of

Solanine

the three phenethylamines. Monoamine oxidase inhibitors are a heterogeneous group of drugs. Clinically-used MAO inhibitors include hydrazine derivatives such as the antidepressent iproniazid. Several phenalkylamines are found in citrus fruits.

Amines may be formed by the metabolic transformation of precursors endogenously present in food of plant origin. Fava beans *(Vicia faba)* contain dihydroxyphenylalanine (DOPA), which may be decarboxylated to dopamine.

Table 2.3 Biogenic amine content of some fruits and vegetables (mg per 100 g fresh weight)

Amines	Avocado	Banana pulp	Eggplant	Orange	Red plum	Tomato (ripe)	Potato
Dopamine	0.4–0.5	66–70		0.1			
Epinephrine		<.25					
Norepinephrine		10.8					0.01–0.02
Serotonin	1.0	2.5–8.0	0.2		1.0	1.2	
Tyramine	2.3	6.5–9.4	0.3	1.0	0.6	0.4	0.1

	Date	Fig	Pawpaw	Pineapple Green	Ripe	Juice	Plantain Green	Ripe	Cooked
Dopamine	<0.08	<0.02	0.1–0.2						
Epinephrine	<0.08	<0.02							
Norepinephrine	<0.08	<0.02					0.2	0.25	
Serotonin	0.9	1.3	0.1–0.2	5.0–6.0	2.0	2.5–3.5	2.0–6.0	4.0–10	4.7

2.2.6   Central stimulants

For most people the everyday diet contains a considerable amount of stimulants. These substances increase the state of activity of the nervous system. A particular class of stimulants is the methylxanthines. They include caffeine, theophylline, and theobromine.

| Caffeine | Theophylline | Theobromine |

Caffeine is found in coffee beans, tea leaves, cocoa beans, and colanuts. In our diet the primary source of caffeine, however, is coffee: one cup of coffee is estimated to contain 100 to 150 mg of caffeine. The caffeine content of cola drinks ranges from 0.1 to 0.15 mg/ml.

In general, methylxanthines cause effects on the peripheral nervous system, but they also induce significant stimulation of the central nervous system. Caffeine is a little more potent than theophylline, and theobromine is relatively inactive.

Further, caffeine has been reported to cause premature aging, a lower growth rate and a lower body weight in experimental animals. In rodents, the oral LD_{50} (see Part 2, Chapter 8, Section 8.9.1) ranges from 127 to 355 mg/kg. Adverse effects of caffeine on cardiac function are questioned, since no relationship has been found between drinking tea and heart disease; the caffeine content of tea is similar to that of coffee. *Theophylline* is present in small amounts in tea. *Theobromine* is the principal alkaloid of the cocoa bean. It is also found in tea leaves and cola nuts.

2.3   Natural contaminants

Natural contaminants can also originate from biological systems different from those in which they occur. There are three important sources. First, raw materials of plant origin can become contaminated if they are mixed with toxic non-nutritive plant species. Secondly, raw materials of animal origin, mainly fish and milk, can also become contaminated if the animal has ingested toxic substances of natural origin. Thirdly, contaminants of natural origin can be the products of microorganisms. This section deals with a number of important examples of contamination with natural toxins.

2.3.1   Mixing of edible plants with toxic plants

Several intoxications have been reported following the consumption of contaminated cereals. The causative agents are pyrrolizidine alkaloids, produced by the genera *Senecio*, *Crotalaria* and *Heliotropium*.

Pyrrolizidine alkaloids can be the cause of acute liver damage and vein lesions. These substances may also contribute to the liver cancer incidence in humans.

Epidemics of pyrrolizidine intoxication have been reported in India and Afghanistan in 1973 and 1976.

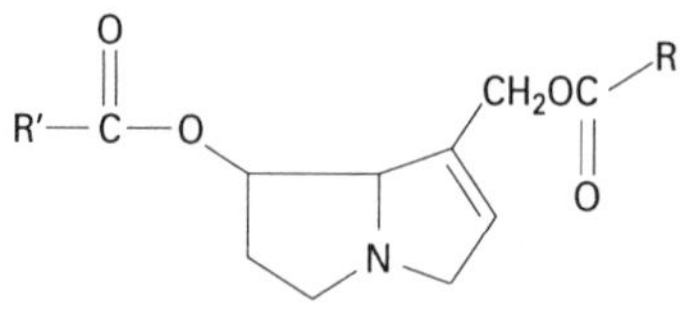

Pyrrolizidine alkaloids

In India, millet, the principal cereal in the diet, appeared to be heavily contaminated with *Crotalaria* seeds. The alkaloid content of the seeds was estimated at 5.3 mg/g, while the precentage of *Crotalaria* seeds in millet varied from 0.0 to 0.34% in unaffected households and 0.0 to 1.9% in affected households.

In Afghanistan, the consumption of wheat bread heavily contaminated with *Heliotropium* seeds was found to be the cause of the intoxication. In this epidemic, the minimum daily consumption per person during 2 years was estimated at 2 mg. The disease had been observed in preceding years, but worsened after the occurrence of a severe drought which caused the wheat fields to become heavily infested with *Heliotropium*.

2.3.2   Contamination resulting from the intake of toxic substances by animals

Contamination of meat with toxic substances of plant origin rarely occurs. Only in a few cases the intoxication appeared to be related to the consumption of wild animals which had ingested highly toxic plant material shortly before they were consumed. Toxic contaminants in milk and aquatic organisms can originate from feed.

2.3.2.1   Contamination of milk with plant toxins

Many foreign substances have been detected in milk. Milk is readily contaminated when lactating animals or women ingest toxins. Contamination of milk with plant toxins has been observed in the US in rural areas, where the inhabitants depend on the local milk supply. The toxin originated from either white snakeroot (*Polygonum*), or the rayless goldenrod (*Solidago*). Especially during periods of drought, when feed plants are scarce and the weeds are in flower, the milk may contain sufficient toxin to give rise to outbreaks of "milk sickness." In this case, the major toxic component appeared to be tremetone.

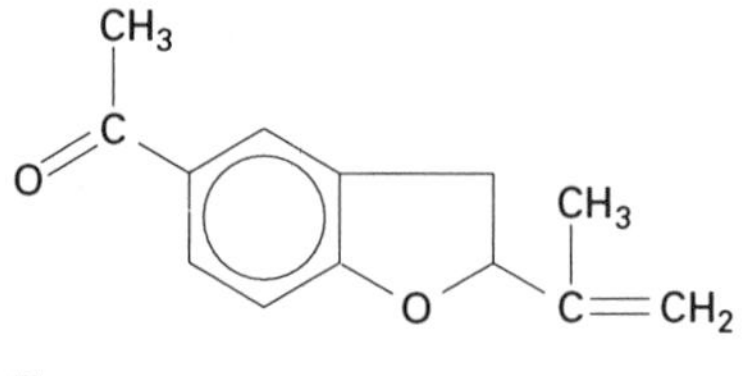

Tremetone

The symptoms were weakness, followed by anorexia, abdominal pain, vomiting, muscle tremor, delirium and coma, and eventually death. A characteristic accompanying phenomenon is the expiration of acetone. The mortality rate was between 10 and 25%.

2.3.2.2   Natural toxins in aquatic organisms

Paralytic shellfish poisoning is attributed to the consumption of shellfish that have become contaminated with a toxin or group of toxins from the ingestion of toxic plankton, in particular toxic dinoflagellates. The shellfish involved are pelecypods, a family of mollusks, including mussels and clams. The dinoflagellates produce a complex mixture of toxins. One of the components has been identified as saxitoxin.

Saxitoxin

Shellfish poisoning symptoms include tingling and burning in face, lips, tongue, and ultimately the whole body, and parathesia followed by numbness, general motor incoordination, confusion, and headache. These symptoms develop within 30 minutes after ingestion. Death, preceded by respiratory paralysis, occurs within 12 hours. The chance of contamination and poisoning is highest during a so-called red tide. In many parts of the world, the sea sometimes suddenly becomes colored, as a result of dinoflagellate bloom. The phenomenon is referred to as red tide, although the bloom may also be yellowish, brownish, greenish, and bluish in color. The red color is probably due to the xanthophyll peridinin.

In spite of the frequent occurrence of red tide and the high toxicity of the paralytic shellfish poisons, intoxication rarely occurs. This is largely due to strict regulations set by many countries and the awareness in coastal areas of the risks associated with eating shellfish during red tides. Although ordinary cooking destroys up to 70% of the toxin(s) and pan-frying destroys even more, there may be sufficient toxin left in the mollusks to cause serious poisoning.

2.3.3 Microbial toxins

2.3.3.1 Introduction

Section 2.3.3 deals with the way in which toxic substances produced in food and feed by microorganisms enter the pathway from raw material to consumer.

Microorganisms are ubiquitous. Any environment supporting higher organisms contains microorganisms too, while the converse is not true. Absence of microorganisms in an environment indicates that special or unusual conditions have occurred, such as heating and filtration for sterilization or preservation.

During food production, raw food materials of plant or animal origin are exposed to soil, water, air, machinery parts, packaging materials, human hands, etc. As these invariably carry microorganisms, all raw food materials have in principle been inoculated with a variety of microbes. The opportunity for these microorganisms to grow is determined by the food environment. Major environmental factors include availability of water (referred to as water activity or a_w) and nutrients, temperature, pH, and presence or absence of atmospheric oxygen. Growth also depends very heavily on how long suitable environmental conditions prevail. The majority of naturally occurring microbial contaminators are unable to multiply, or succumb to other microbes in a food environment. However, even if an infective microorganism remains alive without multiplying, the food may serve as a vehicle to transfer it to the human body and cause illness. Microorganisms which multiply usually degrade the food components enzymatically and excrete their metabolites. In many cases, the resulting loss of structure, or formation of off-smells is regarded as spoilage. However, a wide variety of fermented foods are manufactured of which the desirable taste, flavor, and other properties are especially due to microorganisms and their metabolic activity.

Table 2.4 Food hazards: perception of the consumer versus epidemiological data

Cause	Perception[1]	Relative importance[2]
Microbial contamination	22	49.9
Nutritional imbalance		49.9
Environmental contaminants	31	0.05
Natural toxins	10	0.05
Food additives	30	0.0005
Others, e.g., packaging materials	7	
	100%	100%

[1] Survey held in the Netherlands, 1990.

[2] Ranking based on objective scientific criteria including the severity, incidence, and onset of biological symptoms.

Table 2.5 Food-borne bacterial pathogens and associated diseases

Organism	Pathogenicity	Incubation time (hours)	Duration of disease (days)
Salmonella	infection	6–36	1–7
Shigella	infection	6–12	2–3
Escherichia coli	infection	12–72	1–7
Yersinia enterocolitica	infection	24–36	3–5
Campylobacter jejuni	infection	3–5 (days)	5–7
Listeria monocytogenes	infection	variable	—[a]
Vibrio parahemolyticus	infection	2–48	2–5
Aeromonas hydrophila	infection	2–48	2–7
Staphylococcus aureus	toxin in food	2–6	≤1
Clostridium botulinum	toxin in food	12–96	1–8[b]
Clostridium perfringens	toxin in intestine	8–22	1–2
Bacillus cereus[c]	toxin in food	1–5	≤1
Bacillus cereus[d]	toxin in intestine	8–16	>1

[a] Affects people with a predisposing factor; high mortality rate.

[b] High mortality rate; complete convalescence takes 6–8 months.

[c] Emetic type.

[d] Diarrheal type.

Section 2.3.3 deals with some harmful aspects of microbial food contamination, namely the production of toxic substances causing food-borne disease.

2.3.3.2   Food-borne diseases

Epidemiological evidence has shown that microbial contamination is a major risk factor associated with food consumption. However, the average consumer does not always realize this, and is, for example, more concerned about environmental contaminants in food. This discrepancy between the incidence of food-borne diseases and the perception of the consumer is illustrated in Table 2.4.

Food-borne diseases can be either food-borne infections or food-borne intoxications, depending on whether the pathogen itself or its toxic product (a microbial toxin or toxic metabolite, produced in the food) is the causal agent. Table 2.5 lists the most important bacterial food-borne pathogens.

Of all reported food-borne diseases with microbiological etiology which occurred in Canada in 1984, infections with *Salmonella* and *Campylobacter* spp. constituted 67% and 8%, and intoxications originating from *Clostridium perfringens*, *Staphylococcus aureus* and *Bacillus cereus* 16%, 7% and 1%, respectively.

Other microbial agents causing food-borne intoxications include toxins produced by fungi (mycotoxins) and by algae, and toxic metabolites such as biogenic amines and ethyl carbamate produced by bacteria and yeasts. The various causative factors of food-borne diseases are summarized in Figure 2.3. In Section 2.3.3.3 the major food-borne toxins will be discussed. Although intoxications by biogenic amines and ethyl carbamate are of microbial origin, they can also be regarded as chemical poisonings.

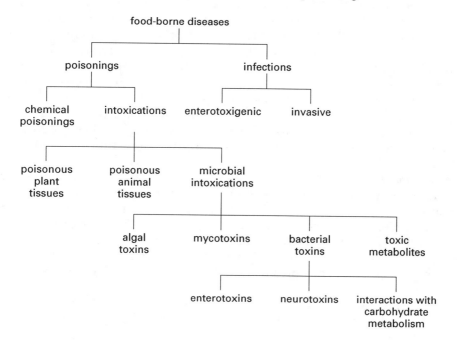

Figure 2.3 Classification of food-borne diseases.

2.3.3.3 Bacterial toxins

According to the mechanisms underlying the effects of bacterial toxins, they can be classified as follows:

- sub-unit toxins (e.g., *Clostridium botulinum* toxins, see Subsection 2.3.3.3.1)
- membrane-affecting toxins (e.g., *Staphylococcus aureus* toxins, see Subsection 2.3.3.3.2)
- lesion-causing toxins (e.g., *Clostridium perfringens* and *Bacillus cereus* toxins, see Subsection 2.3.3.3.3)
- immuno-active endotoxins (e.g., Gram-negative bacteria toxins, see Subsection 2.3.3.3.4).

These toxins will be discussed in relation to the properties of the causative bacteria, the conditions favoring toxin production, as well as the structure and stability of the toxin. Their toxicity will be discussed in detail in Part 2 of this book.

2.3.3.3.1 Sub-unit bacterial toxins.

To this group belong the toxins produced by *Clostridium botulinum*. *C. botulinum* are motile, Gram-positive rod-shaped spore-forming anaerobic bacteria. *C. botulinum* is not one species, but a group of bacteria which are all capable of producing neurotoxins. Biochemically, *C. botulinum* is very similar to *Clostridium sporogenes* and *Cl. novii*. However, the latter do not produce toxins and are therefore not relevant here. According to the toxin they produce, there are eight types of *C. botulinum*: A, B, C1, C2, D, E, F, and G.

Toxicity and symptoms. Botulism, caused by the ingestion of food containing the neurotoxin, is the most severe bacterial food-borne intoxication known. The type A toxin is the most lethal. Types A, B, E, and F are toxic to humans; types B, C, and D to cattle; and types C and E to birds.

After an incubation period of 12 to 72 hours, symptoms may start with nausea and vomiting, followed by tiredness, headache, muscular paralysis, double vision, and respiratory problems, often with fatal results. The duration of botulism is 1 to 10 days, mortality is relatively high (30 to 65%). In most foods, botulinum spores are of no consequence unless they are able to germinate and produce the toxin. The exception is infant foods in which botulinum spores are potentially infective and may give rise to toxicogenesis in the infant intestine. A good example of this is infant botulism caused by contaminated honey.

Recent outbreaks involved yogurt with hazelnut (UK 1989: 27 cases, 1 death; type B), fermented seal oil (Canada 1989: 4 cases, 2 deaths; type E), white fish (1989: 8 cases, 1 death; type E), traditional Eskimo fish product (1984, 1989: type E), and infant botulism (1987–1989: 68 cases).

Chemical properties (structure and stability) of botulinum toxin. *C. botulinum* produces an intracellular protoxin consisting of a non-toxic progenitor toxin (a hemagglutinin with molecular mass approximately 500,000) and a highly toxic neurotoxin (molecular mass approximately 150,000). The protoxin is released upon lysis of the vegetative bacterial cell. The neurotoxin is formed by proteolytic degradation of the protoxin. This proteolysis is caused by *C. botulinum* (type A, and some B and F) proteolytic enzymes, or by exogenous proteases e.g., trypsin when non-proteolytic *C. botulinum* (type C, D, E, and some B and F) are involved.

Botulinum toxin is heat-sensitive (inactivated at 80°C for 10 minutes or 100°C for a few minutes). It is acid-resistant and survives the gastric passage. Botulinum toxin is an exotoxin: it is excreted by the cell, but most of it is released upon lysis of the cell after sporulation.

Environmental conditions. *C. botulinum* grows best at pH >4.6 at temperatures of approximately 37°C (type E at 30°C). The minimum temperature for growth is 12.5°C (type E at 3.5°C).

Type of food involved; prevention. At particularly risk is food of low to neutral pH (>4.5) which has undergone inadequate heating. Examples include home-preserved vegetables which carry soil-borne *C. botulinum*, but also meat and fish which are contaminated during slaughtering with *C. botulinum* originating from the intestines. Of increasing importance are chilled vacuum-packed foods which usually have had minimal heat treatment, and contain no preservatives other than any naturally occurring antimicrobial substances, and are not reheated or only mildly heated prior to consumption.

Preventive measures include adequate heat processing to reduce the number of *C. botulinum* spores with a factor 10^{12} (the "botulinum cook" or "12-D concept", with D being the time required for a tenfold reduction in the population density at a given temperature). The heat-resistance of the spores varies: D-values of 1 minute at 80°C (type E), 100°C (type C), or 113°C (type A and D). Germination of spores surviving the heat treatment can be prevented by the addition of nitrite, lowering the pH or the a_w, addition of salt, thorough heating of food prior to consumption (the toxin is heat-labile) and refrigerated storage (less adequate for type E).

2.3.3.3.2 Membrane-affecting bacterial toxins. A well-known example of a bacterium-producing membrane-affecting toxins is *Staphylococcus aureus*. *S. aureus* are non-motile, non-sporeforming, Gram-positive bacteria which can excrete enterotoxins in food. These enterotoxins show clear antigenic activity, and based on their antigenic properties, they are differentiated in A, B, C1, C2, C3, D and E. Most enterotoxicoses are caused by toxins A, or A and D. A characteristic of *S. aureus* is the formation of the enzyme coagulase which can cause the clumping (coagulation) of blood serum. However, coagulase-negative staphylococci have also been incriminated in food-borne gastroenteritis.

Toxicity and symptoms. A quantity of 1 to 25 µg of the enterotoxin is required to cause sickness in adult humans. After a very short incubation period ($^1/_2$ to 6 hours), symptoms of staphylo-enterotoxicose include violent vomiting and diarrhea, sometimes followed by shock but no fever. Serious dehydration may result from the diarrhea. The duration of the illness is 24 to 72 hours, and the mortality is very low. Everyone who consumes the poisoned food, becomes ill ("maladie du banquet": buffet disease).

Chemical properties (structure and stability) of the enterotoxin. The extracellular enterotoxins are a heterogenous group of globular proteins consisting of linear peptide chains. The molecular mass ranges from approximately 30,000 to 235,000. The enterotoxin is heat-resistant (it withstands boiling at 100°C for more than 1 hour).

Environmental conditions. Growth is possible at temperatures between 7 and 46°C (optimum is 37°C), pH 4 to 9 (optimum is pH 7), a_w ≥0.86, and salt (NaCl) concentrations up to 15%. *S. aureus* is a facultative anaerobe, but grows better under aerobic conditions. It is a poor competitor: it hardly grows in the presence of competitive microflora. About 70% of *S. aureus* of human origin are able to produce enterotoxins. The environmental conditions required for toxin production include: temperature ≥12°C, a_w ≥0.90, pH ≥4.6, aerobic conditions, and little microbial competition.

Type of food involved; prevention. *S. aureus* is a very common microorganism. About 30 to 50% of humans carry the organism in the mucous membrane of the nose and throat, or on the skin. Animals also carry *S. aureus*. Particularly with this microorganism, the human factor plays a very important role in the transfer to food. For instance, sneezing behind the hand increased the *S. aureus* load of a test surface from about 100 to ≥5000 per 25 cm^2. The types of food favoring enterotoxin production include dairy cream, ice cream, cured meats (ham, sausages, meat pies), and opened canned foods (in which fast growth is possible without competition). See also Section 2.4.3 for this aspect. *S. aureus* growth and toxin production can be prevented by proper storage (refrigerated, or too hot for growth), heating (this is not of help if the toxin has already been produced), adequate personal hygiene, cleanliness, and good disinfection practice.

2.3.3.3.3 Lesion-causing bacterial toxins. Two examples of this type of toxin-producing bacteria will be discussed: *Clostridium perfringens* and *Bacillus cereus*.

Clostridium perfringens. *Clostridium perfringens* are Gram-positive, anaerobic (aerotolerant) spore-forming rod-shaped bacteria. Several serotypes are distinguished (A, B, C, D, E, F) which produce different enterotoxins. Particularly, serotype A is associated with food-borne intoxications.

Toxicity and symptoms. Although enterotoxin formation in food (i.e., meat and poultry) may occur, still the incidence of *C. perfringens* food poisoning due to preformed enterotoxin in the food is rare. (Therefore, in Table 2.5, *C. perfringens* itself is listed as the causal agent.) A large number (>10^8) of vegetative *C. perfringens* cells need to be consumed to release sufficient enterotoxin. After an incubation period of 8 to 24 hours, abdominal cramps (much gas produced) and diarrhea with nausea but rarely vomiting can last for 24 hours. A number of enterotoxins have been found to damage the intestinal wall; the glucose resorption is inhibited and the bowel movement is stimulated. The mortality of serotype A poisoning (mainly in the US) is 3 to 4%; serotype C (Europe) is rarely fatal.

Chemical properties (structure and stability) of the enterotoxin. The toxins are protein-type enterotoxins (molecular mass approximately 34,000 dalton). Release of the enterotoxins in the intestine occurs during sporulation and lysis of the *C. perfringens* cells. The proteinous nature of the enterotoxin makes it rather heat-sensitive.

Environmental conditions. Growth can take place at 15 to 50°C (optimum 40°C), pH 5 to 8 and $a_w \geq 0.93$.

Type of food involved; prevention. *C. perfringens* causes mostly problems in meats. In the live animal, the microorganism can penetrate into the body through the intestinal wall. The thermal resistance of the spores varies from heat-labile (D-value 0.3 minutes at 100°C) to relatively heat-resistant (D-value 17.6 minutes at 100°C). The heat resistance also depends on the composition of the food. When contaminated meat receives inadequate heating (e.g., in the center of large pieces of roasted meat) or when the cooked meat is not sufficiently cooled prior to storage, germination of surviving *C. perfringens* spores may occur. Prevention measures include good hygiene, adequate meat heating ($\geq 65°C$ at the center) followed by refrigerated storage ($\leq 7°C$).

Bacillus cereus. *Bacillus cereus* are Gram-positive spore-forming aerobic rod-shaped bacteria. They produce enterotoxins as well as several enzymes of pathogenic relevance, including lecithinase and hemolysin. Two different enterotoxins are known: type I and type II.

Toxicity and symptoms. Type I diarrheagenic enterotoxin occurs most frequently and is mildly toxic. After an incubation period of 8 to 16 hours, 50 to 80% of the consumers develop abdominal cramps and diarrhea which may last for 24 hours. Type II emetic enterotoxin is less common. After a short incubation period of 1 to 6 hours, violent vomiting occurs. Symptoms may last for 8 to 10 hours.

Chemical properties (structure and stability) of the enterotoxin. Type I is a proteinous enterotoxin (with molecular mass approximately 50,000). It is formed in the intestine (relatively long incubation period; large number of cells $\approx 10^6$ required). This enterotoxin is heat-sensitive and, being a protein, undergoes degradation by trypsin. Type II is a toxin with molecular mass ≤ 5000. It is formed in the food during the logarithmic phase of bacterial growth. Type II is stable at pH 10 and is heat-resistant.

Environmental conditions. Growth can take place at 10 to 50°C (optimum 37°C), pH 5 to 9.

Type of food involved; prevention. Particularly cereal products contain *B. cereus* spores. There is no evidence that human factors are involved in the contamination. Cooking with cereal containing dishes followed by inadequate cooling enables germination of the spores that survived the heating. Type I toxin is associated with sauces, pastries, etc.; type II toxin is associated with cooked or fried rice. The main prevention measure is adequate and immediate cooling after cooking. This should be carried out in shallow layers enabling fast heat transfer; storage should be at $\leq 10°C$.

2.3.3.3.4 Immuno-active bacterial endotoxins. Endotoxins are found in the cell wall of Gram-negative bacteria. Examples of bacteria with active endotoxins are *Salmonella abortus equi* and *Escherichia coli*. The endotoxins can be released upon lysis of the vegetative cells.

Toxicity and symptoms. Endotoxins are capable of stimulating the immune system in a non-specific way, and causing inflammations. Symptoms of intoxication include fever, shivering, painful joints, and influenza-like complaints, lasting for approximately 24 hours.

Chemical properties (structure and stability) of the endotoxin. Immuno-active endotoxins consist of lipopolysaccharides (LPS) bound covalently to protein and lipid fractions (Figure 2.4). The polysaccharide part consists of a lipid A fraction and a long polysaccharide chain. The lipid A fraction is identical in almost all bacteria. In the polysaccharide chain, a central part and an O-chain are distinguished. The central part has a similar structure in many bacteria, but the O-chain is rather characteristic.

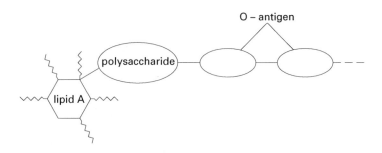

Figure 2.4   General structure of endotoxins.

Little is known about the covalently-bound protein. It is assumed that it is bound to the lipid A and is thus referred to as lipid A associated protein (LAP) or endotoxin protein (EP). The biological activity of the endotoxin is associated with its LPS part. LAP or EP appear to play a minor role. This might explain the different activities of endotoxins of various bacteria.

Environmental conditions. Endotoxins are released at the end of the growth curve, i.e., after death of the bacteria. Favorable conditions for growth of Gram-negative bacteria include pH 4.5, a_w >0.99, and temperatures ranging from 15 to 40°C.

Type of food involved; prevention. In principle, any type of food can serve as a vehicle. The release of endotoxins may take place in the intestine as a result of a food infection. Preventive measures against food infections include avoidance of cross-contamination of cooked food with raw foods, adequate heating, refrigerated storage, and adequate personal hygiene.

2.3.3.4   *Mycotoxins*

2.3.3.4.1 General.   Mycotoxins are secondary metabolites of fungi which can induce acute as well as chronic toxic effects (i.e., carcinogenicity, mutagenicity, teratogenicity, and estrogenic effects) in animals and man. Currently, a few hundred mycotoxins are known, often produced by the genera *Aspergillus, Penicillium,* and *Fusarium.*

Toxicity and symptoms. Toxic syndromes resulting from the intake of mycotoxins by man and animals are known as mycotoxicoses. Although mycotoxicoses have been known for a long time ("Holy Fire" in the Middle Ages in Europe caused by the mold *Claviceps purpurea;* "Alimentary Toxic Aleukia" in the Soviet Union in 1940 caused by *Fusarium* spp.; "Yellowed Rice Disease" in Japan caused by *Penicillium* spp.), the mycotoxin-induced disorders remained the neglected diseases until the early 1960s, when the aflatoxins were discovered. This discovery was followed by much scientific research on mycotoxins.

Chemical properties (structure and stability). The chemical structures of some important mycotoxins are shown in Figure 2.5. The mycotoxins that will be discussed below are chemically stable and resistant to cooking. Several other mycotoxins have been shown to be unstable in foods. As this strongly reduces their toxicity, these will not be discussed here.

Environmental conditions. Mycotoxin contamination of food and feed highly depends on the environmental conditions that lead to mold growth and toxin production. The detectable presence of live molds in food, therefore, does not automatically indicate that mycotoxins have been formed. On the other hand, the absence of viable molds in foods does not necessarily mean there are no mycotoxins. The latter could have been formed at an earlier stage prior to food processing. Because of their chemical stability, several mycotoxins persist during food processing, while the molds are killed.

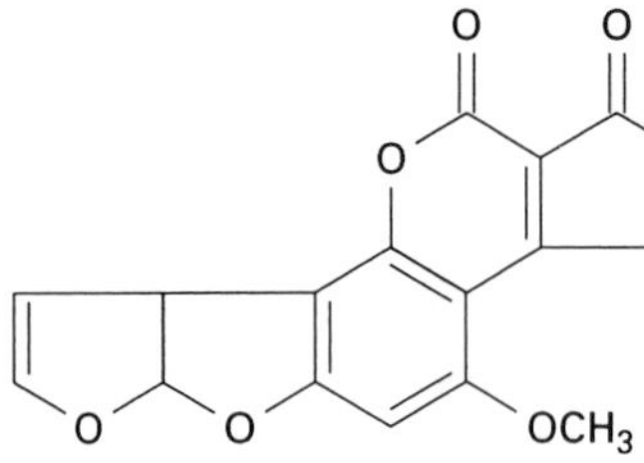

Aflatoxin B1

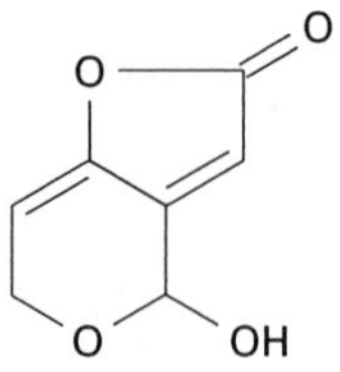

Patulin

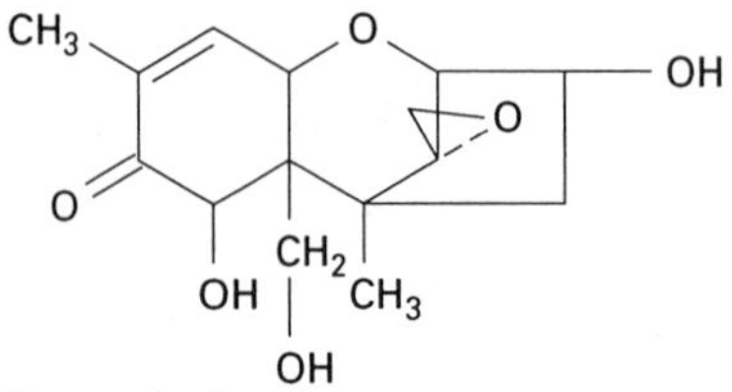

Deoxynivalenol

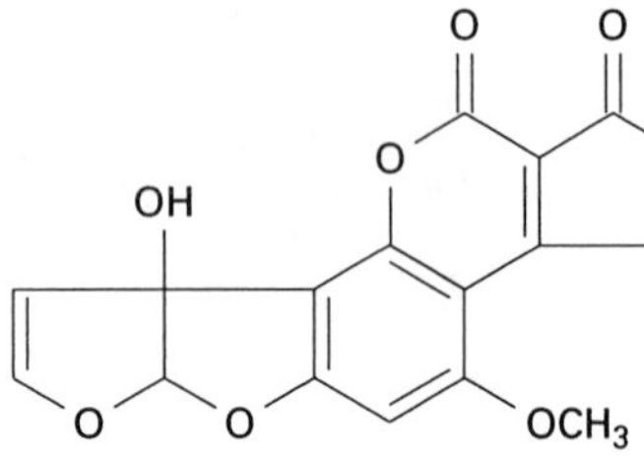

Aflatoxin M 1

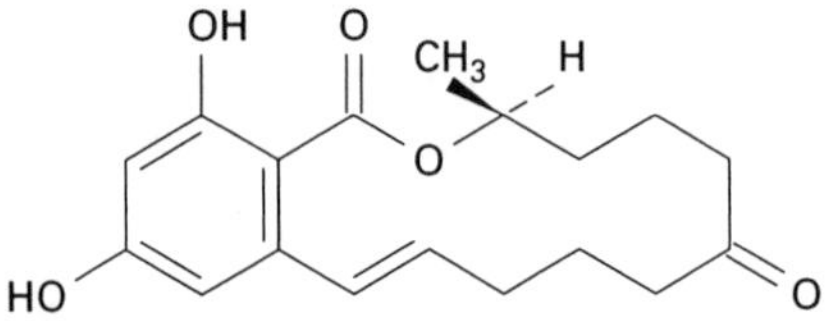

Zearalenone

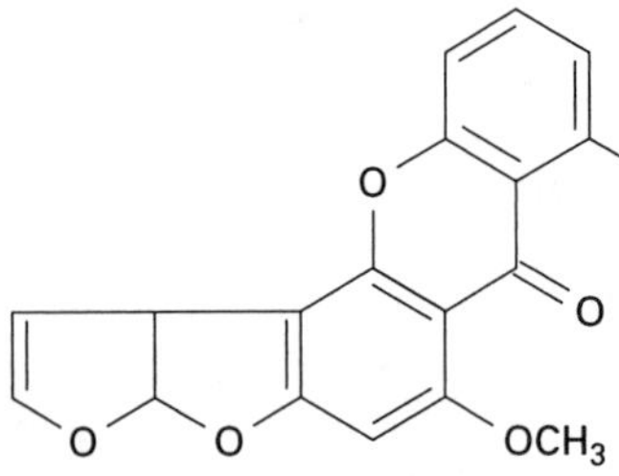

Sterigmatocystin

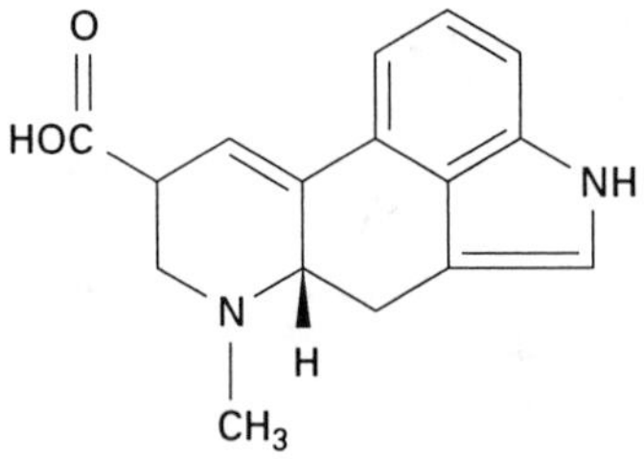

Lysergic acid

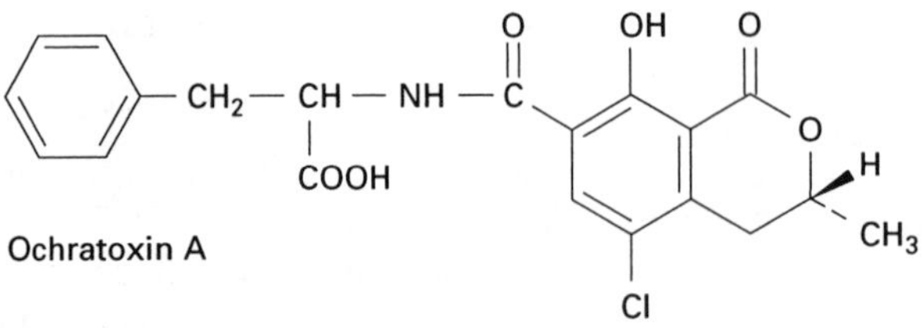

Ochratoxin A

Figure 2.5 Some important mycotoxins.

Type of food involved; prevention. Many foodstuffs and ingredients may become contaminated with mycotoxins. The occurrence of various mycotoxins in foods and feeds has often been reported. Since the discovery of the aflatoxins, probably no commodity can be regarded as absolutely free from mycotoxins. Also, mycotoxin production can occur in the field, during harvest, processing, storage, and shipment of a given commodity.

2.3.3.4.2 Aflatoxins. The aflatoxins are the most important mycotoxins. They are produced by the molds *Aspergillus flavus* and *Aspergillus parasiticus*.

Toxicity and symptoms. Aflatoxins are potent toxins. They are well-known for their carcinogenicity. In view of occurrence and toxicity, aflatoxin B1 is the most important of them, followed by G1 > B2 > G2. Aflatoxin B1 is a very potent hepatocarcinogen in various experimental animal species including rodents, birds, fish, and monkeys. It appears that the aflatoxins themselves are not carcinogenic but rather some of their metabolites. Primary liver cancer is one of the most prevalent human cancers in the developing countries. Epidemiological studies carried out in the 1970s provide statistical support for the association of food consumption, contamination with aflatoxins, and incidence of hepatocellular carcinoma. It is now believed that there are combined actions of aflatoxins and hepatitis B virus infection leading to primary liver cancer. Due to worldwide commercial activities, the threat of aflatoxins to human health is not limited to those countries where the mycotoxins are produced. Moreover, the international trade in animal feed ingredients has contributed to the potential hazard for public health, because milk and dairy products may become contaminated with aflatoxin M1 (see Figure 2.5), the 4-hydroxy derivative of aflatoxin B1 formed in cows after ingestion of aflatoxin B1 with their feed. Aflatoxin M1 is also a suspect carcinogen, although its carcinogenic potency is probably less than that of aflatoxin B1.

Chemical properties (structure and stability). Aflatoxins are derivatives of coumarin. (The structure of coumarin can be found in Section 2.2.1.3.)

The most important types of aflatoxins are B1, B2, G1, and G2 (Figure 2.6). Aflatoxins are heat-stable and are hard to transform to non-toxic products. However, two methods of detoxication should be mentioned. First, the fate of aflatoxin B1 during food fermentation has been investigated in a variety of products. It appeared that fungi involved in food fermentation, for instance *Rhizopus oryzae* and *R. oligosporus*, are capable of reducing the cyclopentanone moiety, resulting in the formation of aflatoxicol A (Figure 2.7). This reaction is reversible. Under suitable environmental conditions (e.g., presence of organic acids), aflatoxicol A is irreversibly converted to its stereoisomer aflatoxicol B. Aflatoxicol A is approximately 18 times less toxic than aflatoxin B1.

In lactic fermentations at pH ≤4.0, aflatoxin B1 is readily converted to aflatoxin B2a (Figure 2.7) which is also less toxic. Both transformations thus reduce the toxicity, but the detoxication is not complete unless the lactone ring of the aflatoxin molecule is opened (Figure 2.8).

This would correspond to the loss of fluorescence at 366 nm. It has been found that loss of fluorescence correlates with reduced mutagenicity. Screening fungi for their ability to reduce the fluorescence of aflatoxin B1 solutions revealed that certain *Rhizopus* spp. were able to transform 87% of aflatoxin B1 into non-fluorescent substances of as yet unknown nature. A similar detoxication by opening of the lactone ring is achieved by treatment with ammonia (NH_4OH) at elevated temperature and pressure, which is applied at industrial scale to detoxicate animal feed ingredients, e.g., groundnut press-cake. At high pH the lactone ring of the aflatoxin molecule is hydrolyzed.

B1

G1

B2

G2

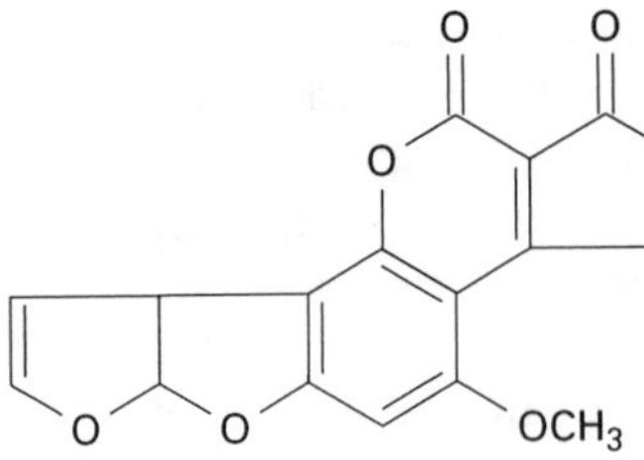

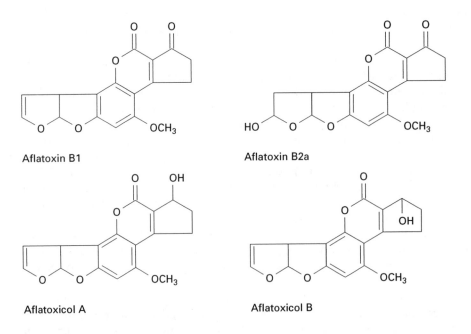

Figure 2.6 Structures of the major aflatoxins.

Aflatoxin B1

Aflatoxin B2a

Aflatoxicol A

Aflatoxicol B

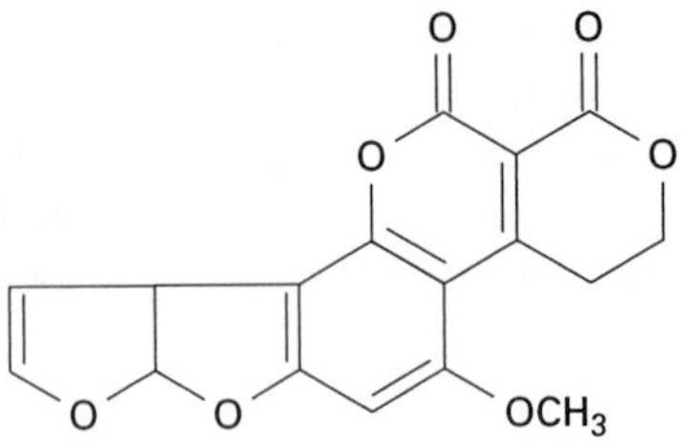

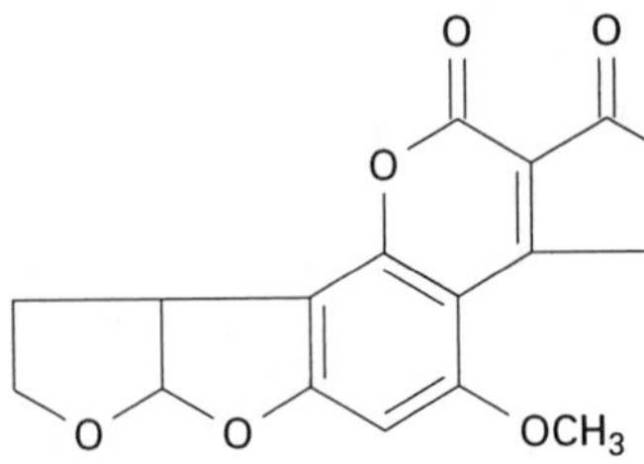

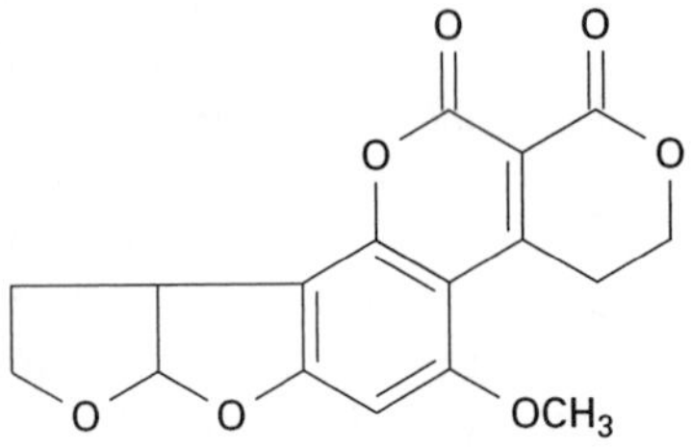

Figure 2.7 Detoxication of aflatoxin B1.

Environmental conditions. The fungi grow best at approximately 25°C at high relative air humidity (≥80%). Aflatoxins are produced both pre- and post-harvest, at relatively high moisture contents and relatively high temperatures.

Type of food involved; prevention. Aflatoxins can occur on various products, such as oilseeds (groundnuts), grains (maize) and figs. Problems with aflatoxin contamination

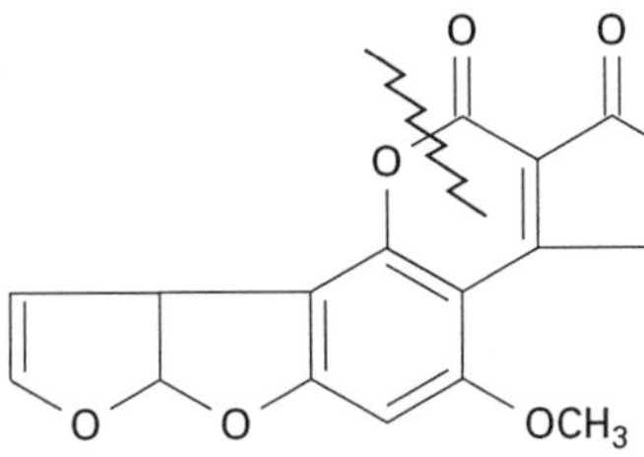

Figure 2.8 Detoxication of aflatoxin B1 by opening the lactone ring.

occur in industrialized countries (US) as well as in the developing countries in Latin America, Asia, and Africa. Aflatoxin M1 can be detected in low concentrations in milk samples from around the world, because of the high sensitivity of the current analytical methods. Prevention of aflatoxin contamination is achieved by discouraging fungal growth. Particularly, adequate post-harvest crop-drying is essential to reduce the chance of fungal growth.

2.3.3.4.3 Deoxynivalenol. Deoxynivalenol (DON) is a mycotoxin belonging to the group of trichothecenes (see Figure 2.5). It is produced by *Fusarium graminearum*.

Toxicity and symptoms. The trichothecenes, including T-2 toxin, HT-2 toxin, diacetoxyscirpenol, neosolaniol, fusarenon-X, nivalenol, and DON, induce a wide variety of toxic effects in experimental animals: diarrhea, severe hemorrhages, and immunotoxic effects. DON occurs worldwide. The toxin is of particular interest in the zootechnic sector, because feeding pigs with DON may lead to economic loss due to refusal of the feed and vomiting.

Chemical properties (structure and stability). DON is quite resistant to conventional food processing conditions.

Environmental conditions. The *Fusarium* producer strains prefer high relative air humidity at moderate temperatures (10 to 30°C).

Type of food involved; prevention. Fusarium species particularly occur on grains, e.g., maize, wheat and rye, in the moderate climate zones.

2.3.3.4.4 Ergot alkaloids. Ergot alkaloids are produced by *Claviceps purpurea*, which grows in the ears of grasses and cereals. The fungus forms sclerotia (2 to 4 cm large ergot kernels), which are the hibernation stage. During the harvest the sclerotia may end up between the cereal grains.

Toxicity and symptoms. Ergot alkaloids act particularly on the smooth muscles. Severe poisoning leads to constriction of the peripheral arteries, followed by dry gangrene of tissue and loss of extremities. Neurological disorders (itching, severe muscle cramps, spasms and convulsions, and psychological disorders) may also occur.

Chemical properties (structure and stability). The sclerotia contain derivatives of lysergic acid (Figure 2.5), the ergot alkaloids. Figure 2.9 illustrates the basic structure of these substances, taking ergotamine as an example. Ergometrine, ergotamine, and ergocristine are among the most important.

Environmental conditions. Ergot formation is favored especially in pre-harvest rye by high relative air humidity and temperatures of 10 to 30°C.

Type of food involved; prevention. Claviceps pupurea is common in pre-harvest grains. Consequently, a strict quality control of grain before milling is required. Taking into account the present-day grain quality assurance systems and its relatively high no-effect level, ergot is not considered a serious threat to human or animal health.

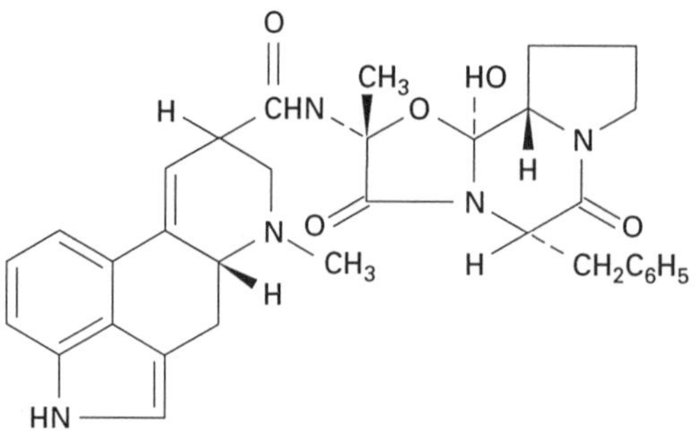

Ergotamine

Figure 2.9 Structure of ergotamine.

2.3.3.4.5 Patulin. Patulin is mainly produced by *Penicillium expansum, Penicillium patulinum* and *Byssochlamys nivea.*

Toxicity and symptoms. Patulin causes hemorrhages, formation of edema, and dilatation of the intestinal tract in experimental animals. In subchronic studies, hyperemia of the epithelium of the duodenum and kidney function impairment were observed as main effects.

Chemical properties (structure and stability). The structure of patulin is shown in Figure 2.5. It is stable under conditions required for fruit juice production and preservation (see below).

Environmental conditions. Moderate temperatures, high moisture content, and relatively low pH (3 to 5) favor the growth of the fungi involved and patulin formation.

Type of food involved; prevention. The toxin occurs in vegetables and fruits (apples). Patulin is an indicator of bad manufacturing practice (use of moldy raw material) rather than a serious threat to human and animal health, as recent subacute and chronic toxicity studies have revealed. Thus, regulatory action based on safety evaluation would not be necessary.

2.3.3.4.6 Sterigmatocystin. Sterigmatocystin is produced by *Aspergillus versicolor* and *Aspergillus nidulans.*

Toxicity and symptoms. Sterigmatocystin is considered to be a carcinogen. Experiments with animals have shown that it causes liver and lung tumors in rats and mice. In comparison to the doses that induce tumors in rats, sterigmatocystin appeared to be a less potent carcinogen than the very potent aflatoxin B1.

Chemical properties (structure and stability). Sterigmatocystin is structurally related to the aflatoxins (Figure 2.5) and is equally stable.

Environmental conditions. Among the factors stimulating fungal growth and toxin production on cheese are lactose, fat, and some fat hydrolysis products.

Type of food involved; prevention. The natural occurrence of sterigmatocystin in food is probably limited. However, investigations on the occurrence of sterigmatocystin in food are, as yet, also limited. Sterigmatocystin occurs occasionally in grains and the outer layer of hard cheeses, when these have been colonized by *Aspergillus versicolor.* The concentration of sterigmatocystin in the outer layer of contaminated cheeses decreases rapidly from outside to inside. Insufficient data are available on the occurrence of sterigmatocystin, for example, in grated cheese to allow an evaluation of the health hazard caused by this product.

2.3.3.4.7 Zearalenone. Zearalenone is produced by some *Fusarium* species, i.e., *Fusarium roseum* and *Fusarium graminearum*.

Toxicity and symptoms. Zearalenone has estrogenic and anabolic properties. Pigs are among the most sensitive animals. The International Agency for Research on Cancer has placed zearalenone in the category "limited evidence of carcinogenicity."

Chemical properties (structure and stability). Zearalenone (Figure 2.5) is structurally related to the anabolic zeranol. Few data are available on its stability.

Environmental conditions. The conditions favoring zearalenone production are similar to those favoring DON formation, i.e., high relative air humidity at moderate temperatures.

Type of food involved; prevention. Zearalenone often co-occurs with DON in various grains, in particular maize and wheat. A risk assessment study on zearalenone carried out in Canada revealed that currently no adverse health effects are anticipated from zearalenone due to the intake of maize products. Other food sources such as wheat, flour, or milk may also contribute to the exposure to zearalenone. For the time being, no regulatory action has been recommended.

2.3.3.4.8 Ochratoxin A. Ochratoxin A can be produced by both *Aspergillus ochraceus* and *Penicillium viridicatum*.

Toxicity and symptoms. Ochratoxin A is a potent nephrotoxin in birds, fish, and mammals. Ochratoxin A is also teratogenic in mice, rats, hamsters, and chickens. The primary target organ is the developing central nervous system. There is a hypothesis that a renal disease observed in some areas of the Balkan countries is associated with exposure to ochratoxin A.

Chemical properties (structure and stability). The structure of ochratoxin A is shown in Figure 2.5. It is a fairly stable substance which is not easily metabolized.

Environmental conditions. Ochratoxin A production in cereals is favored under humid conditions at moderate temperatures.

Type of food involved; prevention. Ochratoxin A occurs in grains and, following transfer, in the organs and blood of a number of animals, especially pigs. Recently, a risk assessment study on ochratoxin A has been published. (Limited) Canadian data on estimated human intakes indicate that the tolerable daily intakes, estimated from carcinogenicity data of ochratoxin A, have been exceeded occasionally. More data are required to estimate the dietary exposure to ochratoxin A and to assess the need for regulatory controls or other control mechanisms. The current concern about ochratoxin A has led the International Union of Pure and Applied Chemistry (IUPAC) to the recent launching of a project in which the worldwide occurrence of ochratoxin A in food and animal feed will be mapped.

2.3.3.5 Toxic microbial metabolites

2.3.3.5.1 Biogenic amines. The main producers of biogenic amines in foods are Enterobacteriaceae and Enterococci. Most lactic acid bacteria which are used to produce fermented foods do not produce significant levels of biogenic amines.

Toxicity and symptoms. Biogenic amines have a stimulatory or toxic effect on the consumer. The symptoms of intoxication, persisting for several hours, include burning throat, flushing, headache, nausea, hypertension, numbness and tingling of the lips, rapid pulse, and vomiting. Especially, histamine has been indicated as the causative agent in several outbreaks of food intoxication. A level of approximately 1000 ppm of total biogenic amines in food is supposed to elicit toxicity, but from a Good Manufacturing Practice (GMP) point of view, levels in food of 50 to 100 ppm, 100 to 200 ppm and 30 ppm for histamine, tyramine, and phenylethylamine, respectively, or a total of 100 to 200 ppm are acceptable. The toxicity of histamine appears to be enhanced by the presence of other biogenic amines found in foods that can inhibit histamine-metabolizing enzymes in the small intestine. Estimating the frequency of histamine poisoning is difficult because most countries have no regulations for histamine levels in foods, nor do they request notification of histamine poisoning. Also, because histamine poisoning closely resembles food allergy, it may often be misdiagnosed.

Chemical properties (structure and stability). Biogenic amines are a group of moderately toxic substances which can be formed in fermented foods, mainly by decarboxylation of amino acids (Figure 2.10).

Environmental conditions. The levels of biogenic amines increase with the presence of free amino acids (precursors), low pH of the product, high NaCl concentrations, and microbial decarboxylase activity.

Type of food involved; prevention. Biogenic amines are especially associated with lactic fermented products, particularly wine, cheese, fish, and meat. Also, very low levels occur in fermented vegetables (Figure 2.11).

Biogenic amines also occur naturally in fruits, vegetables, and fish; they may be produced by microbial decarboxylase activity. For instance, fresh fish (mackerel, tuna, skipjack) contain high levels of histidine which is readily decarboxylated to histamine by Gram-negative bacteria, e.g., *Proteus morganii.*

Amine	Formula	Precursor
Ethylamine C_2H_7N	$CH_3CH_2NH_2$	Alanine
Putrescine $C_4H_{12}N_2$	$H_2N\ (CH_2)_4\ NH_2$	Ornithine
Histamine $C_5H_9N_3$	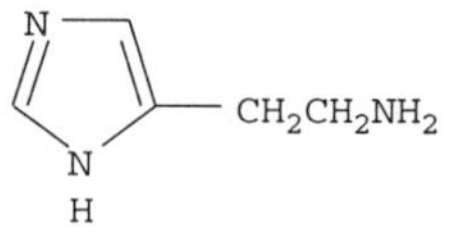 imidazole ring — $CH_2CH_2NH_2$	Histidine
Cadaverine $C_5H_{14}N_2$	$H_2N\ (CH_2)_5\ NH_2$	Lysine
Tyramine $C_8H_{11}NO$	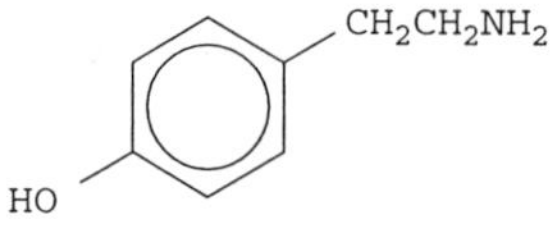 HO—phenyl— $CH_2CH_2NH_2$	Tyrosine
Phenylethylamine $C_8H_{11}N$	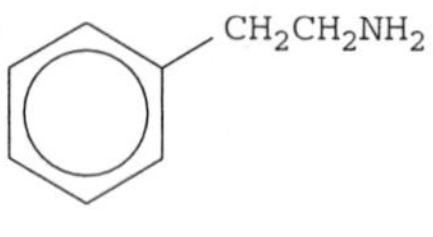 phenyl— $CH_2CH_2NH_2$	Phenylalanine
Tryptamine $C_{10}H_{12}N_2$	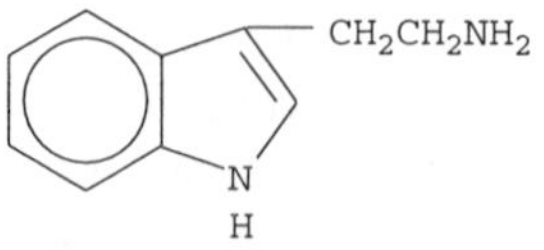 indole— $CH_2CH_2NH_2$	Tryptophan

Figure 2.10 Major biogenic amines.

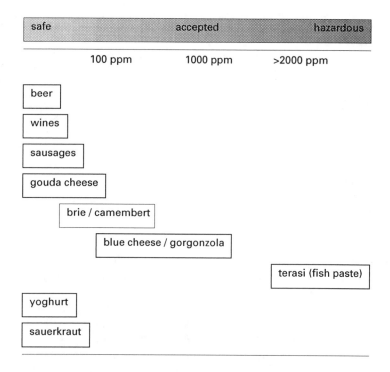

Figure 2.11 Presence of biogenic amines in fermented foods in relation to health hazards.

In meat products, species of Enterobacteriaceae have been found to be associated with cadaverine formation, and lactobacilli with tyramine formation. Also, sauerkraut may contain varying levels of biogenic amines, due to the large variations in the naturally selected microflora. In cheese, Enterobacteriaceae, heterofermentative lactobacilli, and *Enterococcus faecalis* were shown to be associated with the production of up to 600 ppm of biogenic amines including phenylethylamines. *Lactobacillus buchneri* has been shown to be involved in cheese-related outbreaks of histamine poisoning. Pasteurization of cheese milk, good hygienic practice, and selection of starters with low decarboxylase activity are measures to prevent the accumulation of these undesirable products.

2.3.3.5.2 Ethyl carbamate. Ethyl carbamate (urethane) is associated with yeast fermented foods and beverages.

Toxicity and symptoms. Ethyl carbamate is a mutagen as well as a carcinogen.

Chemical properties. The structure of the carbamic acid moiety of ethyl carbamate may originate from several substances including naturally occurring citrulline, and urea and carbamyl phosphate resulting from the metabolism of L-arginine and L-asparagine by yeast. In addition, vicinal diketones and HCN released from cyanogenic glycosides can act as precursors. Ethanol (the other precursor) is formed as a result of alcoholic fermentation by yeasts, or as one of the products of heterofermentative lactic acid fermentation.

$$CH_3 - CH_2 - O - C - NH_2$$
$$\overset{\|}{O}$$

Ethyl carbamate (Urethane)

Table 2.6 Occurrence of ethyl carbamate in fermented foods and beverages[1]

Product	Number of samples	Average level (ppb)	Range (ppb)
Yogurt	12	0.4	ND–4
Cider	8	0.6	ND–4
Bread	30	1.7	ND–8
Malt beverages	69	1.8	ND–13
Bread, toasted	9	5.2	2–14
Soya sauce	12	18	ND–84
Wine	6	18	7–40
Sake	11	52	3–116

Note: ND = not detectable.

[1] Data found in literature.

Environmental conditions. Heat and light enhance the formation of ethyl carbamate.

Type of food involved; prevention. Ethyl carbamate occurs in a variety of fermented foods and beverages (Table 2.6).

In most countries there is no legislative limit value, but the Food and Agriculture Organization World Health Organization (FAO/WHO) suggest a level of 10 ppb for softdrinks, and the Canadian Government recommends 30 to 400 ppb for various alcoholic beverages. Research on wine and stone fruit (cherry, plum) fermentations indicate that reduction of the levels of the precursors by enzymatic treatment, selection of yeast strains, control of fermentation conditions, and treatment of the pH-adjusted fermented pulp with $CuSO_4$ may be useful in keeping the ethyl carbamate levels at a minimum.

2.4   Recent developments in food safety assurance

2.4.1   Good manufacturing practice

In principle, prevention of food-associated intoxications starts at the level of primary, i.e., pre-harvest plant and animal production. However, this is very difficult to achieve on a large scale and can only be considered as a long-term objective.

Consequently, it is important to prevent bacteria and mold spores from starting to grow during food and feed processing and storage. The establishment of, and strict adherence to hygiene guidelines and rules for Good Manufacturing Practice contribute to the systematic microbiological control of industrial processes. The main techniques to achieve growth prevention include: drying (reduction of water activity), control of keeping and storage temperatures (the lower the better) and modified atmosphere storage (CO_2 levels in the gas phase exceeding 35% v/v inhibit microbial growth). Other techniques include the application of gamma irradiation, or fungicides to kill fungal spores. However, some of the methods may have the disadvantage of not being fully effective and of leaving chemical residues in the food product.

2.4.2   Consumer education

It is also important that the consumer should protect him/herself by safe handling of food. Surveys have shown that most food intoxications originate from inadequately refrigerated storage, or use of left-overs which were not or inadequately re-heated. Also, foods of animal origin should not be consumed raw. Cross-contamination of cooked food with raw food must be avoided by keeping raw and cooked foods separated.

Table 2.7 Critical control points in the manufacturing process of sweetened concentrated milk with regard to growth of and enterotoxin production by *Staphylococcus aureus*

Processing stage	Can contamination with *S.aureus* occur?	Can *S.aureus* grow or produce enterotoxins?	Can *S.aureus* survive?
Raw milk	yes	no, if temperature <10°C	yes
Pasteurization	no, if overpressure is maintained		no, if adequate time/temperature combination is used
Concentration	no	no, if temperature >45°C	yes
Seeding with lactose crystals	yes, if not done aseptically		yes
Bottling or can filling	yes, if not done aseptically		yes
Storage	no	growth if $a_w > 0.86$; no toxin formed	yes
Home use	yes, after opening	yes, after dilution if temperature > 15°C	yes

2.4.3   Hazard analysis at critical control points

The introduction in the 1970s of the Hazard Analysis at Critical Control Points (HACCP) (see Part 3, Chapter 21, Section 21.3) concept has marked a change in the philosophy with regard to the microbiological quality assurance of food. This concept provides a means for identifying the microbiologically important stages in food processing and the means to control them. Introduction of this system starts with a detailed analysis of the hazards associated with the manufacture, distribution, and use of the food, and leads to the identification of the critical control points. Systematic and frequent monitoring and control are carried out at these points. In applying these principles, greater assurance of product safety is achieved than would be possible with the traditional procedures.

The above is illustrated by the production of sweetened concentrated milk. In this product, *Staphylococcus aureus* can grow and produce enterotoxins. The stages in the manufacturing process which are of importance from a microbiological point of view are summarized in Table 2.7.

At each stage, the chance of contamination with *S. aureus* is assessed, and the conditions determining growth and toxin production are given. As can be seen, the safety of the process can be monitored comfortably and quickly by regular measurements of temperature and pressure. In addition, maintenance of asepsis during the process and adequate instructions for use and storage at the consumer level are key factors to reduce the risk of intoxication.

2.5   Summary

Some of the many thousands of natural substances present in food have been found to induce toxic effects. Usually, natural toxins are not acutely toxic, except in a few cases in animals. Particularly, those natural toxins occurring in plant-derived foods may induce adverse effects only after chronic ingestion or by allergic reactions.

In this chapter, the natural toxins are divided into endogenous toxins of plant origin and contaminants of natural origin. Endogenous toxins of plant origin comprise many different types of substances. There is no simple way of classifying this group of toxic food components. The way they are dealt with here is based on a classification according to common functional groups (toxic phenolic substances, cyanogenic glycosides, and glucosinolates), the physiological action (acetylcholinesterase inhibitors), and the type of toxic effect induced (biogenic amines and central stimulants). Toxins in food can also be contaminants of natural origin. There are three important sources of this group of natural toxins. First, raw materials of plant origin may be mixed with toxic non-nutritive plant species, e.g., cereals have been reported to be contaminated by pyrrolizidine alkaloids. Secondly, raw materials of animal origin, mainly fish and milk, can also be contaminated if the animal has ingested toxic substances of natural origin. A well-known case is that of paralytic shellfish poisoning. This is attributed to the consumption of shellfish that have become contaminated with a toxin (saxitoxin) on ingestion of toxic plankton. Thirdly, contaminants of natural origin can be products of microorganisms. Several microorganisms, including bacteria and fungi, can cause food-borne diseases in this way. The most important bacterial toxins, their chemical properties, environmental conditions required for their formation, type of food involved, and prevention measures are discussed. In particular, the bacteria *Clostridium botulinum* (botulinum toxin), *Staphylococcus aureus*, *Clostridium perfringens*, *Bacillus cereus*, and endotoxin-forming Gram-negative bacteria are of importance as far as food safety is concerned. Toxins of fungal origin, the so-called mycotoxins, are produced by the genera *Aspergillus*, *Penicillium*, and *Fusarium*. The most important mycotoxins are the aflatoxins (*Aspergillus flavus*). Another major fungal toxin is ochratoxin A (*Aspergillus ochraceus* and *Penicillium vindicatum*). A special group of microbial toxins are metabolites of microorganisms. Important examples are the biogenic amines (formed by decarboxylation of free amino acids during spoilage and in some fermentations) and ethyl carbamate (occurring in yeast-fermented foods and beverages).

Reference and reading list

Culliney, T.W., D. Pimentel, M.H., Pimentel, Pesticides and natural toxicants in foods, *Agric. Ecosys. Environ.*, 41, 297–320, 1992.

Davidek, J., (Ed.), *Natural toxic compounds of foods. Formation and change during food processing and storage*. Boca Raton, CRC Press Inc., 1995.

Egmond, H.P. van, G.J.A., Speijers, Survey of data on the incidence and levels of ochratoxin A in food and animal feed worldwide, *J. Natural Toxins*. 3, 125–144, 1994.

Hardegree, M.C. and A.T. Tu, Eds., Bacterial toxins, Vol. 4 in: *Handbook of natural toxins*. New York, Marcel Dekker Inc., 1988.

Doyle, M.P., (Ed.), *Foodborne Bacterial Pathogens*. New York, Marcel Dekker Inc., 1989.

Hauschild, A.H., K.L. Dodds, (Eds.), *Clostridium botulinum: Ecology and Control in Foods*. New York, Marcel Dekker Inc., 1993.

Hu, Y.H., J.R. Gorham, K.D. Murrell, D.O. Cliver (Eds.), *Foodborne Disease Handbook, Vol. 1, Disease Caused by Bacteria*. New York, Marcel Dekker Inc., 1994.

Keeler R.F. and A.T. Tu, (Eds.), Plant and fungal toxins, Vol. 1 in: *Handbook of natural toxins*. New York, Marcel Dekker Inc., 1983.

Krogh, P. (Ed.), *Mycotoxins in Food*. New York, Academic Press, 1988.

Moy, G., F. Kaeferstein, Y. Metarjemi, Application of HACCP to food manufacturing: some considerations on harmonization through training, *Food Control*. 5, 131–139, 1994.

Sahrma, R.P., D.K. Salunkhe, *Mycotoxins and Phytoalezins*. Boca Raton, CRC Press, 1991.

Salyers, A.A., D.D. Whitt, *Bacterial Pathogenesis*. Washington DC, American Society for Microbiology, 1994.

Todd, E.C.D., Foodborne disease in Canada — a 10-year summary from 1974 to 1984, in: *J. Food Protection*. 55, 123–132, 1992.

Tu, A.T. (Ed.), Marine toxins and venoms, Vol. 3 in: *Handbook of natural toxins*. New York, Marcel Dekker Inc., 1988.

Viviani, R., Butrophication, marine biotoxins, human health, in: *Sci. Total Environ. Suppl.*, 631–632, 1992.

chapter three

Antinutritives

M.M.T. Janssen

3.1 Introduction

It is a well-known fact that food components can also cause toxic effects without being the active agents themselves. The substances in question are known as antinutritives. They induce their toxic effects indirectly, by causing nutritional deficiencies or by interference with the functioning and utilization of nutrients.

Antinutritives can interfere with food components before intake, during digestion in the gastrointestinal tract, and after absorption in the body. The adverse effects of antinutritives usually do not manifest themselves as readily as those of the directly acting toxic food components. The conditions under which antinutritives may have important implications are malnutrition or a marginal nutritional state.

Antinutritives can be of natural as well as synthetic origin. They can be classified as follows:

- type A: substances primarily interfering with the digestion of proteins or the absorption and utilization of amino acids; these are also called *antiproteins*;
- type B: substances interfering with the absorption or metabolic utilization of minerals *(antiminerals)*;
- type C: substances that inactivate or destroy vitamins or otherwise increase the need for vitamins *(antivitamins)*.

Antinutritives are mainly found in plant material. In a number of cases, drugs, antibiotics, and pesticides have been reported to be antinutritive. This chapter deals with food components of natural origin only.

3.2   Type A antinutritives (antiproteins)

Especially people depending on vegetables for their protein supply are in danger of impairment by antiproteins. This is often the case in less developed countries.

3.2.1   Protease inhibitors

Protease inhibitors are proteins which inhibit proteolytic enzymes by binding to the active sites of the enzymes. Their specificities for the different proteases are broadly overlapping. This category of antinutritives occurs in many plants, and in a few animal tissues.

Proteolytic enzyme inhibitors were first found in eggs around the turn of the century. They were later identified as ovomucoid and ovoinhibitor, both of which inactivate trypsin. Also, chymotrypsin inhibitors are found in eggs, especially in the egg white. Other foods in which trypsin and/or chymotrypsin inhibitors are found are legumes (e.g., soybeans), vegetables (e.g., alfalfa), milk, wheat and potato.

The protease inhibitors in soybeans, kidney beans and potatoes also inhibit elastase, a pancreatic enzyme acting on elastin, an insoluble protein in meat. Since protease inhibitors are proteins, they can be expected to be heat labile. Protease inhibitors that are indeed heat labile are particularly sensitive to moist heat, whereas dry heat is less effective. Autoclaving soybeans for 20 min at 115°C or 40 min at 107 to 108°C is necessary for maximum destruction of its inhibitors. Prior soaking in water for 12 to 24 hr makes the heat treatment more effective. Boiling at 100°C for 15 to 30 min is sufficient to improve the nutritional value of soaked soybeans.

However, several protease inhibitors are relatively heat resistant. An example is the trypsin inhibitor in milk. In raw milk, the activity of trypsin can be reduced by 75 to 99%. The inhibitor is unaffected by temperatures up to 70°C. Pasteurization for 40 sec at 72°C destroys only 3 to 4%, heating at 85°C for 3 sec, 44 to 55%, and heating at 95°C for 1 hr 73% of the inhibitor. Other relatively heat-resistant protease inhibitors are the trypsin inhibitor in alfalfa, the chymotrypsin inhibitor in potatoes, and the trypsin inhibitor in lima beans.

3.2.2   Lectins

Lectins is the general term for plant proteins that have highly specific binding sites for carbohydrates. The majority of the lectins are glycoproteins. A carbohydrate-free lectin occurs in jack beans (concanavalin A). The lectin in kidney beans is probably a lipoprotein. The mode of action of lectins may be related to their ability to bind to specific cell receptors in a way comparable to that of antibodies. They can agglutinate red blood cells. Therefore, they are also called hemagglutinins.

This section discusses the origin of lectins in relation to their interference with the absorption of amino acids, fats, vitamins, and thyroxine. It will become clear that the lectins in legumes not only belong to type A antinutritives but also to types B and C.

Lectins occur in plants, especially legumes such as peanut, soybean, lima, kidney, mung, jack, hyacinth, castor and fava bean, and lentil and pea. They are also found in potato, banana, mango, and wheat germ.

Lectins can contribute to a large extent to the protein content of plants. Bean lectins can disturb the absorption of nutrients and other essential substances from the intestines. *In vitro* studies have shown that bean lectins bind to the rat intestinal mucosal cells. In experimental animals fed on a raw soybean diet, the absorption of amino acids, thyroxine, and fats decreased whereas the requirement for the lipophilic vitamins A and D increased significantly. Interference with the absorption of thyroxine could explain the goitrogenic effect of soybeans. However, this effect may also be caused by interference with the absorption of iodine, as iodine supplementation to the diet has a positive effect on soybean goiter.

Among the edible pulses the kidney bean and the hyacinth bean are highly antinutritive. However, one of the most toxic plants is the castor bean. The toxin is ricin which causes intestinal cell necrosis.

The lectins, being proteins, can easily be inactivated by moist heat. Dry heat is ineffective. The hemagglutinin activity of several pea varieties and bean species decreases on germination. Soybeans can lose as much as 92% of their lectin activity during the first day of germination.

3.3 Type B antinutritives (antiminerals)

Substances interfering with the utilization of essential minerals are widely distributed among vegetables, fruits, and cereal grains. The levels of antiminerals in foods seldom cause acute effects if the diet is well-balanced.

3.3.1 Phytic acid

Phytic acid, the hexaphosphoric ester of myo-inositol, is a strong acid. It forms insoluble salts with many types of bivalent and tervalent heavy metal ions. In that way, phytic acid reduces the availability of many minerals and essential trace elements.

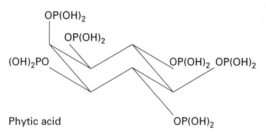

Phytic acid

Phytic acid has been shown to have a negative effect on iron absorption in humans. The absorption of iron depends mainly on the levels in the iron pools, the amount and the chemical form ingested, and the presence of ascorbic acid.

Ferric phytate is least soluble in diluted acid, i.e., it is insoluble in the stomach. At the pH in the duodenum, ferric phytate dissolves in the form of ferric hydroxide.

Phytic acid prevents the complexation between iron and gastroferrium, an iron-binding protein secreted in the stomach. Results from animal experiments and human studies indicate interference of phytic acid with the absorption of magnesium, zinc, copper, and manganese.

An important factor in the precipitation of phytates is the synergistic effect of two or more different cations, which can act together to increase the quantity of phytate that precipitates. For instance, zinc-calcium phytate precipitates maximally at pH 6, which is also the pH of the duodenum, where mainly calcium and trace metals are absorbed. An element makes its deficiency felt as soon as it becomes limiting due to binding to phytic acid.

The results of chemical analysis of the mineral and trace element contents of diets may give a false impression because of the interaction of phytic acid with the elements concerned at the levels of exposure or absorption. The highest levels of phytate occur in grains and some legumes. Generally, grains make up the bulk of the diet especially in developing countries where the diet is often deficient. Therefore, the presence of phytates in grains is a cause for concern. The phytate contents of several foods are listed in Table 3.1.

The phytate of plant seeds is located primarily in the bran and germ. Therefore, brown and wholemeal bread contain more phytate than white bread. In human studies, a diet of

Table 3.1 Phytate contents of selected foods

Food	mg%
Grains	
Wheat	170–280
Rye	247
Maize	146–353
Rice	157–240
Barley	70–300
Oats	208–355
Sorghum	206–280
Buckwheat	322
Millet	83
Wheat bran	1170–1439
Legumes and vegetables	
Green bean (*Phaseolus vulgaris*)	52
Bean (*Phaseolus vulgaris*)	269
Bean (*Phaseolus lunatus*)	152
Soybean	402
Lentil	295
Green pea (*Pisum sativum*)	12
Pea (*Pisum sativum*)	117
Pea (*Lathyrus sativum*)	82
Chick pea	140–354
Vetch	500
Potato	14
Carrot	0–4
Nuts and seeds	
Walnut	120
Hazelnut	104
Almond	189
Peanut	205
Cocoa bean	169
Pistachio nut	176
Rapeseed	795
Cottonseed	368
Spices and flavoring agents	
Millet	83
Caraway	297
Coriander	320
Cumin	153
Mustard	392
Nutmeg	162
Black pepper	115
Pepper	56
Paprika	71

brown bread (containing 214 mg phytic acid per 100 g) resulted in a 33 to 62% decrease in calcium absorption after 3 to 4 weeks, compared to a white bread diet. Addition of bran to white flour gave similar results. Only if the calcium intake was increased to 1.0 to 1.4 g per day, the calcium absorption improved. Calcium absorption is influenced not only by dietary phytate but also by vitamin D and lipids. If vitamin D is limiting in the diet, calcium absorption will be less efficient and the phytate effect will become more pronounced.

In many foodstuffs phytase activity can reduce the phytic acid level. Phytase is an enzyme occurring in plants. It catalyzes the dephosphorylation of phytic acid. Soybeans show weak phytase activity. Rye contains the most active phytase of all cereal grains. The activity of phytase drastically reduces the phytate content of dough during bread-making. Dephosphorylation of phytic acid is facilitated by the increase in acidity of bread dough caused by the reactivity of the yeast. Phytase is added to animal feeds so that no extra phosphate needs to be added. Also, in this way the animals will excrete less phosphate, which may contribute to reduction of environmental pollution.

3.3.2   Oxalic acid

Oxalic acid (HOOC–COOH) can induce toxic as well as antinutritive effects. To humans, it can be acutely toxic. However, it would require massive doses of 4 to 5 g to induce any toxic effect. The oxalic acid levels usually found in food, however, are no cause for concern. This section, discusses the presence of oxalic acid in food in relation to its antinutritive effects. Like phytic acid, oxalic acid reduces the availability of essential bivalent cations. Oxalic acid is a strong acid and, with alkaline earth metal ions and other divalent metal ions, it forms salts that are hardly soluble in water. Calcium oxalate is insoluble in water at neutral or alkaline pH, and dissolves easily in an acid medium. In many animal experiments and human studies, negative effects of oxalate-rich foods have been found, especially on calcium absorption. Vegetable foods such as rhubarb, spinach, and celery, as well as cocoa have been shown to disturb the calcium balance in man.

Negative effects of oxalic acid on calcium absorption can be predicted from the oxalate/calcium ratio of foods. Foods with a ratio higher than 1 may decrease the calcium availability (see Table 3.2). Foods with a ratio of 1 or lower than 1 do not interfere with

Table 3.2 Foods with an oxalate/calcium ratio (meq/meq) >1

Food	Oxalate mg/100 g[a]	Oxalate/Ca. meq/meq
Rhubarb	805	8.5
Common sorrel	500	5.6
Garden sorrel	500	5.0
Spinach	970	4.3
Beet, leaves	610	2.5
Beet, roots	275	5.1
Purslane	1294	4.6
Cocoa	700	2.6
Coffee	100	3.9
Potato	80	1.6
Tea	1150	1.1
New Zeeland spinach	—[b]	3.9
Pig spinach	—	4.9
Orache	900	4.0
Amaranth	—	1.4

[a] Average.

[b] No average available.

calcium absorption. Calcium is irreversibly bound to oxalic acid, so a food with an oxalate/Ca^{2+} ratio of 1 would not be a good calcium source, although it is rich in calcium.

The effects of oxalates may be influenced by the nutritional state of the subjects, the duration of the experiments, and the extent of calcium intake. For example, rats showed no serious effects after a diet containing 2.5% oxalate unless the diet was deficient in calcium, phosphorus, and especially vitamin D. Humans also show a remarkable ability to adapt to a drastic reduction in calcium intake. This can be attributed to the large calcium pool in the form of the skeletal system. Therefore, a decrease in calcium absorption, caused by oxalate, would not make much difference unless the calcium pool is (nearly) depleted. Consumption of foods rich in calcium, such as dairy products and seafood, as well as enhanced vitamin D intake are recommended only if large quantities of foods rich in oxalate are ingested. In this respect, the traditional way of preparing rhubarb is noteworthy. Before cooking the rhubarb , chalk is added to enhance its pallatability. The calcium, of course, is bound by the oxalic acid in the rhubarb, thereby inhibiting the chelation of calcium from other food sources. Today rhubarb varieties are available with lower oxalate levels.

3.3.3   Glucosinolates

A variety of plants contain a third group of antiminerals, the so-called glucosinolates, a class of thioglucosides, whose general structure is shown below.

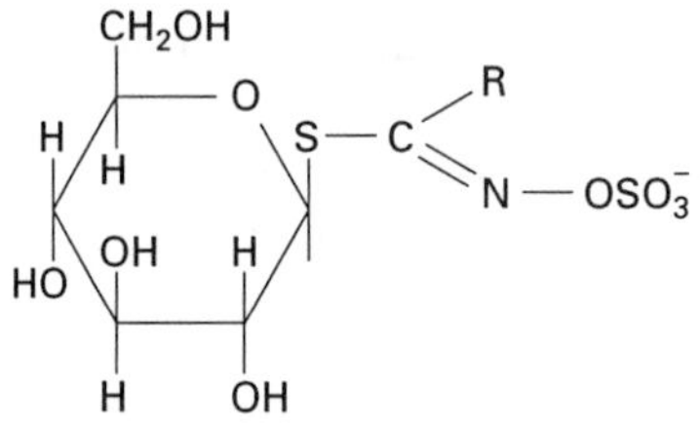

Glucosinolate

Many glucosinolates are goitrogenic. Three types of goiter are distinguished: cabbage goiter (struma), brassica seed goiter, and legume goiter.

Cabbage goiter can be induced by the excessive consumption of cabbage. It seems that cabbage goitrogens inhibit iodine uptake by directly affecting the thyroid gland. Cabbage goiter can be treated by iodine supplementation.

Brassica seed goiter can result from the consumption of the seeds of Brassica plants, such as rutabaga (swede), turnip, cabbage, rape, and mustard, which contain substances that prevent thyroxine synthesis. This type of goiter can only be treated by administration of the thyroid hormone.

Legume goiter is induced by goitrogens in legumes such as soybeans and peanuts. It differs from cabbage goiter in that the thyroid gland is not involved directly. Inhibition of the intestinal absorption of iodine or the reabsorption of thyroxine has been shown in this case. Legume goiter can be treated by iodine therapy.

Fifty types of glucosinolates have been identified. Table 3.3 gives a list of foodstuffs which have been shown to induce goiter, at least in experimental animals.

Rutabaga, turnips, cabbage, peaches, strawberries, spinach, and carrots can cause a significant reduction in the iodine uptake by the human thyroid gland, with rutabaga being the most active.

Table 3.3 Goitrogenic foodstuffs

Food or feedstuff	Glucosinolate[a]	Specific chemical (R) Group
Broccoli (buds)	Glucobrassicin[b]	3-Indolylmethyl
	Gluconapin[c]	3-Butenyl
	Neoglucobrassicin[c]	3-(N-methoxyindolyl)methyl
	Progoitrin[c]	2-Hydroxy-3-butenyl
	Sinigrin[b]	Allyl
Brussel sprouts (head)	As in broccoli	
Cabbage (head)	As in broccoli	
Cauliflower (buds)	As in broccoli	
Charlock (seed)	Sinalbin[c]	*p*-Hydroxybenzyl
Crambe (seed)	Gluconapin[c]	(See broccoli above)
	Gluconasturtlin[c]	2-Phenylethyl
Garden cress (leaves)	Glucotropasolin[c]	Benzyl
Horseradish (roots)	Gluconasturtlin[c]	(See crambe above)
	Sinigrin[b]	
Kale (leaves)	As in broccoli	
Kohlrabi (head)	As in broccoli	
Mustard, black (seed)	Sinigrin[b]	(See broccoli above)
Mustard, white (seed)	Sinalbin[b]	(See charlock above)
Radish (root)		4-Methylthio-3-butenyl
	Glucobrassicin[c]	(See broccoli above)
Rutabaga (root)	Glucobrassicin[c]	(See broccoli above)
	Neoglucobrassicin[c]	(See broccoli above)
	Progoitrin[b]	(See broccoli above)
Turnips (root)	Gluconasturtlin[b]	(See crambe above)
	Progoitrin[c]	(See broccoli above)
		2-Hydroxy-4-pentenyl

[a] Trivial or common name of glucosinolate. The chemical name is formed by the designation of the R group prefixed to the term glucosinolate, e.g., glucobrassicin is 3-indolylmethyl-glucosinolate.

[b] Major glucosinolate.

[c] Minor glucosinolate component.

Hydrolysis, which takes place in damaged plant tissue before or after ingestion, yields the actual goitrogens: thiocyanates, isothiocyanates, cyclic sulfur substances, and nitriles. The thiocyanates and isothiocyanates formed from the glucosinolates are probably the substances responsible for cabbage goiter. As discussed in Chapter 2, goitrogens can also be formed from cyanogens, as their major biotransformation products are thiocyanates. An illustrative example of activation of a glucosinolate by hydrolysis is the potent antithyroid progoitrin occurring in the seeds of Brassica plants and the roots of rutabaga. This substance undergoes hydrolysis as shown in Figure 3.1.

The cyclization product of isothiocyanate III, goitrin (IV), is a powerful goitrogen. The R-enantiomer of goitrin found in crambe seeds is also a strong antithyroid agent. These substances interfere with the iodination of thyroxine precursors so that iodine therapy is not successful. The nitriles I and II obtained from progoitrin are highly toxic, but it is uncertain whether they are goitrogens. Mustard and rapeseed varieties have been bred with low thioglucoside concentration.

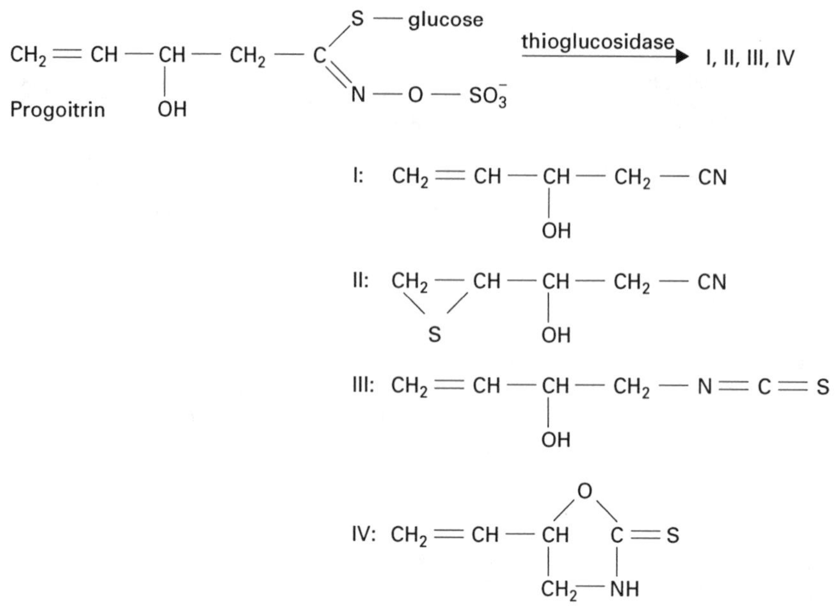

Figure 3.1 Hydrolysis of progoitrin.

3.3.4 Dietary fiber

Dietary fiber is a collective term for all food components derived from plant cell walls that are not digested by the endogenous secretions of the human digestive tract. It has no clearly defined composition. It may differ from foodstuff to foodstuff, and from diet to diet. Dietary fiber consists of pectic substances, hemicelluloses, plant gums and mucilages, algal polysaccharides, celluloses, and lignin. Further, tannins, indigestible proteins, plant pigments, waxes, siliceous materials, and phytic acid can be incorporated in the fiber matrix. These materials give bulk to the fecal matter, not only from their inherent mass, but also by their water-binding capacity. The amount of water bound can be four to six times the dry weight of the fiber.

Around 1970 it was suggested that dietary fiber is a protective factor against many diseases, prevalent in Western communities, e.g., colon cancer. This may be true, but harmful effects of overconsumption of fiber should not be overlooked. The various types of dietary fiber components have many reactive groups, including $-COOH$, $-HPO_3H$, $-OH$, $-SO_3H$ and $-NH_2$, to which metals, amino acids, proteins, and even sugars can be bound.

There are different ways of binding to dietary fiber. First, fiber components of many food products act like ion exchangers. Their binding capacity depends on pH and ionic composition of the bowel contents. Dietary fiber has the capacity to bind various metals, even if the phytic acid is removed. Disturbed Ca^{2+}, Mg^{2+}, Zn^{2+} and P balances have been observed in human subjects using diets rich in fiber in the form of whole wheat bread.

Secondly, amino acids and proteins are bound to dietary fiber. A diet containing 15% cellulose can cause a decrease in nitrogen absorption of as much as 8%. Carrageenans, which are highly indigestible, can cause a decrease in nitrogen absorption of about 16%. The interaction of dietary fiber with sugars does not result in a reduction of sugar absorption, but in a slow release of sugars into the bloodstream.

3.3.5 Gossypol

Gossypol is a plant antinutritive that would probably have remained unnoticed as a food hazard, if the seeds of the plant concerned had not gained importance as a dietary oil and protein source, especially in tropical and subtropical countries. This antinutritive is a yellow pigment present in all parts of the cottonplant. The highest levels are found in cottonseed.

Gossypol exists in three tautomeric forms (Figure 3.2): phenolic quinoid tautomer (I), aldehyde (II), and hemiacetal (III).

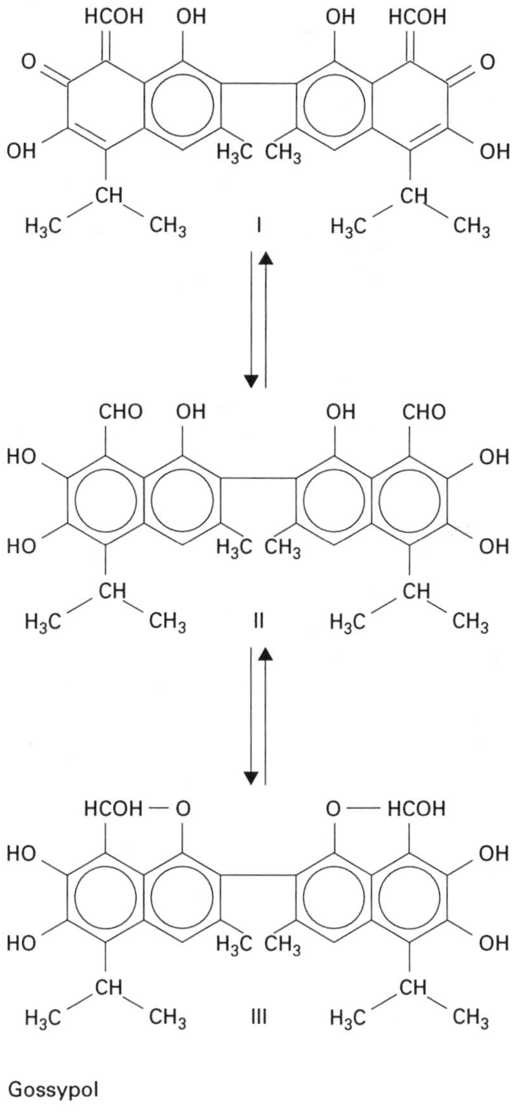

Gossypol

Figure 3.2   The three tautomeric forms of gossypol.

Gossypol is an antimineral as well as an antiprotein. It forms insoluble chelates with many essential metals, such as iron, and binds to amino acid moieties in proteins, especially to lysine. The protein binding suggests that gossypol can reduce the availability of food proteins and inactivate important enzymes.

Processing removes 80 to 99% of the gossypol. The pigment is extracted with the oil and subsequently removed by refining and bleaching. About 0.5 to 1.2% of the total gossypol generally remains in the processed meal. Less than 0.06% is free gossypol. The free gossypol concentration in cottonseed meal can also be lowered considerably by the formation of insoluble metal-gossypol complexes. The use of additives such as $FeSO_4$ and $Ca(OH)_2$ prevents the reaction of gossypol with lysine, even during heat treatment. Breeding of cotton with gossypol-free seeds has been succesful and this cotton variety is now planted commercially. In the US, the maximum allowable gossypol content of cottonseed products for use in human food has been set at 0.045%.

3.4   Type C antinutritives (antivitamins)

As defined in the introduction to this chapter, antivitamins are a group of naturally occurring substances which can decompose vitamins, form unabsorbable complexes with them or interfere with their digestive or metabolic utilization.

Only the more relevant examples of this type of antinutritives are discussed in this section, including ascorbic acid oxidase, antithiamine factors, and antipyridoxine factors.

3.4.1   Ascorbic acid oxidase

Ascorbic acid oxidase is a copper-containing enzyme that mediates the oxidation of free ascorbic acid first to dehydroascorbic acid and next to diketogulonic acid, oxalic acid, and other oxidation products (see Chapter 6, Section 6.6.2.1).

Ascorbic acid oxidase occurs in many fruits and vegetables such as cucumbers, pumpkins, lettuce, cress, peaches, bananas, tomatoes, potatoes, carrots, and green beans. Its activity varies with the type of fruit or vegetable. The enzyme is active between pH 4 and 7. Its optimum temperature is about 38°C. When plant cells are disrupted the compartmentalization of substrate and enzyme is removed. Therefore, if vegetables and fruits are cut, the vitamin C content decreases gradually. In fresh juices, 50% of the vitamin C content is lost in less then one hour. Being an enzyme, ascorbic acid oxidase can be inhibited effectively by blanching of fruits and vegetables.

Ascorbic acid can also be protected against ascorbic acid oxidase by substances of plant origin. Flavonoids, such as the flavonoles quercetin and kempferol, present in vegetables and fruits, strongly inhibit the enzyme. (As far as risk evaluation is concerned, it should be noted that quercetin has also been reported to induce adverse effects: see Part 1, Chapter 2, Section 2.2.1.1).

Kempferol ($R_1 = R_2 = H$)
Quercetin ($R_1 = OH$; $R_2 = H$)

3.4.2 Antithiamine factors

A second group of antivitamins is the antithiamine factors. They interact with vitamin B_1, also known as thiamine. Antithiamine factors can be distinguished as thiaminases, tannins, and catechols. The interaction with vitamin B_1 can lead to serious neurotoxic effects as a result of vitamin B_1 deficiency. Normally, antithiamine factors pose no appreciable risk to humans. They only cause thiamine deficiency in people whose diet is already low in thiamine.

Thiaminases are found in many fish species, freshwater as well as saltwater species, and in certain species of crab and clam. These antithiamine factors are enzymes that split thiamine at the methylene linkage (Figure 3.3).

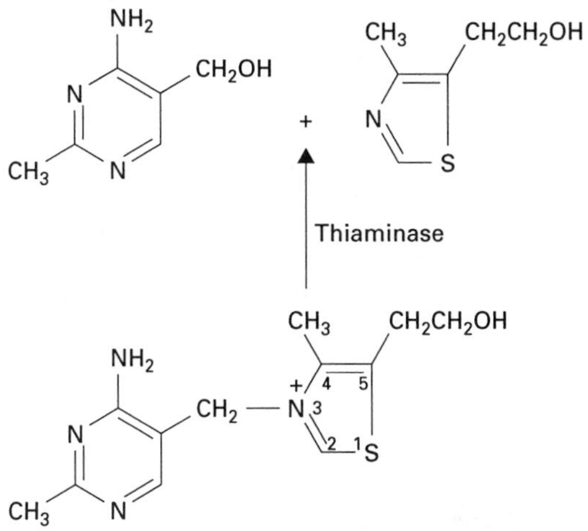

Figure 3.3 Degradation of thiamine by thiaminase.

Thiaminases contain a nonprotein coenzyme, structurally related to hemin, the red pigment component of hemoglobin. The coenzyme is the actual antithiamine factor. Cooking destroys thiaminases in fish and other sources.

Antithiamine factors can also be of plant origin. *Tannins*, occurring in a variety of plants, including tea, are believed to be responsible for inhibition of growth in animals, and for inhibition of digestive enzymes. A study in volunteers on the effects of tannins in tea leaves, tea infusions and betel nuts on thiamine, has shown that the tannins were responsible for thiamine destruction. Tannins are a complex of esters and ethers of various carbohydrates. A component of tannins is gallic acid.

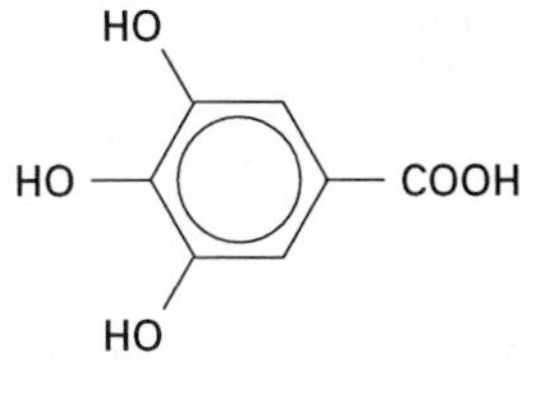

Gallic acid

Gallic acid is obtained by hydrolyzing tannins. The interaction of these substances with thiamine is oxygen-, temperature-, and pH-dependent. It appears to proceed in two phases: a rapid initial phase, which is reversible on addition of reducing agents, such as ascorbic acid, and a slower subsequent phase, which is irreversible.

A variety of antithiamine factors are the *ortho-catechol* derivatives. A well-known example is present in bracken. So-called fern-poisoning in cattle is attributed to this factor. Possibly, there are two types of heat-stable antithiamine factors in this fern, one of which has been identified as caffeic acid (3,4-dihydroxycinnamic acid).

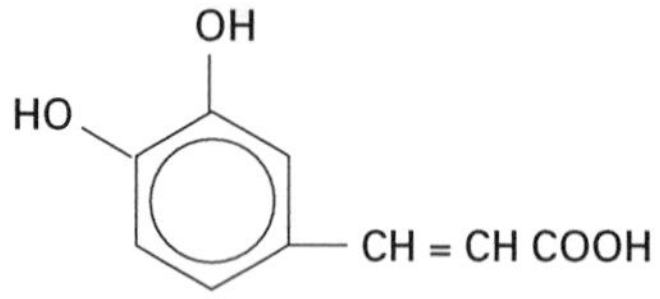

Caffeic acid

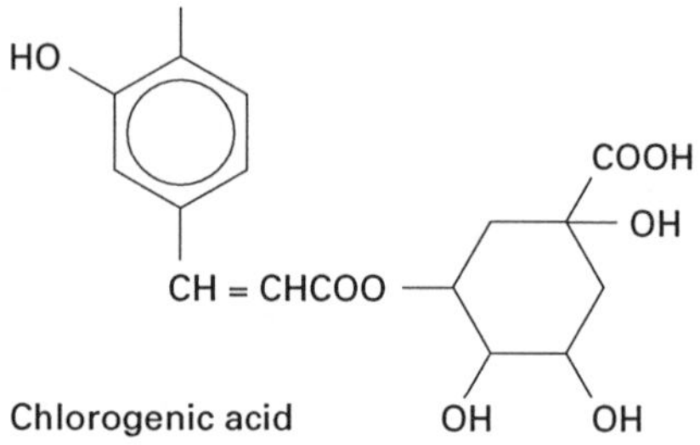

Chlorogenic acid

Caffeic acid can also be formed on hydrolysis of chlorogenic acid by intestinal bacteria. Chlorogenic acid is found in green coffee beans and green apples. Other *ortho-catechols*, such as methylsinapate occurring in mustard or rapeseed, also have antithiamine activity.

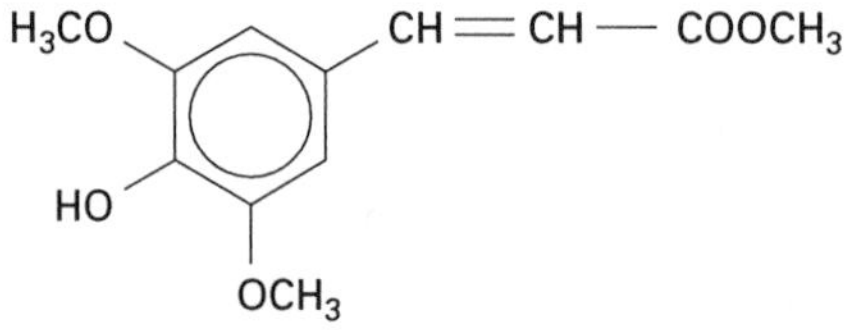

Methylsinapate

3.4.3   Antipyridoxine factors

A variety of plants and mushrooms contain pyridoxine (a form of vitamin B_6) antagonists. The antipyridoxine factors have been identified as hydrazine derivatives.

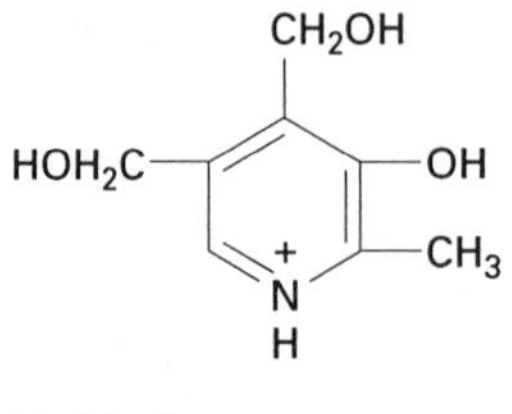

Pyridoxine
(a form of vitamin B_6)

Linseed contains the watersoluble and heat-labile antipyridoxine factor linatine. Linatine is γ-glutamyl-1-amino-D-proline. It readily undergoes hydrolysis to the hydrazine derivative, 1-aminoproline, the actual antipyridoxine factor (Figure 3.4).

Antipyridoxine factors have also been found in wild mushrooms, the common commercial edible mushroom, and the Japanese mushroom shiitake. Commercial and shiitake mushrooms contain agaritine. Agaritine is hydrolyzed in the mushroom by γ-glutamyl-transferase to the active agent 4-hydroxymethylphenylhydrazine (Figure 3.5). The hydrolysis of agaritine is accelerated if the cells of the mushrooms are disrupted. Careful

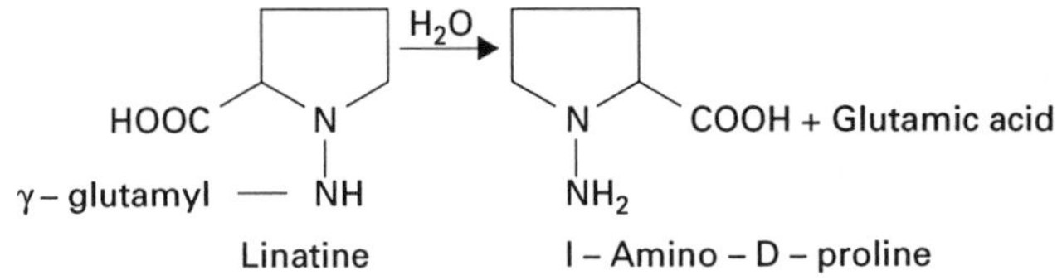

Figure 3.4 Hydrolysis of linatine.

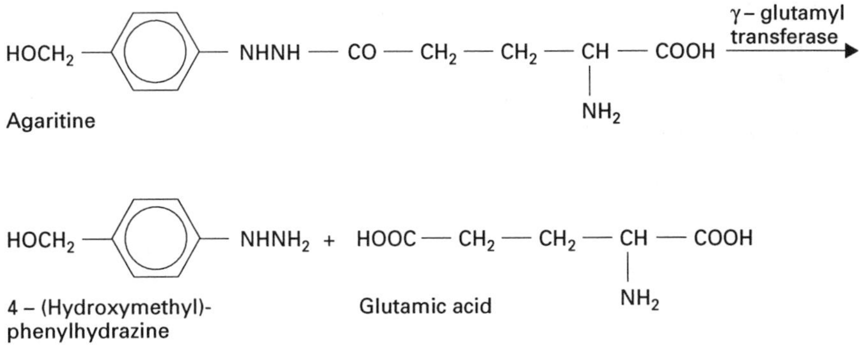

Figure 3.5 Hydrolysis of agaritine.

handling of the mushrooms and immediate blanching after cleaning and cutting can prevent hydrolysis.

The mechanism underlying the antipyridoxine activity is believed to be condensation of the hydrazines with the carbonyl compounds pyridoxal and pyridoxal phosphate — the active form of the vitamin — resulting in the formation of inactive hydrazones.

3.5 Summary

Antinutritives induce their toxic effects indirectly by causing nutritional deficiencies or by obstructing the utilization or functioning of nutrients, mainly proteins, minerals, and vitamins. Especially in the case of marginal nutritional status or malnutrition, the effects of antinutritives can become manifest.

The majority of antinutritives are of natural origin. They are distinguished as three types: antiproteins, antiminerals, and antivitamins. Antiproteins interfere with the digestion of proteins or the absorption and utilization of amino acids. Well-known examples are protease inhibitors and lectins. Both groups are proteins. Protease inhibitors inhibit proteolytic enzymes. Lectins belong not only to the antiproteins but also to the antiminerals and antivitamins. They interfere with the absorption of amino acids as well as that of iodine and vitamins. Antiminerals interfere with the absorption of minerals and essential trace elements, resulting in the reduction of their bioavailability. They include acids, such as phytic and oxalic acid, that form insoluble salts with bivalent and tervalent heavy metal ions. Another group of antiminerals, the glucosinolates, interfere with the absorption of iodine, thus causing goiter. Antivitamins interfere with vitamins in various ways. They can decompose them, form unabsorbable complexes, or disturb their physiological utilization. Relevant examples of this type of antinutritives are ascorbic acid oxidase, antithiamine, and antipyridoxine factors. Many antinutritives can be eliminated from food by various ways of food processing.

Reference and reading list

Belitz, H.-D. and W. Grosch, (Eds.), *Food Chemistry*. Berlin, Springer Verlag, 1987.

Concon, J.M., (Ed.), *Food Toxicology, Part A and Part B*. New York, Marcel Dekker Inc., 1988.

Fennema, O.R., (Ed.), *Food Chemistry*. New York, Marcel Dekker Inc., 1985.

Gibson, G.G. and R. Walker, (Eds.), *Food Toxicology — Real or imaginary problems?*. London, Taylor and Francis, 1985.

Gosting, D.C., (Ed.), *Food safety 1990; an annotated bibliography of the literature*. London, Butterworth-Heinemann, 1991.

Hathcock, J.N., (Ed.), *Nutritional Toxicology, Vol. I*. London, Academic Press, 1982.

Reddy, N.R., M.D. Pierson, Reduction in antinutritional and toxic components in plant foods by fermentation, *Food Res. Int.*, 27, 281–290, 1994.

Tannenbaum, S.R., (Ed.), *Nutritional and safety aspects of food processing*. New York, Marcel Dekker Inc., 1979.

chapter four

Contaminants

M.M.T. Janssen

4.1   Introduction

Food contaminants are substances that are included unintentionally in foods. Some are harmless and others are hazardous because of the toxicological risks from their intake to the consumer. Harmless contaminants may still have the disadvantage of interfering with food processing and causing interactions during storage. Examples are metal ions and plant pigments. This chapter deals with contaminants that are hazardous.

Contamination can occur at every step on the way from raw material to consumer. Raw material of plant origin can be contaminated with environmental pollutants, such as heavy metals, pesticide residues, industrial chemicals, and products from fossil fuels (in the exhaust gases of combustion engines). The sources of contaminants in raw materials of animal origin — mainly fish and milk — are to a large extent comparable to those of raw materials originating from plants. In animal products also, residues of veterinary drugs and growth promoting substances may be present.

During processing, food can be contaminated with processing aids, such as filtering and cleaning agents, and with metals coming from the equipment.

Finally, contaminants can be included in foods during packaging and storage. These can originate from plastics, coatings, and tins.

A number of important examples of hazardous contaminants originating from the above sources are dealt with in this chapter. Toxic contaminants of natural origin have already been discussed in Chapter 2. The formation and inclusion of the well-known environmental pollutants polycyclic aromatic hydrocarbons are dealt with in Chapter 6.

4.2   Contamination with heavy metals

4.2.1   Mercury

The widespread use of mercury and its derivatives in industry and agriculture has resulted in serious environmental pollution. This has led to increased levels of mercury in foods. Fish products in particular can be contaminated with mercury, as methylmercury accumulates extensively in fish. Data on mercury residues in food are shown in Table 4.1.

The toxicity of mercury depends on the chemical form involved: elemental, inorganic, or organic. Exposure to organic mercury compounds, especially methylmercury, is more dangerous than exposure to elemental or inorganic mercury. Organic mercury compounds easily pass across biomembranes and are lipophilic.

The primary target for mercury is the central nervous system. Human response data are available from epidemics of methylmercury poisoning in Japan and Iraq. The first epidemic was caused by consumption of fish from water that was heavily contaminated by industrial waste water. In Iraq, the poisoning appeared to result from the ingestion of wheat treated with a fungicidal mercurial. The total daily intake of mercury per individual in the US and in Western Europe is estimated at 1 to 20 µg. The tolerable weekly intake (TWI) is 300 µg, of which not more than 200 µg should be in the form of methylmercury. The TWI is an estimate of the amount of a contaminant in food or drinking water, which can be ingested weekly over a lifetime by humans without appreciable health risk. Compare to Part 3, Chapter 16, Section 16.3.2.1. The US has set limit values for seafood only.

4.2.2   Lead

In Roman times, extensive lead poisoning occurred as a result of drinking wine treated with lead salts for neutralizing the sour taste. Over a long period of time contamination of

Table 4.1 Levels of mercury residues in food in several countries

Foods	Hg in µg/kg (ppb)		
	United States	United Kingdom	Japan
Cereal (grains)	2–25	5	12–48
Bread and flour		20	
Meats[a]	1–150	10–40	310–360
Fish[b]	0–60	70–80	35–540
Dairy products			
Milk	8	10	3–7
Cheese	80	170	–
Butter	140	10	–
Fruits	4–30	10–40	18
Vegetables (fresh)	0–20	10–25	30–60
Canned	2–7	20[c]	0
Eggs			
White	10	ND[d]	80–125
Yolk	62		330–670
Beer	4		

[a] Includes beef, pork, beef liver, canned meats, and sausages.
[b] Includes canned salmon, shellfish, and whitefish.
[c] Canned peas.
[d] Not detectable.

Table 4.2 Ranges and means of lead content of food

Food	Pb, µg/100 g		Food	Pb, µg/100 g	
	Range	Mean		Range	Mean
Cereal grains	0–62	22	Cider, apple		90 µg/l
Cereal grain products	0–749	10.5	Vinegar, cider		100 µg/l
Seafood, raw	17–250	62	Cola (2 samples)	18–65 µg/l	
Canned	6–30	16	Ginger ale		10 µg/l
Meats	7–37	19	Beer, canned		40 µg/l
Gelatin		57	Wine, red		50 µg/l
Eggs, whole	0–15	7	Sugar, white	0–7	
Vegetables, leafy	0–126	37	granulated		
Legumes, raw, dried	0–16	7	Molasses		53
or frozen			Backing powder		150
Canned	3–11	7	Yeast, dry		117
Apple, raw		38	Black pepper		40
Pear, raw		3	Cinnamon		11
Milk, whole, fresh		0	Nutmeg		41
Skim, dried and packaged		2	Allspice		64
Skim, bulk package		2	Chili powder		18
evaporated	4–5	4.5	Bay leaves		55
Tea, leaves		1.37			
Cocoa, dry		0.10			

food and water with lead occurred in improperly lead-glazed earthenware containers, tins with lead solders, or lead water pipes.

At the moment, the main causes of environmental contamination with lead are industrialization and the use of leaded gasoline. The lead content of food, however, has not significantly increased. The soil retains lead effectively. Nevertheless, the diet, including drinking water, is believed to be the principal source of the total body burden of lead (Table 4.2).

Chronic lead intoxication has been reported to lead to central and peripheral nervous system effects, anemia, and disturbance of renal function and weight loss. Lead intoxications following the intake of contaminated food and water rarely happen. In the majority of cases, lead poisoning of adults is occupational.

Lead compounds are hardly soluble in water. As a result, their absorption is low. Approximately 10% of the ingested lead is absorbed. The levels of lead in bones, hair, and teeth increase with age, suggesting a gradual accumulation of lead in the body. Therefore, contamination of food with lead and the possibility of chronic lead intoxication through the diet needs constant monitoring.

Annual FDA Total Diet Studies have shown that the average lead intake (by adults) in the US has decreased from 90 µg/day in 1974 to 8.1 µg/day in 1989. The TWI through food is 3 mg for adults, and 25 µg per kg body weight for children. The US has set the safety limit for lead in calcium supplements at 5.0 µg/g. Further, the lead content of drinking water and bottled water should not exceed 5 ppb. Decontamination of plant foods by trimming or dehulling is sometimes possible.

4.2.3 Cadmium

Cadmium is widely distributed in the environment, due to extensive industrial use. Sewage sludge, which is used as fertilizer and soil conditioner, is an important source of soil pollution with cadmium. In food, only inorganic cadmium salts are present. Organic

Table 4.3 Cadmium content of foods in the US in μg/100 g

Class of foodstuffs	1972–1973 Range	1972–1973 Average[a]	1973–1974 Range	1973–1974 Average[a]	1974–1975 Range	1974–1975 Average[a]	1975 Range	1976 Average[a]
Dairy products	1–6	trace (5/30)	1–14	1 (4/30)	trace	trace (4/20)	1–2	0.2 (3/20)
Meat, fish, and poultry	1–6	1 (12/30)	1–6	2 (21/30)	trace	trace (11/20)	1–3	1.0 (17/20)
Grains and cereals	2–5	1 (30/30)	2–5	3 (29/30)	5–8	trace (19/20)	2–5	3.0 (20/20)
Potatoes	2–12	5 (30/30)	2–13	5 (29/30)	5–12	4 (20/20)	2–9	5.0 (20/20)
Leafy vegetables	1–28	5 (30/30)	1–14	4 (28/30)	5–14	5 (20/20)	2–10	4.0 (19/20)
Legumes	1–3	trace (10/30)	1–10	1 (8/30)	trace	trace (3/30)	1–7	1.0 (14/20)
Root vegetables	1–6	2 (24/24)	1–31	3 (24/30)	trace	trace (16/20)	1–8	2.7 (19/20)
Garden fruits	1–6	2 (25/25)	1–10	2 (23/30)	trace	trace (17/10)	1–4	2.0 (18/20)
Other fruits	1–2	trace (4/30)	1–6	trace (3/30)	trace	trace (5/20)	1–2	0.3 (5/20)
Oils, fats, shortening	1–6	3 (29/30)	1–7	2 (24/30)	trace	trace (17/20)	1–3	1.6 (18/20)
Sugars and adjuncts	1–6	1 (13/30)	1–9	1 (12/30)	trace	trace (8/20)	1–3	1.1 (14/20)
Beverages	1–8	trace (5/30)	1–3	trace (6/30)	trace	trace (1/20)	0–1	0.2 (3/20)

[a] For numbers in parentheses, numerators represent positive composites; the denominators, the total number of composites analyzed.

cadmium compounds are very unstable. In contrast to lead and mercury ions, cadmium ions are readily absorbed by plants. They are equally distributed over the plant.

Foods of animal origin that can be contaminated with cadmium, include liver, kidney and milk. Table 4.3 shows the cadmium content of foods in the US.

Cadmium accumulates in the human body, especially in the liver and kidney. In experimental animals it can cause anemia, hypertension, and testicular damage. Chronic cadmium intoxication in humans occurred in Japan after the consumption of rice heavily contaminated as a result of environmental pollution. 0.1 to 1 mg/day were ingested for a period of possibly more than 12 years. The painful disease that developed was characterized by skeletal deformation, reduced body height and multiple fractures. Vitamin D deficiency appeared to be a predisposing factor in this case.

The intake of cadmium in the US from 1982 to 1991 ranged from 3.7 to 14.4 μg/day (also determined on the basis of data obtained in FDA Total Diet Studies). The absorption from food varies, depending on genetic factors, age, and nutritional factors. Infants absorb and accumulate more cadmium than adults. Calcium or iron deficiency can increase the absorption of cadmium. Pyridoxine deficiency appears to decrease its absorption. The TWI is 0.4 to 0.5 mg. The US has set a safety limit for drinking water and bottled water: 0.005 mg/l. For foods, no limit values have been set.

4.3   Nitrate

Contamination of the biosphere with nitrogen compounds can result in a nitrate concentration increase in groundwater. This can ultimately lead to increases in the nitrate concentration of drinking water as well as in the nitrate level of food (of plant origin).

Public waterworks use both groundwater and surface water as sources of drinking water. At the moment, it is not common practice to remove nitrate during drinking water production. Where there is no connection with the water system, groundwater is also used as a source of drinking water through private wells.

There are a number of nitrate sources in the soil. For example, nitrate can originate from microbial fixation of nitrogen in symbiotic relationships with leguminous plants, i.e., from the only biological way of binding nitrogen. Other sources are soil pollution caused by the use of fertilizers in agriculture and manure production in cattle breeding and dairy farming. Microbial nitrification is responsible for the conversion of ammonia and urea to nitrate in the soil. The toxicological risks due to intake of nitrate are attributed to its reduction product nitrite.

Nitrite can oxidize hemoglobin to methemoglobin. In acidic environments, it may react with secondary amines under the formation of nitrosamines. Numerous alkyl- or alkylarylnitrosamines are carcinogenic in experimental animals. Nitrite and nitrosamines can be formed *in situ* in food, but also in the body (after ingestion of nitrate). Oral and intestinal bacteria can convert nitrate to nitrite.

Certain vegetables tend to accumulate nitrate: beets, celery, lettuce and spinach. The nitrate levels in some foods are listed in Table 4.4.

The intake of nitrate via food consumption is estimated at 1.4 to 2.5 mg/kg/day, and from water at 0.3 mg/kg/day. The acceptable daily intake (A.D.I.) of nitrate is 3.64 mg/kg/day. (See Part 3, Chapter 17, Sections 17.3.2 and 17.3.3)

4.4   2,3,7,8-Tetrachlorodibenzo-p-dioxin

2,3,7,8-Tetrachlorodibenzo-*p*-dioxin (TCDD) is a well-known environmental pollutant, formed from chlorinated hydrocarbons at the high temperatures reached in incinerators. Further, it is also a contaminant of the herbicide 2,4,5-trichlorophenoxyacetic acid (2,4,5-T).

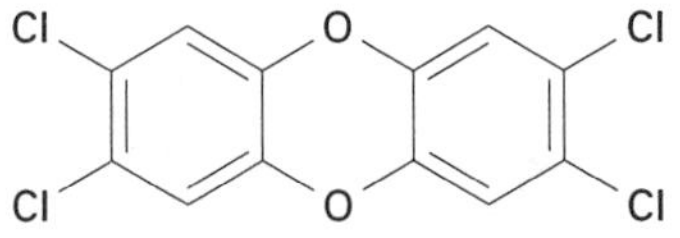

Tetrachlorodibenzo–*p*–dioxin (TCDD)

TCDD is highly toxic on acute exposure. The oral LD_{50} in guinea pigs of either sex is 1 µg/kg. It induces a variety of adverse effects in experimental animals: liver damage, porphyria, teratogenic effects, immune suppression and increased tumor incidence. It also causes enzyme induction. In man, the following effects have been reported (based on occupational exposures and industrial accidents): chloracne, porphyria, liver damage and polyneuropathies.

The main route of exposure to TCDD is dietary intake. TCDD can reach food via:

(a)   spraying of crops with 2,4,5-T;
(b)   ingestion of contaminated feed by livestock;
(c)   magnification via food chains;
(d)   contamination of fruits and vegetables in the proximity of incinerators.

Table 4.4 Average nitrate contents of common foods in the US and per capita daily intake

Food	Nitrate, mg/100 g	
	Content	Ingestion
Total vegetables	1.3–27.6	8609.1
Asparagus	2.1	2.8
Beet	276	546.0
Beans, dry	1.3	10.0
Beans, lima	5.4	6.6
Beans, snap	25.3	258.0
Broccoli	78.3	127.0
Cabbage	63.5	548.0
Carrot	11.9	104.0
Celery	234.0	1600.0
Corn	4.5	77.0
Cucumber	2.4	7.8
Eggplant	30.2	14.8
Lettuce	85.0	1890.0
Melon	43.3	935.0
Onion	13.4	159.0
Peas	2.8	19.8
Pepper, sweet	12.5	33.5
Pickles	5.9	56.0
Potato	11.9	1420.0
Potato, sweet	5.3	26.4
Pumpkin/squash	41.3	38.0
Spinach	186.0	420.0
Sauerkraut	19.1	33.2
Tomato and tomato products	6.2	198.0
Breads	2.2	198.0
All fruits	1.0	130.0
Juices	0.2	10.7
Cured meats	20.8	1554.0
Milk and milk products	0.05	25.0
Water	0.071	71.0

Data on the dietary intake of TCDD are scarce. An indication for the body burden may be the 2 ppt, measured in mother's milk.

The ADI is set at 10 pg/kg/day (by a World Health Organization Expert Committee).

4.5   Pesticide residues

Pesticides are chemicals developed and produced for use in the control of agricultural and public health pests. The main groups of pesticides are insecticides, herbicides, and fungicides. Pesticides are of vital importance in the fight against diseases, e.g., malaria, and for the production and storage of food. In spite of their extensive use, an average of 35% of the produce is lost worldwide.

Common classes of pesticides include organochlorine compounds, organophosphates, and carbamates.

Many members of the various classes are highly toxic. A common misconception is that pesticides have the same mode of action. The ways in which they act are as diverse as their chemistry. Chlorinated cyclodiene insecticides (e.g., aldrin) are neurotoxicants that interfere with γ-aminobutyric acid transmitters in the brain. In humans and experimental animals, seizures have been reported, in addition to symptoms such as nausea, vomiting, and headache. The toxicity mechanism of the chlorophenoxy herbicides 2,4-dichlorophenoxyacetic acid and 2,4,5- trichlorophenoxyacetic acid is poorly understood. They induce their herbicidal effects by acting as growth hormones in plants. However, they do not act as hormones in experimental animals. In animals, effects such as stiffness of the extremities, inability to coordinate muscular movements, paralysis, and eventually coma have been observed.

2, 4 – Dichlorophenoxyacetic acid   2, 4, 5 – Trichlorophenoxyacetic acid

The organophosphorous insecticides (e.g., parathion) inhibit acetylcholinesterase, resulting in symptoms (that mimic the action of acetylcholine) such as lachrymation, pupillary constriction, convulsions, respiratory failure, and coma.

Carbamate herbicides such as propham (isopropyl-N-carbanilate) have relatively low acute toxicities. The oral LD_{50} of propham in rats is 5 g per kg. Herbicidal carbamates are not inhibitors of cholinesterase.

Parathion   Propham (isopropyl carbanilate)

The toxicological risks from residues of s   ynthetic pesticides in foods are minimal because of careful food safety legislation and regulation. Contamination of vegetables may result from treatment as well as from conditions such as improper use of pesticides, residues from preceding treatments in the soil and cross-contamination (particularly during harvesting). Sources of residues in products of animal origin include contaminated water or feed, pesticide-treated housing, and contaminated milk (during weaning).

Table 4.5 lists the pesticide residue levels in food in the US.

Organochlorine insecticides deserve particular attention, as they are very stable and can accumulate in food chains. Products of animal origin as well as mother's milk almost always contain residues of organochlorine compounds. The residue content of mother's milk is 10 to 30 times higher than that of cow's milk.

From May 1990 through July 1991, 806 milk samples from 63 metropolitan areas in the US were collected and analyzed for pesticide residues by the FDA. In the samples from eight of the metropolitan areas, no residues could be detected. Pesticide residues appeared to contaminate 398 milk samples though. The most frequently occurring residues were *p,p'*-DDE (4,4'-dichlorodiphenyltrichloroethane) (in 212 samples) and dieldrin (in 172 samples). The highest residue level measured was 0.02 ppm *p,p'*-DDE (whole milk basis). These chlorinated pesticides have not been registered for agricultural use for about 20 years.

Table 4.5 Pesticide residues in food in the US in 1991

Food	Origin	Number of samples	Samples with residues below permissible level in %	Samples with residues above permissible level in %
Grains/grain products	Domestic	495	40.8	0.8
	Import	396	25.5	2.3
Milk/dairy products/eggs	Domestic	809	12.5	0
Milk/dairy products	Import	216	10.2	0
Fish/shellfish/other meats	Domestic	536	41.6	0.2
Fish/shellfish	Import	611	23.2	0.2
Fruits	Domestic	2168	50.9	0.5
	Import	3481	34.1	1.3
Vegetables	Domestic	3811	30.6	1.3
	Import	4311	28.3	3.3
Other	Domestic	462	19.5	0
	Import	918	17.9	3.5

Source: Food and Drug Administration Pesticide Program, Residue Monitoring 1991 (5th annual report).

The use of organochlorine compounds is decreasing in favor of that of organophosphates and carbamates. Both latter classes of pesticidal chemicals are much more readily degraded, in the environment as well as during processing.

Many of the techniques presently used in food processing give a considerable reduction of pesticide residue levels. Many types of residues are degraded to harmless products during processing due to heat, steam, light, and acid or alkaline conditions. In addition, major reductions of residue levels result from their physical removal by peeling, cleaning or trimming of foods such as vegetables, fruits, meat, fish and poultry.

Table 4.6 lists the results of the Total Diet Study 1991 on the occurrence of pesticides in food. In general, residues present at or above 1 ppb could be measured. Malathion continues to be the residue most frequently found; it is used on a wide variety of crops, including many post-harvest uses on grains. From 1987 to 1991, the occurrence of malathion has decreased from 23 to 18% (see Table 4.6), and that of DDT from 22 to 10%.

4.6   Food contaminants from packaging material

Contact of packaging material with food may result in the transfer of trace quantities of particular chemicals, such as monomers and plasticizers. Well-known chemicals used in the production of polymers are vinyl chloride and styrene. Vinyl chloride is the monomer of polyvinyl chloride, and styrene is used in the manufacturing of a number of plastics. Important plasticizers in polyvinyl chloride plastics are the phthalic acid esters di-(2-ethylhexyl) phthalate (DEHP) and di-*n*-butyl phthalate (DBP).

$CH_2 = CHCl$

Vinyl chloride

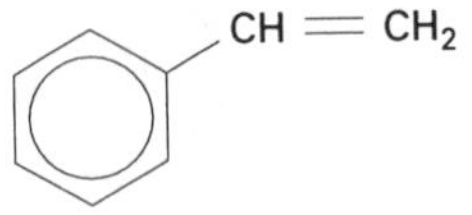

$CH = CH_2$

Styrene

Table 4.6 Occurrence of pesticides in total diet study in 1991

Pesticide[a]	Number of food items contaminated with	Occurrence %[b]
Malathion	167	18
Chlorpyrifos-methyl	97	10
DDT	93	10
Dieldrin	73	8
Endosulfan	67	7
Methamidophos	58	6
Chlorpyrifos	51	5
Dicloran	44	5
Acephate	42	4
Diazinon	42	4
Dimethoate	34	4
Chlorpropham	28	3
Heptachlor	24	3
Lindane	22	2
Omethoate	22	2
Ethion	21	2
Hexachlorobenzene	20	2
Permethrin	16	2
BHC, alpha	13	1
Chlordane	12	1
Parathion	12	1
Quintozene	12	1
Dicofol	10	1

[a] Including parent compounds, isomers, metabolites and related compounds.
[b] On the basis of 936 items. NB: a food item can contain several pesticides.

Dibutyl phthalate   Di(2 – ethylhexyl) phthalate

Vinyl chloride has been identified as a liver carcinogen in animal models as well as in humans. Acute intoxication causes depression of the central nervous system and hepatic damage. Vinyl chloride leaches out of packaging materials into water as well as into fatty material. Mineral water (stored in polyvinyl chloride bottles) has been shown to take up vinyl chloride. After 6 months, a concentration of 170 mg per l was measured. This may lead to a daily intake of 120 ng per person in countries where polyvinyl chloride bottled drinking water is used. In cooking oils, higher concentrations have been found, viz. 14.8 mg/kg.

Styrene-induced toxic effects include renal and hepatic damage, pulmonary edema, and cardiac arrhythmia. The oral LD_{50} in rats is relatively low: 5 g/kg. Styrene appears to leach out of polystyrene packaging material, preferably into the fatty components of food.

Average concentrations of 27 ppb have been measured in high-fat yogurt, 71 ppb in fruit yogurt, 20 to 70 ppb in other desserts, 18 to 180 ppb in meat products and 5 ppb in packed fruit and vegetable salads. For styrene, a provisional ADI of 40 ng per kg has been calculated.

The phthalic acid esters DEHP and DBP have low acute toxicities. The intraperitoneal LD_{50} in mice are 14.2 and 4.0 g/kg, respectively. However, liver or lung damage by the leached plasticizers has been suggested. DEHP and BBP appear to be non-genotoxic carcinogens.

Since they are widely distributed in materials involved in transportation, construction, clothing, medicine, and packaging, the concern about their health effects has increased.

4.7 Summary

Food contaminants are substances unintentionally included in foods. Some are harmless but others may be hazardous. Contamination can occur at every step on the way from raw material to consumer. Raw materials of plant origin can be contaminated with environmental pollutants, such as heavy metals, pesticide residues, industrial chemicals, and products from fossil fuels. The sources of contaminants from raw materials of animal origin are to a large extent comparable with those from raw materials of plant origin. During processing, food can be contaminated with processing aids (filter and cleaning agents) and equipment materials (e.g., metals), and during packaging and storage with components of the packaging material. A number of important examples of hazardous food contaminants originating from the above sources were dealt with, namely heavy metals (mercury, lead, and cadmium), nitrate, 2,3,7,8-tetrachlorodibenzo-*p*-dioxin, pesticides, vinyl chloride, styrene, and plasticizers di-(2-ethylhexyl) phthalate and di-*n*-butyl phthalate.

Reference and reading list

Belitz, H.-D. and W. Grosch, (Eds.), *Food Chemistry*. Berlin, Springer Verlag, 1987.

Concon, J.M., (Ed.), *Food Toxicology, Part A and Part B*. New York, Marcel Dekker Inc., 1988.

Culliney, T.W., D. Pimentel, M.H. Pimentel, Pesticides and natural toxicants in foods, *Agric. Ecosys. Environ.*, 41, 297–320, 1992.

Farris, G.A., P. Cabras, L. Spanedda, Pesticides residues in food processing, *Ital. J. Food Sci.*, 4, 149–159, 1992.

Gilbert, J., The fate of environmental contaminants in the food chain, *Sci. Total Environ.*, 143, 103–111, 1994.

Gormley, T.R., G. Downey and D. O'Beirne, (Eds.), *Food, Health and the Consumer*. Amsterdam, Elseviers Applied Sciences, 1987.

Gosting, D.C., (Ed.), *Food Safety 1990; An Annotated Bibliography of the Literature*. London, Butterworth-Heinemann, 1991.

Hathcock, J.N., (Ed.), *Nutritional Toxicology, Vol. I*. London, Academic Press, 1982.

Hotchkiss, J.H., Pesticides residue controls to ensure food safety, *CRC Crit. Rev. Food Sci. Nutr.*, 31, 191–203, 1992.

Tannenbaum, S.R., (Ed.), *Nutritional and Safety Aspects of Food Processing*. New York, Marcel Dekker Inc., 1979.

Walters, C.L., Reactions of nitrate and nitrite in foods with special reference to the determination of nitroso compounds, *Food Add. Contam.*, 9, 441–447, 1992.

chapter five

Food additives

M.M.T. Janssen

5.1   Introduction

Food additives are those substances that are intentionally added to food for maintaining or improving its appearance, texture, flavor, and nutritional value, as well as for the prevention of microbial spoilage. This includes any substance intended for use in manufacturing, processing, preparing, packaging, transporting, or keeping food. Evaluation of the safety of the intended use of additives in food is extensively provided for by regulation and legislation.

In compliance with safety requirements, and if no other economic or technically practical means are available, the following may serve as justification for the use of additives in food (according to the Joint FAO/WHO Codex Alimentarius Committee):

(a)   preservation of the nutritional value of food. Reduction of the nutritional value of food is justified in the case of consumers with specific dietary needs, and if the food is not an essential component of the diet;
(b)   use in special food for consumers with specific dietary needs;
(c)   improvement of the stability, the organoleptic properties and the nutritional value of food while its nature is not drastically changed;
(d)   use in manufacturing, processing, transport and storage of food, but not with the intention of disguising the use of inferior raw materials, undesirable practices, and techniques.

Five main categories of additives are distinguished:

–   texturizing agents: gelling agents, thickeners and emulsifiers;

- colorings;
- flavoring agents: flavors, flavor enhancers and non-nutritive sweeteners;
- preservatives: antioxidants and antimicrobials;
- miscellaneous additives, such as anticaking agents, catalysts, clarifying agents, filter aids, and solvents.

This chapter considers the use of a number of selected food additives in relation to their safety. The use of the majority of the texturizing agents and miscellaneous additives do not pose toxicological risks.

5.2   Use of food additives in relation to their safety

Food additives have been used since prehistoric times to maintain and improve the quality of food products. Smoke, alcohol, vinegar, oils, and spices have been used for more than 10,000 years to preserve foods. These and a small number of other chemicals, such as salts, copper, and chalk, were the major food additives used until the time of the Industrial Revolution. Since then many changes in food manufacturing and food distribution have taken place as a result of urbanization, the decrease in opportunity for individual families to grow their own food, and the increase in the consumer's demand for a broader food assortment and a higher quality of food. Also, food had to be produced on a larger scale. Distribution over long distances involves a longer span of time between production and consumption. Further, food needs to be stored in warehouses and shops, and also at home. In addition, convenience foods require extensive preservation.

Processed food is more perishable than the individual food components themselves. This can be overcome by the use of food additives. Without these, the food choice would be limited, many food products would be prohibitively expensive, and much food would be wasted. Also, food-related poisonings would occur more often. All these factors together have led to an increased use of additives in food, particularly since the 1950s. More than 2500 different chemicals are now in use. Apart from the consumption of salt and sugar, which are also important preservatives, the yearly additive consumption per capita in the early 1960s was estimated to exceed 3 lbs. However, the demand for new, tasty, convenient and nutritious foods continued to increase. In the US, where this development is most pronounced, the additive consumption per capita has increased from 3 to 9 lbs per year. Besides being beneficial, the use of food additives may also involve adverse health effects which can be either indirect or direct. Indirect effects are concerned with unbalanced diets and direct effects with potential toxicity.

The *indirect* health effects of additives are the opposite of some of their beneficial effects. The use of additives has led to a wider food assortment, but also to an increased availability of food with a low nutrient content. This type of food (so-called junk food) can be (and often is) consumed as dietary substitute for more nutritious food. Obviously, educational programs are needed to alert consumers to the need for a balanced diet.

The *direct* effects include short-term as well as long-term toxic effects. Short-term effects of additives are unlikely because of the low levels at which they are applied. On the other hand, hypersensitivity has been attributed to additives, even if they are used at legally acceptable levels. Further, little or no data are available on the health risks from the daily intake of combinations of additives.

Toxicological problems after long-term consumption of additives are not well-documented. There is no conclusive evidence for the relationship between chronic consumption of food additives and the induction of cancer and teratogenic effects in humans. Results of animal studies, however, have suggested that the use of certain additives involves safety problems. Most of these additives are now banned.

Nowadays, food additives undergo extensive toxicological screening before they are admitted for use. However, the majority of additives already in use are believed to be safe for the consumer at the levels applied in food, even though they have not been examined toxicologically. The substances involved are of natural origin and traditionally have been in use since the early days of food processing. Many additives that are used by the consumer in preparing food in the natural matrix, e.g., pectin as thickener, egg yolk as emulsifier, tomato juice as flavor enhancer, and lemon juice as antioxidant, are used in the food industry in a purified form.

The search for new and safer additives to replace debatable ones, and for processing techniques that require fewer additives, continues.

5.2.1 Colorings

Colorings are used to improve the overall attractiveness of food.

Food colors may be of natural as well as synthetic origin. About 50 colors of natural origin and their derivatives are in use, including chlorophylls (green), carotenoids (yellow, orange, and red) and anthocyanins (purple). They have all been toxicologically evaluated. This section deals with synthetic colorings only.

Synthetic colorings are superior to natural pigments in tinctorial strength, brightness, and stability. After the discovery of the first synthetic dye in 1856, a wide variety of colorings became rapidly available. By the end of the 19th century, 80 colorings were in use. In the first decade of this century, most of these substances were prohibited by law on the basis of their composition and purity.

The toxicology of synthetic food colorings was not given any attention until the early 1930s, when 4-dimethylaminoazobenzene was found to be carcinogenic. This dye was used to color butter and margarine yellow, hence its name "butter yellow." Since then other dyes have proved to be toxic and, as a consequence, have been banned from addition to food. Currently, only 9 synthetic colorings are allowed in the US and 11 in the EU. The majority belong to the class of the azo dyes. A few typical examples are discussed below: *amaranth* and *tartrazine*.

Amaranth, (trisodium 1-(4-sulfo-1-naphthylazo)-2-naphthol-3,6-disulfonic acid) has been approved for use as food color in several countries, including the member states of the EC. It is a water-soluble red dye.

Amaranth

In many long-term studies on carcinogenicity, amaranth has been found to be safe. It is used in food products, such as packaged soup, packaged cake and dessert mix, and canned fruit preserves. In the USA, however, amaranth is no longer in use. The reason for this is the development of tumors in rats fed on a diet containing 3% amaranth.

Tartrazine (5-hydroxyl-1-(*p*-sulfophenyl)-4-(*p*-sulfophenylazo)pyrazole-3-carboxylic acid) is a yellow food coloring.

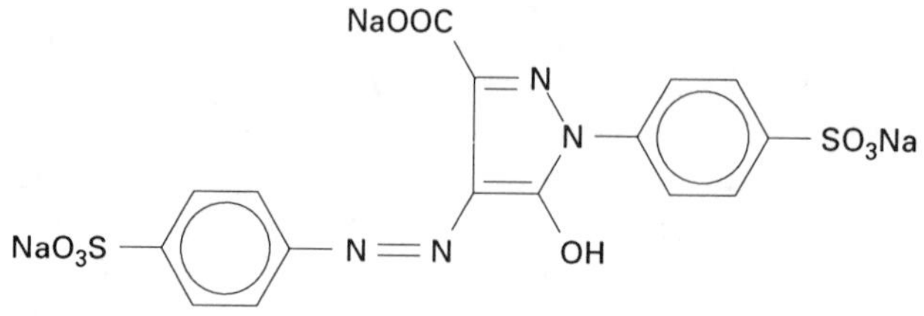

Tartrazine

Tartrazine is widely used in foods, such as the packaged convenience foods mentioned above, smoked fish, chewing gum, sweets, beverages, and canned fruit preserves. The dye has undergone extensive testing, and was found to be harmless in experimental animals. However, various types of allergic reactions are attributed to tartrazine. As little as 0.15 mg can elicit an acute asthmatic attack in sensitive persons. The average daily intake of tartrazine is estimated at 9 mg/kg body weight in the US, while the ADI is 7.5 mg/kg body weight.

5.2.2   *Flavoring agents*

Flavor has a profound influence on the consumption of food. It imparts that quality to products by which they distinguish themselves. Flavoring agents make up the largest number of food additives. There are three types of flavoring additives: flavorings, flavor enhancers, and (non-nutritive) sweeteners. More than 1500 substances are used as *food flavorings*. The majority are of natural origin or are nature-identical, and do not give rise to concern from a safety point of view. Only a few synthetic substances have been approved as food flavoring. Examples are ethylvanillin, ethylmaltol, and anisylacetone.

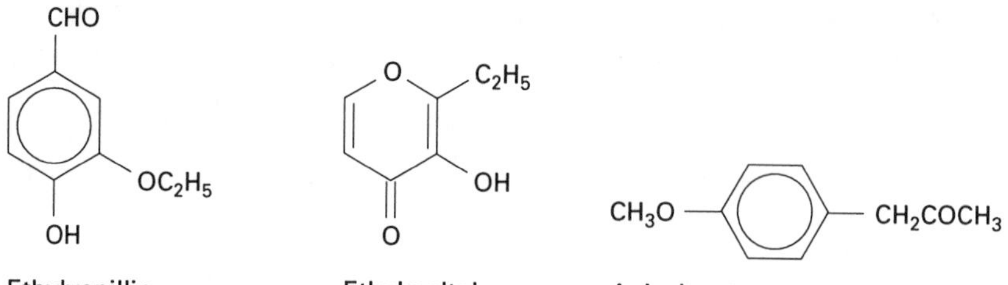

Ethylvanillin   Ethylmaltol   Anisylacetone

Flavor enhancers intensify or modify the flavor of food. They have no taste of their own. They include substances such as monosodium glutamate (MSG) and various nucleotides. These substances are present in Japanese seaweed (traditionally used for seasoning), mushrooms, tomatoes, peas, meat, and cheese. They are often used in soups, sauces and oriental food. No known adverse effects of flavor enhancers have been reported, except for the case of MSG. Humans have been described to be sensitive to food to which MSG had been added. The symptoms include numbness, general weakness, and heart palpitations (see also Part 2, Chapter 2).

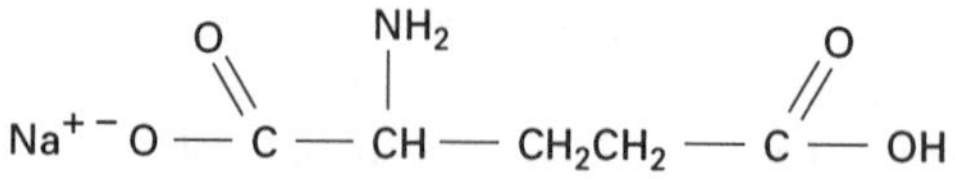

Monosodiumglutamate (MSG)

Sweeteners present the consumer with one of the most important taste sensations. This is reflected by the world production of sugar, which has increased from 8 million tons in 1900 to 70 million tons in 1970. For nutritional and health reasons, however, there is a growing need for sugar substitutes in food that are non-nutritive, i.e., noncaloric, and noncariogenic. Two important noncaloric synthetic sweeteners are saccharin and aspartame.

Saccharin

Aspartame

In the US *saccharin* has been used commercially since 1900. It is 300 times sweeter than saccharose and very stable under almost all food processing conditions. Since World War II the consumption of saccharin has steadily increased even though its safety has been questioned repeatedly. Almost 50% of its use is in soft drinks. Individual use as table top sweetener amounts to approximately 20%. The average consumption of saccharin in the US for the whole population has been estimated at 7.1 mg/day per capita, while the intake by the subpopulation of saccharin consumers was 25 mg/day. In Europe, the average intake has been reported to be 15 mg/day.

Since the beginning of its short commercial history, saccharin has been suspected regarding its safety. In 1912 it was prohibited in the US on the basis of acute toxicity tests. However, the ban was lifted during World War I, as sugar became short in supply. After World War II, numerous studies on the toxicity of saccharin were carried out. Up to now, no mutagenicity has been found. However, long-term animal tests showed a higher incidence of bladder cancer. Although it is difficult to extrapolate from experimental animals to the human situation, it appears unlikely that the intake of saccharin at the present average level involves risks of cancer. Therefore, the use of saccharin in food is still approved in the US and in Europe.

Aspartame was discovered in the early 1960s. In the early 1980s, it was admitted in many countries as a sweetener, in addition to saccharin and cyclamate, another synthetic sweetener whose use in food has now been greatly restricted. Aspartame is a dipeptide, consisting of the amino acids phenylalanine and aspartic acid. It is 200 times sweeter than saccharose and is an excellent sweetener for dry products. At high temperature and low pH, aspartame is gradually hydrolyzed, losing its sweetness. It is suitable as table top sweetener, in chewing gum, in soft drinks, dairy products, ice cream, and dessert mixes. Since aspartame is a dipeptide, it is digested and absorbed by the body. However, the amount necessary for a sweet taste is so small that the energy produced is believed to be irrelevant.

Results from toxicity tests suggest that aspartame has no adverse effects on humans even when extreme amounts of 8 mg/kg body weight are taken in. The ADI for aspartame is 40 mg/kg body weight.

The market for sweeteners is still growing and the situation where the ADI for the known sweeteners is reached, is not inconceivable. There is, however, a need for sweeteners that are stable under specific technological conditions and are less controversial than saccharin. Although since the introduction of aspartame the use of saccharin has slowly declined, aspartame can not replace saccharin completely because of its instability when heated under acidic conditions. Therefore, the search for new sweeteners continues. At present, several non-nutritive sweeteners of natural origin are being investigated. Examples are thaumatin, a macromolecular protein sweetener from an African fruit, 2000 to 3000 times as sweet as saccharose and neohesperidin, present in orange peel, 1500 times as sweet as saccharose. Thaumatin has been admitted in the EU for use in chewing gum and sweets.

5.2.3   Preservatives

Preservatives are added to decrease the degradation rate of foods during processing and storage. They include antioxidants, antimicrobials and antibrowning agents.

5.2.3.1   Antioxidants

Antioxidants primarily prevent or inhibit autoxidation of fatty acids (see also Part I, Chapter 6) in food products and, consequently, the development of rancidity and off-flavor. They are especially useful in preserving dry and frozen foods for long periods of time. The major antioxidants for the protection of dietary fats and oils are phenols. They are either synthetic or natural substances. The synthetic antioxidants include butylated hydroxyanisole (BHA), butylated hydroxytoluene (BHT), *n*-propyl gallate and tertiary-butyl hydroquinone (TBHQ). Important natural food antioxidants are the *tocopherols*. They occur naturally in the majority of fats and oils. The mechanism of the antioxidant action of phenols is shown in Figure 5.1.

$$AH + ROO^{\bullet} \longrightarrow ROOH + A^{\bullet}$$

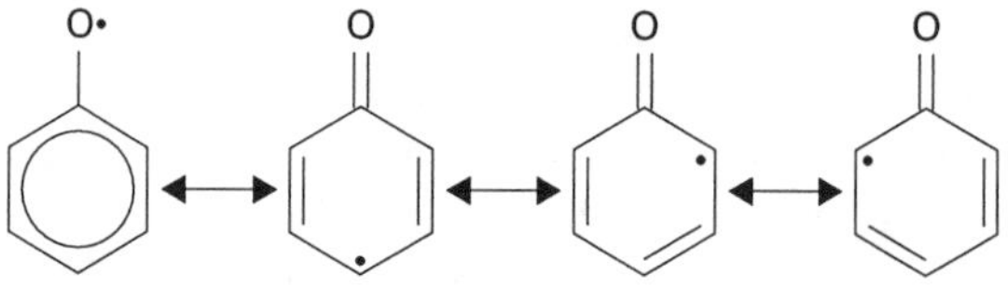

Figure 5.1   Diagrammatic representation of the mechanism underlying the antioxidant action of phenolic antioxidants.

Generally, *BHA* was not believed to be a hazardous substance. However, the results of recent studies in experimental animals suggest that the intake of BHA involves a cancer risk.

The case of *BHT* is more complex. There is evidence that BHT promotes several types of chemical carcinogenesis in a number of experimental animals. Further, liver damage and cytotoxic effects have been found.

Accurate data on the daily intake of BHA and/or BHT by man are not available. Generally, estimates are in the range of 1 to 5 mg/day. Estimated total use in the US of BHA is 143,000 lb/year and of BHT 670,000 lb/year. In the case of *propyl gallate*, no evidence of carcinogenicity, mutagenicity or teratogenicity has been provided. BHA, BHT and propyl gallate are almost universally accepted for use in food since the 1950s.

TBHQ is the most recently developed synthetic food antioxidant. It has been designed especially to protect polyunsaturated oils. In long-term animal feeding tests, no indications of carcinogenicity were obtained. As yet, TBHQ is allowed for use in food in the US and a few other countries, but not in the EU.

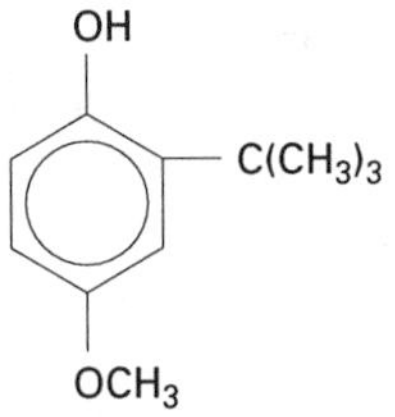

BHA OCH₃

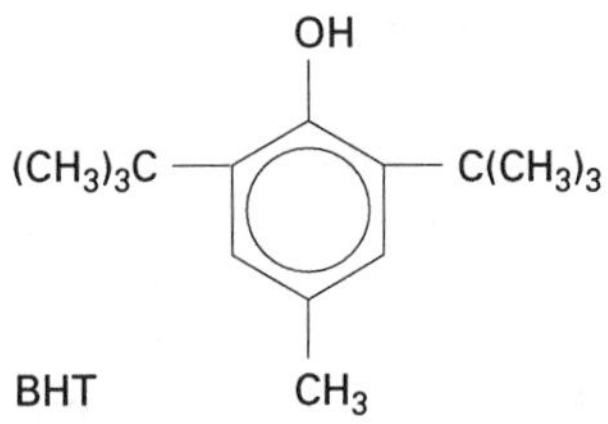

BHT CH₃

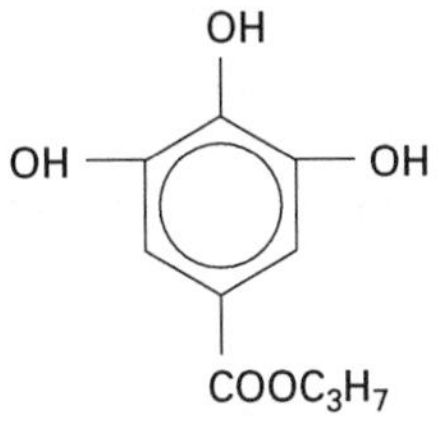

Propyl gallate
(3, 4, 5 – trihydroxybenzoic
acid propyl ester)

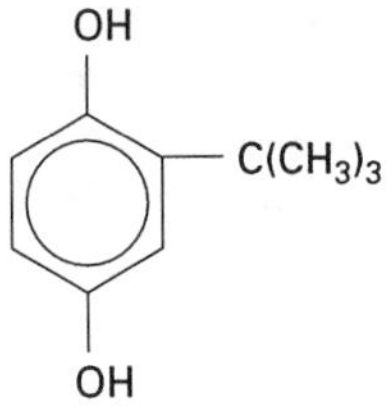

TBHQ

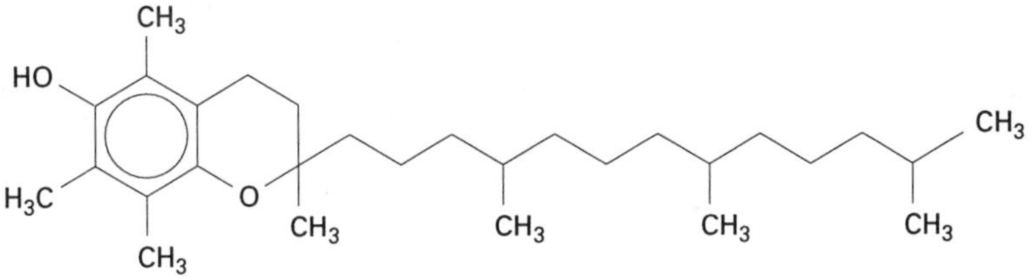

α – Tocopherol (vitamin E)

There are many phenols of natural origin which are strong antioxidants. Sometimes, they are more effective than the major synthetic phenolic antioxidants. At present, *tocopherols* and *rosemary extract* are commercially available.

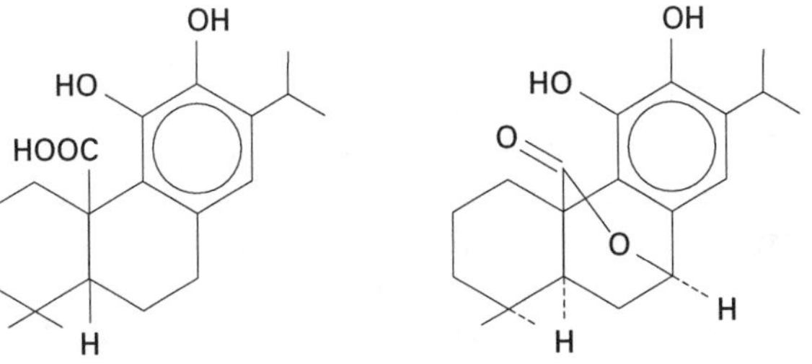

Antioxidants in rosemary extract

5.2.3.2 Antimicrobials

Antimicrobials are used to prevent or inhibit the growth of microorganisms. They play a major role in prolonging the shelf life of foods. Nowadays, consumers expect all foods to be available all year round and to have a fairly long shelf life. In dietary behavior, however, risks from microbial contamination are often overlooked (see Part 1, Chapter 2, Section 2.3.3).

As far as food safety from a microbiological viewpoint is concerned, some advances have been made without calling in the help of additives. These involve the application of certain packaging and processing methods. Nevertheless, the use of chemical antimicrobials is indispensable for safe food handling. Common antimicrobial food additives are benzoic acid and benzoates, sorbic acid and sorbates, short-chain organic acids (acetic acid, lactic acid, propionic acid, citric acid), parabens (alkyl esters of *p*-hydroxybenzoic acid), sulfite, and nitrite. Most of these substances are believed to be safe for application in food. They are easily excreted and metabolized by both animal and man. An exception should be made for one of them, namely nitrite. The intake of nitrite can lead to the formation of nitrosamines, which are well-known carcinogens.

Nitrite has been used as meat preservative for many centuries. It contributes to the development of the characteristic color and flavor, to the improvement of the texture of meat products, such as bacon, ham, frankfurters, fermented sausages, and canned meats, and also of fish and poultry products. Its antimicrobial effect was not recognized until the late 1920s. The primary aim of using nitrite as an antimicrobial is to prevent germination of the spores of *Clostridium botulinum* and hence the production of the botulinum toxin (see also Chapter 2).

Prolonged ingestion of sodium nitrite has been shown to cause methemoglobinemia, especially in infants. The major adverse effect of nitrite intake is the induction of cancer. In many animal species, this is attributed to the formation of nitrosamines in the reaction of nitrite with secondary amines. Nitrosamine formation can take place in the food itself as well as in the body. The normal acidity of the stomach is ideal for nitrosamine formation.

From a food toxicological point of view, three types of nitrosamines are of importance: dialkyl nitrosamines, acylalkylnitrosamines, and nitrosoguanidines. Cyclic nitrosamines are similar to the dialkyl type. The nitrogen atom becomes part of the heterocyclic ring. Nitrosoguanidines are a special class of highly reactive nitrosamides.

general structure of
nitrosamines

general structure of
alkylacylnitrosamines

general structure of
N – alkyl (R) – N′ – alkyl (R′) –
N – nitrosoguanidines

The hazards due to nitrosamines in food depend strongly on the types and levels of precursors present. Precursors can be endogenous substances, products of food components, and endogenous substances, and also contaminants. Tables 5.1 and 5.2 list nitrosamine precursors and the corresponding nitrosamines that can be formed.

Since many nitrosable substances are formed on degradation of proteins and amino acids, nitrosamine formation cannot always entirely be prevented in food and in the body. One of the most effective inhibitors of nitrosation is ascorbic acid. This vitamin reacts rapidly with nitrite to form nitric oxide and dehydroascorbic acid. In that way, it can inhibit the formation of dimethylnitrosamine by more than 90%. Other inhibitors of nitrosation are gallic acid, sodium sulfite, cysteine, and tannins. Nitrosamine levels in food also depend on the temperature at which food is prepared. Table 5.3 gives some examples of nitrosamine levels in foodstuffs. Cooking can increase the nitrosamine level in food. As can be seen from Table 5.3, frying can increase the nitrosamine level in bacon quite considerably. Up to 135°C, cooking or frying does not result in detectable nitrosamine formation. Above 175°C, however, the nitrosamine levels increase rapidly.

Nitrite addition to fresh meat and food products is still under discussion because of the earlier-mentioned toxicological hazards. Up to now, banning of this additive has been blocked by the food industry. It is stressed that so far no other antimicrobial agent has been

Table 5.1 Nitrosamine precursors, endogenous or formed in food

Compound	Food	Nitrosamine formed
Creatine, creatinine	Meats, meat products, milk, vegetables	Nitrososarcosine (NSA)
Trimethylamine oxide	Fish	Dimethylnitrosamine (DMN)
Trimethylamine	Fish	DMN
Dimethylamine	Fish, meat, and meat products, cheese	DMN
Diethylamine	Cheese	Diethylnitrosamine (DEN)
Sarcosine	Meat and meat products, fish	NSA
Choline, lecithin	Eggs, meat and meat products, soybeans, corn	DMN
Proline, hydroxyproline	Meat and meat products, other foodstuffs	Nitrosoproline and nitrosopyrrolidine (NPyr)
Pyrrolidine	Meat and meat products, paprika	NPyr
Piperidine	Meat and meat products, cheese, black pepper	Nitrosopiperidine (NPip)
Methylguanidine	Beef, fish	Methylnitrosourea
Carnitine	Meat and meat products	DMN
Dipropylamine	Cheese	Di-*n*-propylnitrosamine
Dibutylamine	Cheese	Di-*n*-butylnitrosamine

Table 5.2 Nitrosamine Precursors which Contaminate Foodstuffs

Compound	Chemical class	Nitrosamine derivative
Atrazine	Secondary Amine	N-Nitrosoatrazine
Benzthiazuram	Carbamate	N-Nitrosobenzthiazuram
Carbaryl	Carbamate	Nitrosocarbaryl
Fenuron	Carbamate	DMN
Ferbam	Amide	DMN
Morpholine	Secondary Amine	Nitrosomorpholine
Propoxur	Carbamate	Nitrosopropoxur
Simazine	Secondary Amine	Nitrososimazine
Succinic acid 2,2-dimethyl hydrazide	Amide	DMN
Thiram	Amide	DMN
Ziram	Amide	DMN

Note: See Table 5.1 for abbreviations.

found that can provide protection against *Clostridium botulinum* as effectively as nitrite. In some EU countries (but not in Germany and the UK) and the US, nitrite addition to fresh meat is allowed up to a maximum of 200 ppm.

5.2.3.3 *Antibrowning agents*

Antibrowning agents are chemicals used to prevent browning of food, especially dried fruits and vegetables. Browning of food can occur enzymatically as well as non-enzymatically. The latter is dealt with extensively in Chapter 6. Enzymatic browning is mediated by polyphenol oxidase (PPO). This enzyme becomes available for catalysis upon cell disrupture. PPO contains copper and catalyzes two types of reactions (Figure 5.2):

Table 5.3 Nitrosamine levels in food

Food	Nitrosamine	Level (ppb)
Bacon, raw		0
fried	DMN, DEN, NPyr	1–40
	NPip	10–108
	NPyr	11–38
	DMN, NPyr	2–30
Bacon, frying fat	NPyr	10–108
drippings	NPyr	16–39
Luncheon meat	DMN, DEN	1–4
Salami	DMN, DEN	1–4
Danish pork chop	DMN, DEN	1–4
Sausage	DMN	1–3
Sausage, metwurst	NPyr, NPip	13–105
chinese	DMN	0–15
Fish		
raw: sable	DMN	4
salmon	DMN	0
shad	DMN	0
smoked: sable	DMN	4–9
salmon	DMN	0–5
smoked and nitrate- or nitrite-treated:		
salmon, sable, shad	DMN	4–17
salted marine fish	DMN	50–300
smoked and nitrate- or nitrite-treated:	DMN	20–26
Other fish products	DMN	1–9
Fish sauce	DMN	0–2
	NPyr	0–2
Cheese	DMN	1–4
Baby foods	DMN	1–3
Shrimp, dried	DMN	2–10
	NPyr	0–37
Shrimp sauce	DMN	0–10
Squid	NPyr	0–10
	DMN	2–8
	NPyr	0–7
Canned meats (uncooked)	DMN	1–3
Ham and other pork products (uncooked)	DMN	0–5
Beef products (uncooked) (4 days after slaughtering)	DMN	1–2
Wheat flour	DEN	0–10

Note: See Table 5.1 for abbreviations.

- hydroxylation of monophenols to catechols, i.e., *o*-diphenols;
- oxidation of catechols to *ortho*-quinones.

The *ortho*-quinones subsequently undergo a sequence of non-enzymatic reactions to yield brown-black melanin pigments.

Generally used antibrowning agents are vitamin C, citric acid, and sodium sulfite; the latter is also a well-known antimicrobial agent. Usually, antibrowning agents are not hazardous. Sulfite, however, can cause allergic reactions. It is one of the most widely used food additives. It is cheap and can be used efficiently in a variety of applications. Recently,

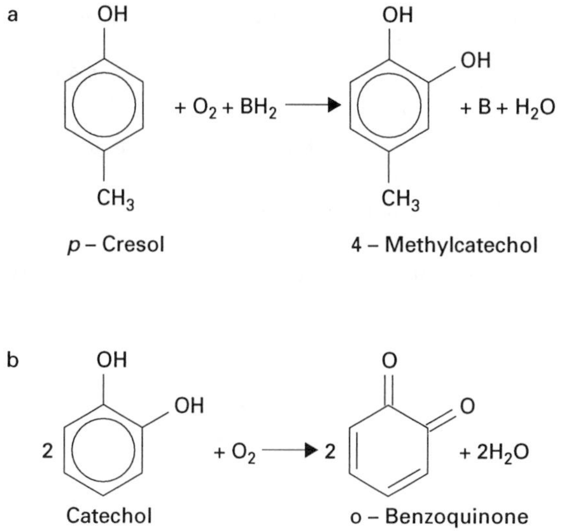

Figure 5.2 Reactions catalyzed by PPO: (a) hydroxylation; (b) oxidation.

attention has been drawn to the reactions between sulfite and nutrients and other food components. Although sulfite itself is considered to be safe for the majority of consumers, there is hardly any information on the nature and toxic effects of its reaction products.

5.3   Summary

Food additives are intentionally added to food not only to prevent microbial spoilage, but also to maintain stability, organoleptic properties, and the nutritional value of foodstuffs. Evaluation of the safety of food additives is extensively provided for by regulation and legislation.

There are five main categories of additives, namely texturizing agents, colorings, flavoring agents, preservatives, and miscellaneous additives. This chapter treats the use of a number of selected food additives in relation to their safety:

- the colorings amaranth and tartrazine;
- the flavorings monosodium glutamate and saccharin;
- the preservatives BHA, BHT, nitrite, and sulfite.

The majority of the texturizing agents and miscellaneous additives do not pose toxicological hazards.

Reference and reading list

Aroma, D. and B. Halliwell, (Eds.), *Free radicals and food additives*. London, Taylor and Francis, 1991.

Belitz, H.-D. and W. Grosch, (Eds.), *Food Chemistry*. Berlin, Springer Verlag, 1987.

Birch, G.G., (Ed.), *Food for the 90's*. Amsterdam, Elsevier Applied Sciences, 1990.

Branen, A.L., P.M. Davidson and S. Salminen, (Eds.), *Food Additives*. New York, Marcel Dekker Inc., 1990.

Concon, J.M., (Ed.), *Food Toxicology, Part A and Part B*. New York, Marcel Dekker Inc., 1988.

Gibson, G.G. and R. Walker, (Eds.), *Food Toxicology — Real or imaginary problems?* London, Taylor and Francis, 1985.

Gormley, T.R., G. Downey, and D. O'Beirne, (Eds.), *Food, health and the consumer.* Amsterdam, Elseviers Applied Sciences, 1987.

Gosting, D.C., (Ed.), *Food safety 1990; an annotated bibliography of the literature.* London, Butterworth-Heinemann, 1991.

Parke, D.V., D.F.V. Lewis, Safety aspects of food preservatives, *Food Add. Contam.*, 9, 561–577, 1992.

Walters, C.L., Reactions of nitrate and nitrite in foods with special reference to the determination of nitroso compounds, *Food Add. Contam.*, 9, 441–447, 1992.

chapter six

Nutrients

M.M.T. Janssen

6.1 Introduction

Nutrients are necessary for growth, maintenance, and reproduction of living organisms, and foods are their vehicles. Individual foodstuffs may contain a smaller or larger number of nutrients. This means that no individual food meets every physiological needs. The main categories of nutrients are carbohydrates, fats, proteins, vitamins and minerals.

Together, the former three are known as *macronutrients*. They are the major sources of energy and building materials for the organism. Vitamins and minerals are the so-called *micronutrients*, as they are only needed in small amounts.

Living organisms have a complex metabolic system at their disposal to maintain the concentrations of nutrients and their metabolites at proper physiological levels. If the metabolic capacity of an organism is exceeded, physiological homeostases may be disturbed, ultimately leading to adverse effects. This may be of particular importance in cases of metabolic disorder (e.g., lactose-intolerance and phenylketonuria), infection, specific physiological state (e.g., pregnancy) or drug treatment. The treatment of a metabolic disorder usually includes a diet with a low level of the nutrient concerned.

Considering the pathway from raw material to consumer in relation to the manifestation of adverse effects on human health after intake of nutrients, the following questions arise:

(a)   Which components, that belong to the nutrients and are hazardous to man, occur in raw materials?
(b)   Which harmful changes can nutrients undergo during the storage of raw materials?
(c)   Which harmful changes can nutrients undergo during the processing of raw materials?
(d)   Which harmful changes can nutrients undergo during the manufacturing, preparation, and storage of food?

In answering these questions, distinctions are made between macronutrients and micronutrients on the one hand, and between types of macronutrients on the other hand. When looking at the macronutrients, there is a difference in relevance between the four questions. The last question will be answered for all three categories of macronutrients, while the questions a, b and c are only relevant to fats. For the micronutrients, all questions will be answered.

The reactions taking place during industrial food manufacture are essentially the same as in food preparation at home. This means that for both cases the risks of harmful changes in food may be the same. However, industrial processes are evaluated and controlled extensively to prevent any possible harmful effect to human health, while inappropiate handling and preparation at home may escape notice.

In the following sections, the above questions will be answered. Section 6.2 concerns the macronutrients, whereas Section 6.3 deals with the micronutrients. Interference with the utilization and functioning of nutrients, so-called antinutritive effects, has already been discussed in Chapter 3.

6.2   Macronutrients

There are three categories of macronutrients: fats, carbohydrates, and proteins. The fats will be discussed first, as only for this category all of the above questions are relevant.

6.2.1   Fats

Dietary fats serve various needs of the living cell. First, they are concentrated energy sources. Further, they provide building blocks, i.e., polyunsaturated fatty acids, for many biological membranes. Third, oxygenation products of the essential fatty acids serve as mediators in the communication between the various cells of the organism. Fats are important to the physiology of the consumer as well as food technology and food safety.

The texture and taste of foods depend on their composition, crystal structure, melting behavior, and association with non-lipid molecules. In baking and frying, fats act as heat-transferring media. Lipids in foods contribute either directly or indirectly to the palatability by the products formed during heating. Fats may affect the quality and safety of food, even if they are present in small quantities only, because of their high reactivity.

6.2.1.1 Undesirable fat components in raw materials

The dietary fatty acids of nutritional value have (saturated or unsaturated) linear chains. The double bonds in unsaturated fatty acids follow specific patterns. The double bond(s) in fatty acids of plant origin is (are) at position 9 in the monoenes (e.g., oleic acid and palmitoleic acid), at positions 9 and 12 in the dienes (e.g., linoleic acid), at positions 6, 9, and 12 in the trienes and at positions 5, 8, 11, and 14 in the tetraenes (e.g., arachidonic acid).

In polyunsaturated fatty acids of marine origin (e.g., fish oil), the double bonds are at positions 9, 12, and 15 in the fatty acids with 18 carbon atoms (e.g., linolenic acid), at positions 5, 8, 11, 14, and 17 in the fatty acids with 20 carbon atoms, and at positions 4, 7, 10, 13, 16, and 19 in the fatty acids with 22 carbon atoms.

Unsaturated fatty acids

Structure	Name
$CH_3(CH_2)_5CH = CH(CH_2)_7COOH$	Palmitoleic
$CH_3(CH_2)_7CH = CH(CH_2)_7COOH$	Oleic
$CH_3(CH_2)_4CH = CHCH_2CH = CH(CH_2)_7COOH$	Linoleic
$CH_3CH_2CH = CHCH_2CH = CHCH_2CH = CH(CH_2)_7COOH$	Linolenic
$CH_3(CH_2)_4(CH = CHCH_2)_3CH = CH(CH_2)_3COOH$	Arachidonic

The specificity of the structures of dietary fatty acids suggests that any structural deviation may result in adverse effects, unless the organism can succesfully eliminate these fatty acids.

First, there is the distinction between fatty acids of the *n*-6 type (with the last double bond 6 carbon atoms from the methyl end, e.g., arachidonic acid) and fatty acids of the *n*-3 type (with the last double bond 3 carbons from the methyl end, e.g., linolenic acid). Overproduction of particular oxidation products of the *n*-6 fatty acid, arachidonic acid, the so-called eicosanoids, may lead to pathophysiological events, such as cardiovascular disorders. Acids of the *n*-3 type have been reported to inhibit the formation of *n*-6 eicosanoids. Epidemiological data on Eskimo populations, which consume large amounts of seafood (rich in *n*-3 types), suggest a reduction in cardiovascular disorders.

Secondly, several unusual fatty acids have been shown to be toxic. These include erucic acid (cis-13-docosenoic acid), sterculic acid and malvalic acid (cyclopropene fatty acids), and cetoleic acid (cis-11-docosenoic acid).

Some unusual unsaturated fatty acids

$CH_3(CH_2)_7CH = CH(CH_2)_{11}COOH$

Erucic acid

$$CH_3(CH_2)_7C \overset{\overset{\displaystyle CH_2}{\diagup\diagdown}}{=\!=} C(CH_2)_nCOOH$$

n = 6: Malvalic acid
n = 7: Sterculic acid

$CH_3(CH_2)_9CH = CH(CH_2)_9COOH$

Cetoleic acid

Erucic acid occurs mainly in the plant family of the Cruciferae, notably in *Brassica*. The oils from rapeseeds *(B. campestris)* and mustardseeds (*B. hirta* and *B. juncea*) contain particularly high levels of erucic acid: 20 to 55%. The toxic effects are fat accumulation in the heart muscle, growth retardation, and liver damage.

Rapeseed and mustardseed oils are commercially valuable in many parts of the world, but their potential toxicity is a reason for concern. Oil refining procedures can reduce the erucic acid content to 1.4%. Selective breeding has resulted in rape varieties that produce oil with an erucic acid content of approximately 0.5%.

The cyclopropene fatty acids *malvalic acid* and *sterculic acid* are found in all plants belonging to the order of the Malvales, except *Theobroma cocoa*. Crude cottonseed oil may contain 0.6 to 1.2% cyclopropene fatty acid in the form of sterculic acid and malvalic acid. For both acids, the reproductive organs are the primary targets. Further, these substances have been found to be carcinogenic. Processing can reduce the cyclopropene fatty acid levels to 0.1 to 0.5%. Residue levels in cottonseed feed may be about 0.01%. Hydrogenation of the ring can lead to the disappearance of some of the biological effects of these acids. *Cetoleic acid* occurs in herring oil. Its toxic effects are similar to those of erucic acid.

6.2.1.2 Changes in dietary fats during storage and processing of raw materials, and during manufacturing, preparation and storage of food

6.2.1.2.1 Rancidity. Spoilage of fat- or oil-containing material is mainly characterized by rancidity, the deterioration of fats and oils in foods. It is accompanied by an unpleasant odor and taste. There are two types of rancidity, hydrolytic and oxidative. Both can occur in all steps along the way from raw material to consumer. Oxidative rancidity can lead to the formation of harmful components.

Hydrolytic rancidity results from the hydrolysis of glycerides to fatty acids and glycerol. This process may be catalyzed by lipases (enzymes present in foods or originating from microorganisms), alkali, or acid. From a food safety point of view, hydrolytic rancidity has no important direct implications. Indirectly, however, it may be involved in combined actions. In addition, foods (e.g., milk) containing short-chain fatty acids such as butyric acid, caproic acid, caprylic acid, and capric acid, may become inedible or organoleptically unacceptable because of the strong off-flavor of the free fatty acids. Sometimes, however, hydrolytic rancidity is considered desirable, for example in strong tasting cheeses. The hydrolysis of glycerides can be minimized by cold storage, proper transportation, careful packaging and sterilization.

Oxidative rancidity is caused by reactions of fatty acids with atmospheric oxygen. Oxidation of fats and oils usually results in the formation of a variety of toxic substances. Therefore, oxidative rancidity will be dealt with separately in more detail.

6.2.1.2.2 Oxidation of fats and oils, and adverse health consequences. At various stages on the way from raw material to consumer, circumstances can arise in which oxidation is either extensified or prevented, and in which the oxidation products can be removed. For the sake of quality and safety, those circumstances are taken into consideration as much as possible in modern fat and oil production.

For clarity's sake, this subsection continues with a discussion of the various oxidation, reactions that fats and oils can undergo: autoxidation, photo-oxidation, and enzymatic oxidation. But first a detailed description of the processing of dietary fats and oils will be given in the following intermezzo.

Intermezzo

Processing of edible fats and oils. Oils and fats are particularly used for frying and baking. Some oils, such as olive oil, are appreciated for their distinctive flavor. Generally, however, fats and oils are expected to be rather tasteless and odorless. In addition, solid fats are expected to have well-defined physical properties and to give reproducible results. The consumer relies on the producer of oils and fats for specific properties and no impurities. It is well known that a relationship exists between certain processing techniques and the quality and safety of their products. Generally, the primary products of pressing and extraction are not suitable for use in food. Taste, color, general appearance, and maintenance of quality of crude oils and fats do not meet the required specifications. They may still contain impurities (see Table 6.1).

The impurities may originate from several sources. First, they may already have been in the crude product and left behind. Secondly, undesirable material may be formed during processing. The oxidation of lipids has been reported to lead to the formation of unwanted polymeric material. A third source of impurities may be the solvent in solvent extraction. The majority of the vegetable oils are obtained by solvent extraction with volatile petroleum ether hydrocarbons. It has been suggested that traces of carcinogenic aromatic hydrocarbons are passed from such solvents to the oils. However, this has not been confirmed. Moreover, should contamination be the case, these substances will certainly be removed during deodorization. The removal of impurities from crude oils and fats or the reduction of impurity concentrations in crude oils and fats to acceptable levels is realized by a sequence of refining processes. These are listed in Table 6.2.

It should be noted that during refining, the resistance of oils to oxidation may be reduced.

Bleaching with bleaching earth is not simply a matter of adsorption; free radicals are produced and oxidation takes place. During the process, small quantities of positional and geometric isomers of fatty acids as well as of a variety of hydroxy, keto, and epoxy derivatives are formed. Bleaching improves the quality of oils at the expense of their stability.

Deodorization is carried out under anaerobic conditions and no oxidation takes place. During this process less volatile substances are removed, including tocopherols and plant sterols. The physical properties of naturally occurring oils and fats are often not optimal for specific applications. However, in most cases they can be altered. This is especially so for melting behavior (see Table 6.3).

Saturating double bonds with hydrogen, so-called *hydrogenation*, converts oils into fats. This process requires a catalyst. Saturation of double bonds reduces the sensitivity to oxidation. The double bonds may be eliminated. Also, their position and their geometry

Table 6.1 Some impurities in crude dietary oils and fats

Impurity	Origin
Free fatty acids	Hydrolysis of triglycerides
Partial glycerides	Hydrolysis of triglycerides
Gums/Lecithin	Oil seeds such as soybean, rapeseed, sunflowerseed
Waxes	Oil seed coat, particularly sunflowerseed
Colored compounds	Oil seeds and fruit, e.g., chlorophyll in rapeseed and β-carotene in palm fruit
Insoluble materials	Oil seed fragments
Pollutants/Pesticides	Environmental pollution and pesticides used for oil seed crops
Volatile components	Breakdown of triglycerides and oxidized triglycerides
Polar components	Breakdown of triglycerides and oxidized triglycerides
Trace metals	Particularly iron and copper from processing machinery and equipment

(cis-trans) may be altered. In this way, fats with specific physical and chemical properties can be produced. For example, the combination of catalyst and reaction conditions can lead to the formation of trans-fatty acids. These are absorbed and metabolized in the same way as saturated fatty acids. Trans-fatty acid intake is 5 to 10% of the total fat intake. The trans-fatty acid content of shortenings, margarines and salad oils, made from partially hydrogenated oils, is 14 to 60%, 16 to 70% and 8 to 17%, respectively.

· Usually, thorough mixing of fats and oils is not sufficient to obtain homogeneous products. The triglycerides retain their physical properties. However, *reesterification* under the influence of a specific catalyst may result in a rapid interchange of fatty acid moieties with the formation of triglycerides of random fatty acid composition. Reesterification may lead to a small reduction in stability.

Of the three types of oxidation of fats and oils, photo-oxidation and enzymatic oxidation are less important; they are dealt with in the Intermezzo on page 82.

The reactions of organic compounds with elemental oxygen under mild conditions are generally referred to as *autoxidations* (see Figure 6.1). Such oxidations often take place by themselves, if the (slightly contaminated) substrate is exposed to air. Compounds of many types, including hydrocarbons, such as (poly)unsaturated fatty acids, alcohols, phenols, and amines may undergo autoxidation. Free fatty acids are more susceptible to oxidation than the fatty acid moieties in glycerides. Hydrolysis of glycerides is catalyzed by alkali, acid, and enzymes.

The attack of oxygen on hydrocarbons, i.e., (poly)unsaturated fatty acids, may be initiated by a radical derived from an outside source (generally peroxides, see Figure 6.1). The reaction of a C–H bond with a radical proceeds more readily if the carbon is tertiary or secondary, than if it is primary, and still easier if the carbon is allylic, as in (poly)unsaturated fatty acids.

(Hydro)peroxides, as is to be expected, may act as free radical initiators. The free radicals may be produced from (hydro)peroxides in two ways:

– in reactions with metals that have at least two readily accessible oxidation states (e.g., Fe^{2+}/Fe^{3+}, Cu^+/Cu^{2+}):

$$Fe^{2+} + ROOH \rightarrow Fe(OH)^{2+} + RO^{\cdot}$$

– on heating or by the action of visible or ultraviolet light:

$$ROOR \xrightarrow{\Delta \text{ or } h\nu} RO^{\cdot} + {\cdot}OR$$

Table 6.2 Removal of impurities from crude oils and fats

Refining process	Impurity to be removed	Processing technique
Degumming	Gums, phosphoglycerides (mainly lecithin)	Hot water rinsing
Neutralization	Traces of gums and phospho-glycerides, free fatty acids, mono- and diglycerides	Treatment with aqueous alkali solutions
Bleaching	Colored components and polar substances	Treatment with bleaching earth
Filtration	Bleaching earth and insoluble material	Filtering under pressure or using a mesh filter
Deodorization	Volatile substances, pesticides and pollutants	Vacuum steam distillation

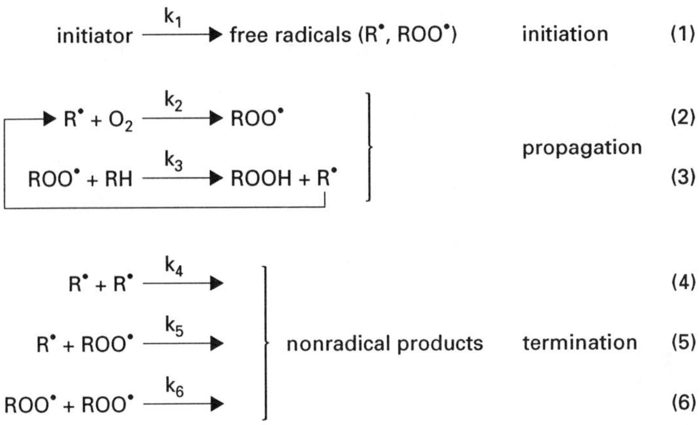

$$\text{initiator} \xrightarrow{\ k_1\ } \text{free radicals (R}^\bullet\text{, ROO}^\bullet\text{)} \qquad \text{initiation} \qquad (1)$$

$$R^\bullet + O_2 \xrightarrow{\ k_2\ } ROO^\bullet \qquad (2)$$
$$\qquad\qquad\qquad\qquad \text{propagation}$$
$$ROO^\bullet + RH \xrightarrow{\ k_3\ } ROOH + R^\bullet \qquad (3)$$

$$R^\bullet + R^\bullet \xrightarrow{\ k_4\ } \qquad (4)$$

$$R^\bullet + ROO^\bullet \xrightarrow{\ k_5\ } \quad\Big\}\ \text{nonradical products} \quad \text{termination} \quad (5)$$

$$ROO^\bullet + ROO^\bullet \xrightarrow{\ k_6\ } \qquad (6)$$

Figure 6.1 Diagrammatic representation of autoxidation.

Table 6.3 Alteration of physical properties of crude oils and fats

Process resulting in alteration	Result of processing
Hydrogenation	Rearrangement of double bonds; conversion of double bonds to single bonds
Fractionation	Separation of solid and liquid triglycerides
Reesterification	Random distribution of fatty acid moieties over the triglycerides

In the case of polyunsaturated fatty acids, the initiation is followed by a rearrangement of the double bonds (in the fatty acid radical), resulting in the formation of so-called conjugated diene radicals (see Figure 6.2). Further propagation is simple. The conjugated diene radicals may be oxygenated to peroxide radicals. The latter, in turn, may abstract hydrogen atoms again under the formation of hydroperoxides and radicals of the parent compound (see Figures 6.1 and 6.2).

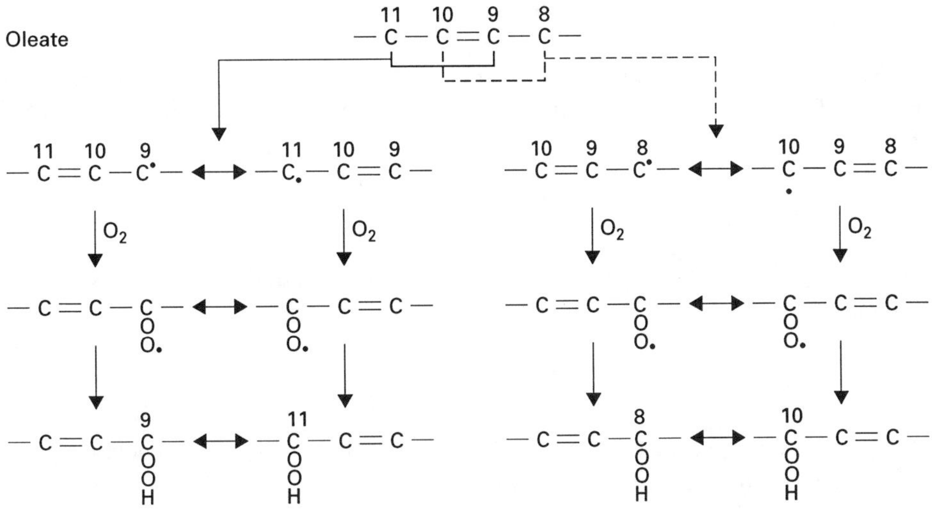

Figure 6.2 Formation of hydroperoxides.

One of the termination reactions is undoubtedly radical coupling:

$$R^{\cdot} + {}_{.}R \rightarrow R\text{--}R$$

The chain reaction may also be terminated by radical scavenging:

$$R^{\cdot} + S\text{--}OH \rightarrow RH + S\text{--}O^{\cdot}$$
$$(\text{e.g. vitamin E})$$

Intermezzo

Photo-oxidation and enzymatic oxidation. (Poly)unsaturated fatty acids may also undergo oxidative degradation in photoreactions (photo-oxidation) and enzymatic processes (enzymatic oxidation). The *photo-oxidations* that (poly)unsaturated fatty acids can undergo are of two types:
1.) Free-radical chain reactions, which start from the excited state of another molecule:

$$A \xrightarrow{\text{hv}} A^{*}$$

A = molecule of another food component, e.g., riboflavin (in milk), and A* = excited state of A.
A* may abstract an electron or a hydrogen atom from the substrate, RH:

$$RH + A^{*} \rightarrow R^{\cdot} + AH^{\cdot}$$

RH = e.g., linoleic acid.
These radicals can undergo further reactions in the presence of oxygen to form hydroperoxides:

$$R^{\cdot} + O_2 \rightarrow R\text{--}OO^{\cdot} \xrightarrow{\text{RH}} R\text{--}OOH$$

2.) Singlet oxygen ($^1O_2$) reactions in which the absorption of photons by molecules of another food component is followed by energy transfer to ground-state oxygen, leading to the formation of singlet oxygen:

$$A \xrightarrow{\text{hv}} A^{*}$$

A = e.g., protoporphyrin (occurring in hemoglobin, myoglobin and most of the cytochromes).

$$O_2 + A^{*} \rightarrow {}^1O_2 + A$$

Singlet oxygen can attack the double bonds in the unsaturated fatty acids (e.g., linoleic acid), yielding hydroperoxides:

$$2 \quad \diagup\diagdown\diagup\diagup\diagdown\diagup\diagup\diagdown\diagup\diagup \quad + 2\ {}^1O_2 \rightarrow \quad \overset{\text{OOH}}{\diagup\diagdown\diagup\diagdown\diagup\diagdown\diagup\diagup} \quad + \quad \overset{\text{OOH}}{\diagup\diagdown\diagup\diagdown\diagup\diagup}$$

As against autoxidation, tocopherols (e.g., vitamin E) can provide protection against sensitized photo-oxidation by acting as quenchers of singlet oxygen. Photodegradation of foods can be further prevented by using packaging materials that absorb the photochemically active light, and by removing endogenous photosensitizers and oxygen from the food.

Oxidative degradation of fats and oils may also be *enzyme*-mediated. The oxidation of fats and oils of plant origin may be catalyzed by lipoxygenase. The lipoxygenase-mediated oxidation is a hydroperoxide-initiated free radical chain reaction. Enzymatic oxidation also leads to the formation of hydroperoxides. Lipoxygenases can be inactivated by heat treatment.

In the three routes of oxidative degradation described above — autoxidation, photo-oxidation and enzymatic oxidation — a large variety of products is formed from fats and oils (see Figures 6.3 and 6.4).

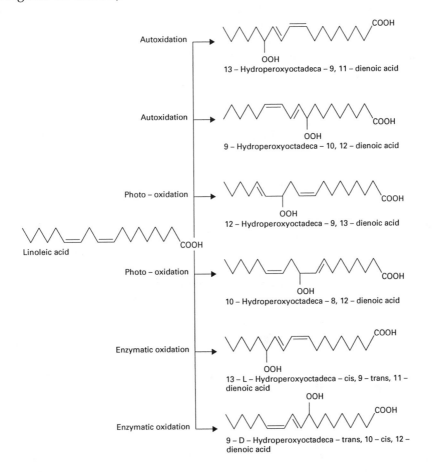

Figure 6.3 Hydroperoxide formation in autoxidation, photo-oxidation and enzymatic oxidation of linoleic acid.

The initially-formed lipid hydroperoxides are unstable. They degrade further in metal–ion catalyzed reactions to compounds such as alkanes (e.g., ethane and pentane) and (unsaturated) aldehydes (e.g., hydroxynoneal), and ketones (e.g., acetone):

Pentane

Unsaturated aldehydes, in particular α,β-unsaturated carbonyl compounds, may undergo toxic conjugations (Michael additions) with biologically essential nucleophiles such as sulfhydryl compounds and DNA bases. Hydroxynonenal is known to form adducts with DNA. Further, while the hydroperoxides are relatively unvolatile, tasteless, and odorless, the products formed in the secondary degradation reactions are volatile, and play a role in the development of off-flavors.

Hydroperoxides can also react with other nutritive components, such as amino acids and carotenoids. For example, methionine is oxidized to the sulfoxide, and lysine degrades to diaminopentane, aspartic acid, glycine, alanine, α-amino adipic acid and many other substances.

In all three of the oxidative degradations, termination of the chain reaction may result in dimerization of oxygen-centered radicals:

$$R-O^{\cdot} + .O-R \rightarrow R-OO-R$$

The formation of these peroxidic dimers may lead to polymerization reactions. This can be explained by both their instability and the large variety of functional groups they may contain. The dimers readily undergo homolysis, even at low temperatures. Further, homolysis may be followed by several rearrangements.

Another well-known end product of the peroxidation of polyunsaturated fatty acids such as linoleic acid and arachidonic acid is malondialdehyde:

$$\overset{H}{O = C} - CH_2 - \overset{H}{C = O}$$

Malondialdehyde is capable of cross-linking to primary amino groups, forming a conjugated Schiff base with the general structure:

$$R - N = \overset{H}{C} - \overset{H}{C} = \overset{H}{C} - \overset{H}{N} - R$$

where R may be free amino acids, proteins or nucleic acids.

Enzymes may be inactivated as a result of cross-linking, either directly by a reaction or indirectly by alteration of the membrane structure. Furthermore, malondialdehyde has been demonstrated to be carcinogenic in experimental animals and mutagenic in the Ames

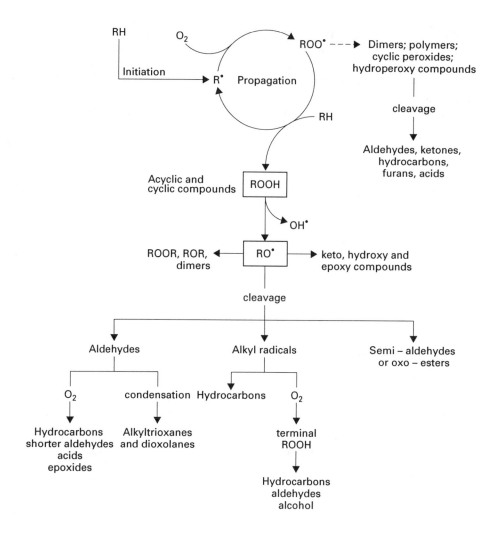

Figure 6.4 Diagrammatic representation of the degradation and polymerization reactions following autoxidation of unsaturated fatty acids.

test. The above data emphasize that peroxidation of food components may have negative effects on its nutritional value, sensoric quality and safety.

6.2.1.2.3 Effects of processing techniques on the oxidation of dietary fats and oils. The effects of various processing techniques used in food manufacture on the oxidation of dietary fats and oils can be largely predicted on the basis of tissue damage, exposure to oxygen, presence of metal ions, time, and temperature range involved.

Processing may already cause problems in the early stages of food manufacture. *Rapid freezing* of raw plant material may be accompanied by lipoxygenase-mediated oxidation. This phenomenon depends on the extent of tissue damage, and on storage temperature and time. *Blanching* and storage at low temperature may inhibit peroxidaton, but cannot prevent it.

During *dehydration* and *freeze-drying*, food lipids are extensively exposed to air as a thin film, thereby promoting autoxidation. The water content of a foodstuff is critical for the autoxidation rate. Excess water prevents extensive contact of lipids with oxygen. Further, storage temperature and time are important factors in determining the extent of oxidative deterioration. In the case of dehydration, the detection of oxidatively developed off-flavors may be facilitated, while the natural flavors and odors may disappear.

During *baking* lipids may be spread in thin films over large surfaces. Since baked products are usually consumed fairly soon after production, autoxidation will only occur to a limited extent.

Some types of *fermentation* are used for the production of substances that are undesirable in other products. Examples are the formation of short-chain acids and carbonyl compounds in cheeses and the high rancidity of a number of traditional Asian fermented fish and soy products.

Minor effects of the above processing techniques on product quality, shelf life, and vitamin content have been reported. Nutritional value and food safety do not appear to be much affected, not even under extreme conditions.

Deep-frying of foods in oils gives more rise to concern. As dealt with before, saturated and unsaturated fatty acids may undergo decomposition upon heating in the presence of oxygen. A diagrammatic summary of the thermolytic and oxidative mechanisms involved is shown in Figure 6.5.

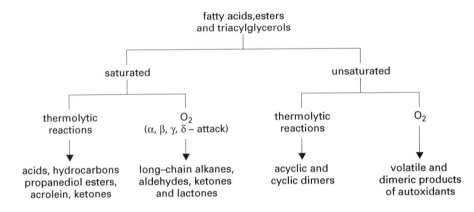

Figure 6.5 Diagrammatic representation of thermal and oxidative decompositions of lipids.

Frying of food under normal conditions may result in the formation of small amounts of stable peroxides. During industrial processing under vacuum, dimers, polymers, and cyclic products may be formed. The products that are formed when cooking oils are heated may be taken up by the fried food products. Meat, deep-fried in rapeseed oil, appeared to contain 0.63 to 1.1% of the nonvolatile oxidation products of the oil. French fried potatoes have been shown to contain secondary oxidation products of the cooking oil of high molecular mass. If oil is used in discontinuous batch-type operations, as in restaurants and at home, it is eventually discarded because of either high viscosity or excessive foaming. Discarded oils are found to contain approximately 25% polymers. Stable foams are formed if the polymeric oxidation product content is about 9%.

Generally, under normal frying conditions, oxidation of fats and oils has no harmful consequences. However, it should be noted that inappropriate heating and storage of fats and oils may lead to the formation of harmful substances at toxic levels. From the viewpoint of food safety, the conditions under which moderately rancidifying fats and oils are handled in the consumer's kitchen deserve particular attention.

6.2.2 Carbohydrates

In the human diet, carbohydrates are mainly present as starch. Well-known sources are cereals, potatoes and pulses.

Celluloses and other polysaccharides in the plant cell wall do not serve as energy sources. They cannot be digested by humans and contribute mainly to the dietary fiber intake. Fiber appears to play an important role in the maintenance of gastro-intestinal function, metabolism and health. Carbohydrates are useful food components because of their sweetness, solubility, cristallization behavior, water activity, hygroscopic behavior and rheological properties.

6.2.2.1 Changes in dietary carbohydrates during manufacturing and storage of food

Introduction. Reducing sugars may undergo a well-known (non-enzymatic) browning reaction, the so-called *Maillard reaction* in which sugars condensate with amino acids. Pentoses are usually more reactive than hexoses. The mechanism underlying this reaction has not yet been fully elucidated. The Maillard reaction is a sequence of reactions, resulting in the formation of a mixture of insoluble dark-brown polymeric pigments, known as melanoidins. In the early steps of the reaction, a complex mixture of carbonyl compounds and aromatic substances is formed. These products are water-soluble and mostly colorless. They are called *premelanoidins*.

Initial steps of the Maillard reaction. The first step in the Maillard reaction is the condensation of sugars with amino groups of amino acid moieties. The initial products, glycosylamines, are quite unstable and undergo Amadori rearrangement (Figure 6.6). The course of the condensation depends on the water content of the food. Further, these reactions proceed more rapidly upon heating, especially under neutral to alkaline conditions.

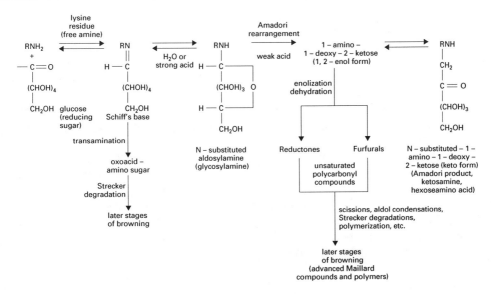

Figure 6.6 Condensation of glucose with lysine, followed by Amadori rearrangement.

The classic example of a foodstuff in which the initial steps may occur is drum-dried milk powder in which a 10 to 40% decrease in availability of lysine has been reported. If spray-drying is used, the availability of lysine does not decrease.

Subsequent steps of the Maillard reaction. In the following steps, reactive unsaturated polycarbonyl compounds such as reductones, and heterocyclic compounds such as pyrazines, are formed. These compounds bind to α-terminal, ε-amino and other amino groups of different polypeptide chains to form colored, high-molecular mass, highly cross-linked carbohydrate–protein polymers of low solubility, low digestibility, and low nutritional value. These steps may be followed by breakage of the polypeptide chains, and decarboxylation and ultimately deamination of the amino acid moieties (Strecker degradation, Figure 6.7).

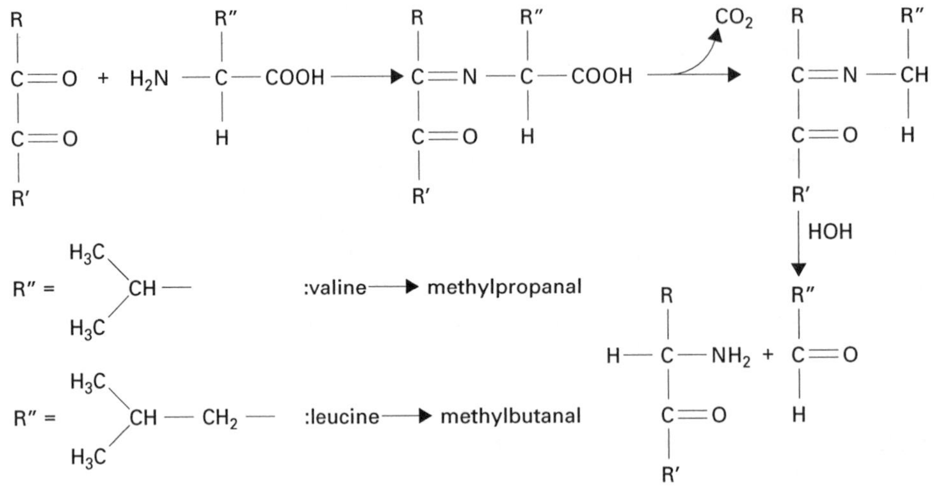

$$
\begin{array}{l}
\text{R} \\
| \\
\text{C}{=}\text{O} \quad + \quad \text{H}_2\text{N}-\overset{\displaystyle \overset{\text{R}''}{|}}{\underset{\displaystyle \underset{\text{H}}{|}}{\text{C}}}-\text{COOH} \quad \longrightarrow \quad \overset{\displaystyle \overset{\text{R}}{|}}{\text{C}}{=}\text{N}-\overset{\displaystyle \overset{\text{R}''}{|}}{\underset{\displaystyle \underset{\text{H}}{|}}{\text{C}}}-\text{COOH} \quad \xrightarrow{\;CO_2\;} \quad \overset{\displaystyle \overset{\text{R}}{|}}{\text{C}}{=}\text{N}-\overset{\displaystyle \overset{\text{R}''}{|}}{\underset{\displaystyle \underset{\text{H}}{|}}{\text{CH}}} \\
| \\
\text{C}{=}\text{O} \\
| \\
\text{R}'
\end{array}
$$

R'' = $\overset{\text{H}_3\text{C}}{\underset{\text{H}_3\text{C}}{\diagup\!\!\diagdown}}$CH —   :valine ⟶ methylpropanal

R'' = $\overset{\text{H}_3\text{C}}{\underset{\text{H}_3\text{C}}{\diagup\!\!\diagdown}}$CH — CH$_2$ —   :leucine ⟶ methylbutanal

After hydrolysis (HOH):

$$
\text{H}-\overset{\displaystyle \overset{\text{R}}{|}}{\underset{\displaystyle \underset{\text{C}{=}\text{O}}{|}\underset{|}{}\underset{\text{R}'}{}}{\text{C}}-\text{NH}_2 \quad + \quad \overset{\displaystyle \overset{\text{R}''}{|}}{\text{C}}{=}\text{O}
$$

Figure 6.7 Strecker degradation of amino acids.

The last steps of the Maillard reaction may take place to a considerable extent on heating of the food, especially chocolate and baked products such as bread, biscuits, and cakes. The initial steps of the Maillard reaction do not cause marked changes in the color and flavor of the foodstuff. However, as mentioned above, the course of the Maillard reaction depends on the water content of the food. As a result, the deterioration may be intensified by concentration and dehydration of protein-containing food, e.g., concentration and drum-drying of milk, dehydration of egg white, and drying of oilseed products. The last steps are largely responsible for the desired color and flavor of baked products.

Animal studies have indicated that premelanoidins inhibit growth, disturb reproduction and cause liver damage. Further, certain types of allergic reactions have been attributed to Maillard reaction products. Maillard reactions can be prevented by using the additive power of the carbonyl group in reducing sugars. A reaction characteristic of aldehydes and some ketones is addition of sodium bisulfite. The addition products are crystalline salts, very soluble in water. Further, Maillard reactions may be inhibited by regulating the temperature, pH, and water content.

6.2.3 Proteins

Proteins are the source of essential amino acids. They are the only dietary source of nitrogen for protein synthesis. Denaturation of protein occurs when food is heated during preparation. This is a desirable effect, as denatured protein is more readily digested. Apart from their nutritional value, proteins are important to the physical properties of foods. Their solubility and dispersability, hygroscopic behavior, viscosity, and stabilizing properties determine the structure and texture of foods.

6.2.3.1 Changes in proteins during processing of raw materials, and during manufacturing, preparation and storage of food

A technique that is increasingly used in the processing of proteins is treatment with alkali. It involves solubilization and purification of proteins. Protein concentrates are prepared this way. Alkali cooking of maize is a traditional Mexican technique to increase its digestibility. Treatment with sodium hydroxide is advocated for the peeling of grain. The use of gaseous ammonia has been proposed to free peanut and cottonseed products from aflatoxin.

Intensive treatment of proteins with alkali is known to result in advanced degradation of several amino acids, cystine, arginine, threonine, and serine being the most sensitive. Under mild alkaline conditions and at moderate temperatures, products may be formed that have been found to be nephrotoxic in rats, e.g., lysinoalanine (LAL), ornithinoalanine (OAL) and lanthionine. LAL is the condensation product of dehydroalanine and lysine (Figure 6.8). Dehydroalanine is formed on alkaline desulfurization of cystine. Sulfur is released as H_2S. Treatment of serine and phosphoserine moieties with alkali also yields dehydroalanine.

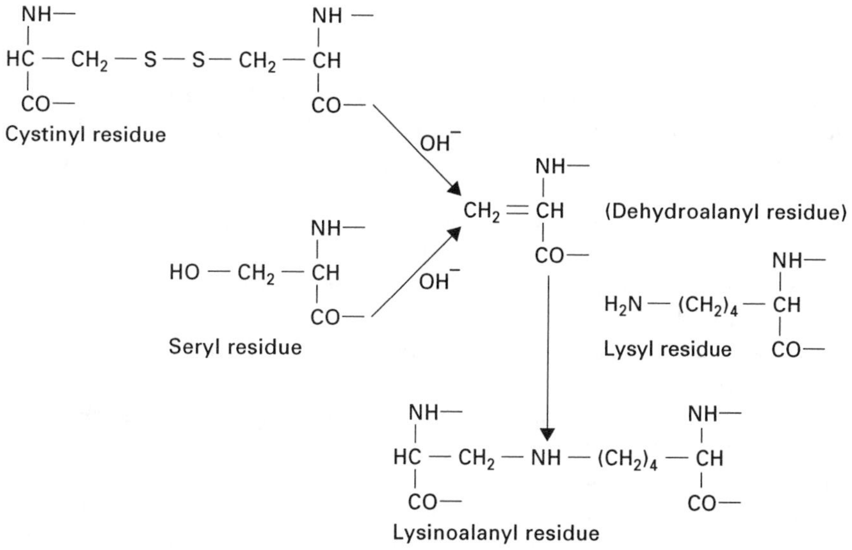

Figure 6.8 Formation of lysinoalanine (LAL).

The extent of LAL formation depends on the nature of the protein, the duration and temperature of the reaction, and the alkali concentration. Different legume proteins from mung beans, cow peas, and peanuts have been shown to produce considerable amounts of LAL when treated with 0.05 to 0.075 N NaOH at 20°C for 30 min. The LAL content may range from 200 to 800 mg/100 g of protein. Some legumes, such as kidney beans, lima beans, and vetch were stable under these conditions. After treatment with alkali at 80°C, however, all legumes contained LAL. Their LAL content decreased on prolonged treatment at higher temperatures. In several other foods, LAL is formed during cooking in the absence of alkali (Table 6.4). Chicken meat which was free from LAL before cooking, contained 200 µg/g after cooking in a microwave oven. Egg white free from LAL when fresh, contained 270 to 370 µg/g after boiling for 10 to 30 min, and 1.1 mg/g if pan-fried at 150°C for 30 min.

Table 6.4 Lysinoalanine content of heated proteins and some protein-containing food products

Protein or food	LAL (μg/g)
Sausage after boiling in water for 10 min	50
Corn chips	390
Pretzels	500
Tortillas	200
Evaporated milk	590–860
Simulated cheese	1070
Egg white solids	160–1820
Hydrolyzed vegetable proteins	40–500
Whipping agent	6500–50,000
Soya protein isolates	0–370

The reactions of proteins with oxidation products of oils and fats and with reducing sugars (Maillard reaction) have already been discussed in Sections 6.2.1.2.2 and 6.2.2.1 respectively.

6.2.4   Pyrolysis products occurring in food

Decomposition of a compound into smaller, more reactive structures by the action of heat alone is known as *pyrolysis*. The fragmentation is usually followed by combination of the smaller structures to more stable compounds, provided the conditions do not allow the conversion to CO and CO_2.

Since pyrolysis products may occur in all three of the macronutrient categories and the formation of these products proceeds largely by the same mechanism, this subsection deals with all three categories. The formation of pyrolysis products depends on the type of parent compound and the temperature. In the case of food, hazardous compounds are formed from about 300°C. In the following paragraphs, some well-known types of pyrolysis products occurring in food are discussed.

Polycyclic Aromatic Hydrocarbons. Polycyclic aromatic hydrocarbons (PAHs) are likely to be formed from degradation products consisting of two- or four-carbon units, such as ethylene and butadiene radicals (Figure 6.9).

The most potent carcinogenic PAH is benzo[a]pyrene (3,4-benzpyrene). Benzo[a]pyrene has been demonstrated in pyrolysis products of food. It has been identified in the charred crusts of biscuits and bread, in broiled and barbecued meat, in broiled mackerel and in industrially roasted coffees. The levels found in broiled meat ranged from 0.17 to 10.5 ppb. Fat is an important "precursor" for the formation of PAHs (in meat and fish). Broiling of high-fat hamburgers led to the production of 43 ppb of PAHs, of which 2.6 ppb was benzo[a]pyrene. In the lean product, only 2.8 ppb of PAH were found, and no benzo[a]pyrene.

Starch may also undergo pyrolysis. On heating of starch from 370 to 390°C, 0.7 ppb benzo[a]pyrene was formed. The range of 370 to 390°C is readily reached in cooking procedures, e.g., at the surface of bread during baking and in boiling cooking fats.

The conditions during the preparation of food may affect the levels of PAH formed. In T-bone steaks cooked close to the charcoal and relatively long, benzo[a]pyrene levels up to 50 ppb have been detected. These levels could be considerably reduced by cooking at a larger distance from the charcoal. PAHs are abundantly found in smoked food (Table 6.5). They originate from the combustion of wood and other fuels.

Heterocyclic pyrolysis products from amino acids. Some potent mutagens are produced on pyrolysis of amino acids. Tryptophan has been shown to be the "precursor" of the

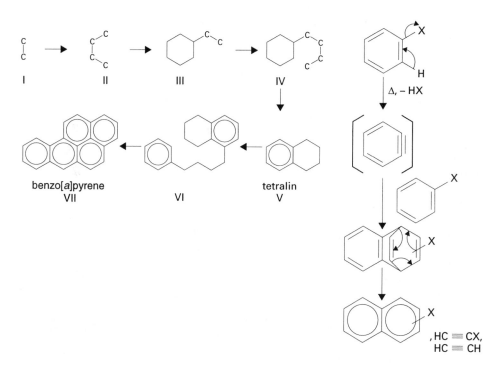

Figure 6.9 Proposed routes for the formation of PAH.

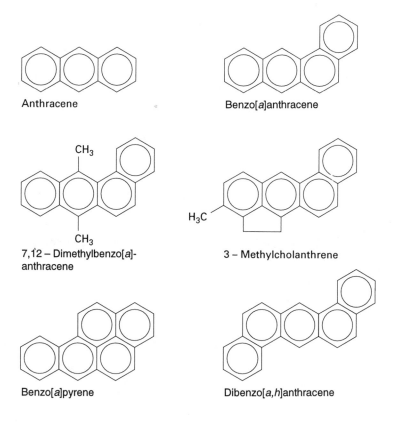

Table 6.5 Benzo[*a*]pyrene content of smoked and other foods

Food	Benzo[*a*]-pyrene (ppb)	Food	Benzo[*a*]-pyrene (ppb)
Smoked fish		Barbecued meats	
Eel	1.0	(Charcoal broiled)	
Herring	1.0	Hamburgers	11.2
Sturgeon	0.8	Pork chop	7.9
Chubs	1.3	Chicken	3.7
White fish	6.6	Sirloin steak	11.1
Kippered cod	4.5	T-bone steak	57.4
Smoked meats		T-bone steak	4.4
Ham	0.7–55.0	(flame broiled)	
Mutton		Ribs	10.5
close to stove	107.0	Other steaks	5.8–8.0
distant from stove	21.0	Other foods	
Lamb	23.0	Spinach	7.4
Sausage (with skin)		Kale	12.6–48.1
cold smoked	2.9	Yeast	1.8–40.4
hot smoked	0.7	Tea	3.9–21.3
Salami	0.8	Coffee	0–15.0
Bacon	3.6	Cereals	0.2–4.1
		Soybean	3.1

mutagens 3-amino-1,4-dimethyl-5H-pyrido[4-*b*]indole and 3-amino-1-methyl-5H-pyrido[4,3-*b*]indole. Pyrolysis of phenylalanine may lead to the formation of the mutagenic substance 2-amino-5-phenylpyridine.

3 – Amino – 1 – methyl –
5H – pyrido (4,3 – *b*) indole

These mutagens and several structurally related substances have been isolated from the surface of protein-containing food cooked at 250°C and higher. Mutagens of the aminoimidazoazoarene type have also been isolated from different types of protein-rich foods heated at about 200°C. They include, for example, quinolines and quinoxalines:

2 – Amino – 3 – methylimidazo
(4,5 – *f*) quinoline

2 – Amino – 3,8 – dimethylimidazole
(4,5 – *f*) quinoxaline

6.3 Micronutrients

This section discusses various relevant aspects of the micronutrients *vitamins* and *minerals*. There are 13 known vitamins. They are usually divided into two groups: lipophilic and hydrophilic vitamins. Generally, vitamins play no part in food technology, with the exception of the vitamins C and E which can function as antioxidants.

The minerals may be divided into two groups based on the levels at which they occur in the body. The elements that are present in considerable quantities are sodium, potassium, calcium, chlorine, magnesium, and phosphorus. Their combined mass is about 3 kg in adult men. The second group comprises the so-called trace elements. They include sulfur, iron, fluorine, iodine, zinc, copper, selenium, manganese, molybdenum, chromium, and cobalt. They are required in very small amounts only. Their combined content is roughly 30 g. Minerals are important in food processing because of their effects on the course of enzymatic as well as non-enzymatic processes. Further, they can affect the texture of foods by reacting with polysaccharides (gel formation). Minerals are also important flavorings.

Health risks due to micronutrients are usually associated with deficiencies in the diet. This has already been mentioned in Chapter 3. Dietary excess of micronutrients does not cause great problems because the minerals and the majority of the vitamins are water-soluble and are readily eliminated by excretion as well as metabolism. Only the intake of the lipid-soluble vitamins A and D may lead to toxic effects, as they accumulate easily in the body. Intake of vitamins in excess of the required amounts results in the toxic syndrome of hypervitaminosis. Selenium is of special importance because the margin between the physiological need and the toxic dose is very small. Dietary intake is estimated at 56 mg/day in regions poor in selenium and at 326 mg/day in regions rich in selenium. Selenium is the only mineral that accumulates in plants in considerable amounts. It is incorporated in amino acids by replacing sulfur. This has been reported to lead to intoxications in animals feeding on selenium-rich plants.

In view of the large intake level range for micronutrients indicated above, the nutritional state of an individual can affect cellular processes other than the specific metabolic reactions in which the vitamins and minerals play essential physiological roles. Micronutrients may influence cellular differentiation, hormone metabolism and regulation, immunological control, and metabolic activation and inactivation of protoxins. Vitamins A, B_2 and C have been found to protect against carcinogenesis in a number of human as well as animal studies. Copper, zinc, and especially selenium can act in a dualistic way. These minerals may cause enhancement of and provide protection against carcinogenesis, depending on the toxic substance involved.

6.3.1 Hypervitaminosis

6.3.1.1 Vitamin A

The major dietary sources of vitamin A are carotenoids, particularly β-carotene, occurring in red, orange, and green plants, and retinyl esters, present in animal tissues. The active form of vitamin A is retinol. Carotenoids act as provitamins. They are converted to retinol in the intestinal mucosa. After absorption in the intestines, retinol is transported to the liver bound to plasma proteins, where it is stored as retinyl fatty acid esters.

Generally, carotenoids are not considered to be toxic, as their conversion to retinol is rather inefficient. On very high intake, carotene accumulates in the body. The toxic symptoms of high vitamin A intake are yellow pigmentation of the skin, headache, dizziness, vomiting, and diarrhea followed by swelling of the skin, which eventually cracks and peels. In most cases, these symptoms disappear within a few days after termination of the high intake.

Table 6.6 Effect of the degree of maturity on the ascorbic acid content of a tomato variety

Weeks after bloom	Average weight (g)	Color	Ascorbic acid (mg %)
2	33.4	green	10.7
3	57.2	green	7.6
4	102.5	green-yellow	10.9
5	145.7	yellow-red	20.7
6	159.9	red	14.6
7	167.6	red	10.1

The natural foods that contain sufficient retinol to induce toxic effects in man are the livers of animals at the top of long food chains, such as marine fish and polar bears. Cases of acute toxicity have been reported by researchers and fishermen in polar regions after eating generous liver portions, containing up to 100,000 IU (1.0 IU equals 0.3 µg retinol) of vitamin A per gram. The toxic intake by adults is estimated at 2 to 5 million IU. Single intakes of 30 million IU have been reported.

An insidious cause of vitamin A intoxication today is the chronic consumption of vitamin supplements. Capsules containing retinol doses of 25,000 IU are easily obtainable, while the recommended daily intake is 5000 IU. Daily intakes of 20,000 to 40,000 IU for a number of years have been found to lead to toxic symptoms.

6.3.1.2   Vitamin D

Vitamin D can be formed from 7-dehydrocholesterol in the skin under the influence of ultraviolet light. The product is cholecalciferol (vitamin D_3). This passes to the liver where it is hydroxylated to 25-hydroxycholecalciferol (25-OHD). After transport to the kidney, 25-OHD is metabolized to 1,25-dihydroxycholecalciferol, a more potent form. The synthesis of vitamin D in the skin is carefully regulated. However, high intakes of this vitamin may result in hypercalcemia. This may be accompanied by tissue injury and hypersensitivity. Daily intakes in excess of 50,000 IU (1.25 mg) have been found to cause toxic effects. The normal dietary intake is estimated at 1000 IU per day.

6.3.2   Changes in raw materials during storage and processing, and in food during manufacture, preparation and storage

For the micronutrients, no changes leading to the formation of harmful substances are known. On the other hand, all foods lose vitamins and minerals to some extent when stored and processed, either industrially or at home. The food industry tries to minimize these losses by careful regulation of the various processing steps.

Apart from the above, the micronutrient content of food largely depends on a number of (pre-storage and pre-processing) factors, including genetic variation, degree of maturity, soil conditions, use and type of fertilizer, climate, availability of water, light (day length and intensity), and post-harvest or post-mortem handling. Data on these factors, however, are scarce. As an example, Table 6.6 shows the effect of maturing of tomatoes on their ascorbic acid content. The maximum vitamin content is already reached before maturity.

6.3.2.1   Vitamins

Vitamins are reactive substances. They may be sensitive to light, heat, moisture, oxidizing and reducing agents, acid, alkali and (traces of) heavy metals. It should be noted that each vitamin reacts in a different way, as is shown in Table 6.7.

Effects of food handling on vitamin content will be illustrated for vitamin C (ascorbic acid). This vitamin is probably the most extensively investigated with regard to its behav-

Table 6.7 Sensitivity of vitamins to chemical and physical factors

Type of vitamin	Light	Heat	Humidity	Oxidizing agents	Reducing agents	Acids	Bases
Vitamin A	XX	X	0	XX	0	X	0
Vitamin D	XX	X	0	XX	0	X	X
Vitamin E	X	X	0	X	0	0	X
Vitamin B$_1$	X	XX	X	XX	0	0	XX
Vitamin B$_2$	XX	0	0	0	X	0	XX
Vitamin B$_6$	X	0	0	0	X	X	X
Vitamin B$_{12}$	X	0	0	0	X	X	XX
Panthotenic acid	0	X	X	0	0	XX	XX
Folic acid	X	0	0	XX	XX	X	X
Vitamin C	0	X	X	XX	0	X	XX

Note: 0 = hardly or not sensitive; X = sensitive; XX = very sensitive.

ior during food processing. Ascorbic acid has an enediol structure, in which the double bond is conjugated with a carbonyl group. Due to this structure, it has both acidic and (strong) reducing properties. The natural form is the L-isomer. The D-isomer has about 10% of the activity of the L-isomer. The latter is used as food antioxidant.

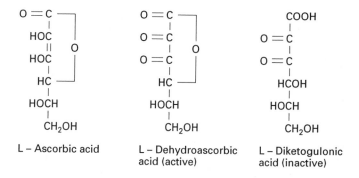

L – Ascorbic acid   L – Dehydroascorbic acid (active)   L – Diketogulonic acid (inactive)

Ascorbic acid is easily and reversibly oxidized to dehydroascorbic acid. The oxidation is catalyzed by metal ions. Subsequent hydrolysis of dehydroascorbic acid to 2,3-diketogulonic acid results in irreversible loss of vitamin acitivity. The latter compound may undergo anaerobic degradation. The primary degradation products are xylosone and 4-desoxypentosone. These in turn may produce ethyl glyoxal, reductones, furfural, and 2,5-dihydrofuroic acid (Figure 6.10).

The presence of ascorbic acid, dehydroascorbic acid and the degradation products in amino group-containing food can give rise to Maillard-type browning reactions.

Ascorbic acid is readily lost by leakage from cut or bruised surfaces of foods. Loss by leakage is most prominent during cooking (Figure 6.11). In processed foods, the most considerable losses result from oxidative degradation, especially upon irradiation with sunlight. This is most frequently seen in foods which are high in ascorbic acid, such as fresh fruits. Warm foods should not be left in open metal containers before eating in order to prevent oxidative degradation catalyzed by the metal of the container.

6.3.2.2 *Minerals*

Losses of minerals, result from physical removal or interaction with other food components rather than from chemical degradation. Primarily responsible for loss are processing

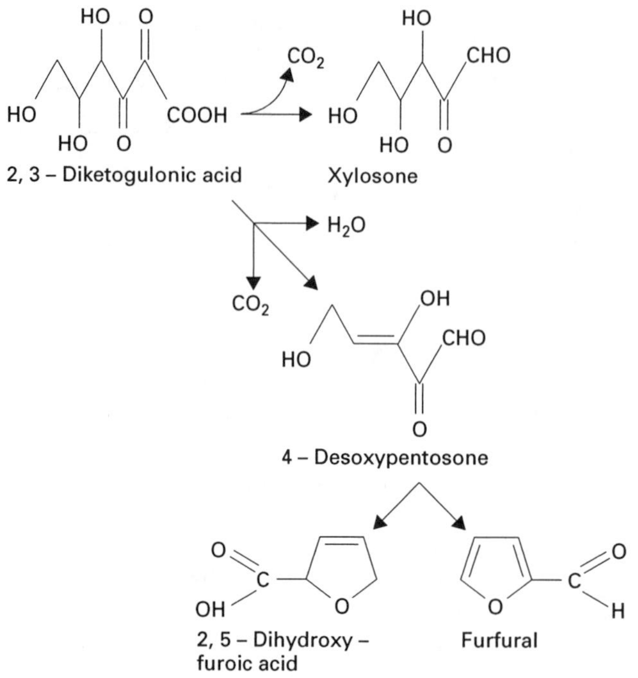

Figure 6.10   Degradation reactions of 2,3-diketogulonic acid.

techniques in which contact with water is possible. These include leaching (of water-soluble materials), blanching, and cooking. Table 6.8 shows the effect of blanching on mineral loss from spinach.

As can be seen from Table 6.8, the differences in loss between the minerals can be attributed to differences in water solubility.

In the case of cereals, milling appears to be a major cause of mineral loss. In general, the overall loss of minerals during storage and processing of foodstuffs has no negative effect on the dietary mineral intake levels. A varied diet still provides the proper amounts of minerals.

6.4   Summary

Nutrients are the major food components. They are necessary for growth, maintenance, and reproduction of living organisms. The main categories of nutrients are carbohydrates, fats, proteins, vitamins, and minerals. The first three categories, the so-called macronutrients, are

Table 6.8 Effect of Blanching on the Mineral Loss from Spinach

Mineral	Unblanched g/100 g	Blanched g/100 g	Loss (%)
Potassium	6.9	3.0	56
Sodium	0.5	0.3	43
Calcium	2.2	2.3	0
Magnesium	0.3	0.2	36
Phosphorus	0.6	0.4	36
Nitrate	2.5	0.8	70

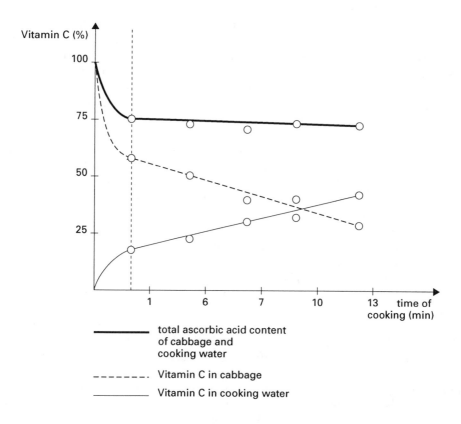

Figure 6.11   Ascorbic acid loss during cooking of cabbage.

the major sources of energy and building materials to the organism. Vitamins and minerals are known as micronutrients, since they are only needed in small amounts. Micronutrients play essential roles in specific metabolic reactions.

Living organisms have a complex metabolic system at their disposal to maintain the concentrations of nutrients and their metabolites at physiological levels. If the metabolic capacity of an organism is exceeded, physiological homeostases may be disturbed, ultimately leading to adverse effects.

On the way from raw material to consumer, however, conditions may be encountered under which nutrients can undergo harmful changes, i.e., during storage and processing of the raw materials, and during manufacturing, preparation, and storage of the actual good food.

Fats are highly reactive. Their reaction products can affect both the quality and the safety of food. Only a few unusual fatty acids have been shown to be toxic themselves, e.g., erucic acid and sterculic acid. Deterioration of fats is mainly associated with hydrolytic and oxidative rancidity. The former has no important consequences from a food safety point of view. Oxidation of fats and oils usually leads to the formation of a variety of toxic substances. Three types of oxidation can be distinguished: autoxidation, photo-oxidation, and enzymatic oxidation. Toxic oxidation products are hydroperoxides, unsaturated aldehydes (e.g., hydroxynonenal), and malondialdehyde. Hydroxynonenal is known to form adducts with DNA. Malondialdehyde has been demonstrated to be carcinogenic in experimental animals and to be mutagenic in the Ames test. Oxidative degradation of fats and fat-containing foods during storage and processing depends on the extent to which the structures of the raw materials and foodstuffs have been damaged on exposure to oxygen and light, and on the time and temperatures involved.

Carbohydrates may undergo a well-known (non-enzymatic) browning reaction, the Maillard reaction. They condensate with amino acids. The Maillard reaction is a sequence of reactions, ultimately leading to the formation of a mixture of insoluble dark-brown polymeric pigments, the so-called melanoidins. In the early steps of the reaction, a complex mixture of carbonyl compounds and aromatic substances is formed. These are known as premelanoidins. They have been found to inhibit growth, to disturb reproduction and to cause liver damage. Further, certain types of allergic reactions have been attributed to products of the Maillard reaction.

A technique that is increasingly used in the processing of proteins is treatment with alkali. Severe treatment with alkali can result in advanced degradation of amino acids. Under mild alkaline conditions and at moderate temperatures, products may be formed that have been found to be nephrotoxic in rats, e.g., lysinoalanine.

All three macronutrient categories can undergo pyrolysis during food preparation. Well-known types of toxic pyrolysis products are polycyclic aromatic hydrocarbons and heterocyclic compounds. The former are likely to be formed from degradation products consisting of two- or four-carbon units. A highly potent carcinogenic of this type of pyrolysis products in food is benz[*o*]pyrene. Heterocyclic products are formed by pyrolysis of amino acids. Pyrolysis of tryptophan appears to lead to the formation of the mutagen 3-amino-1-methyl-5H-pyrido[4,3-*b*]indole.

Vitamins and minerals are essential dietary components. Health risks due to micronutrients are usually associated with deficiencies in the diet. Only intake of the lipid-soluble vitamins A and D may lead to toxic effects, as they accumulate easily in the body. Intake of vitamins in excess of the required amounts results in the toxic syndrome called hypervitaminosis. During storage and processing of raw materials and food, micronutrients are lost to some extent, either in industry or at home.

Reference and reading list

Belitz, H.-D. und W. Grosch, (Eds.), *Food Chemistry*. Berlin, Springer Verlag, 1987.

Birch, G.G., (Eds.), *Food for the 90's*. Amsterdam, Elsevier Applied Sciences, 1990.

Concon, J.M., (Ed.), *Food Toxicology, Part A and Part B*. New York, Marcel Dekker Inc., 1988.

Davidek, J., (Ed.), *Natural toxic compounds. Formation and change during food processing and storage*, Boca Raton, CRC Press Inc., 1995.

Fennema, O.R., (Ed.), *Food Chemistry*. New York, Marcel Dekker Inc., 1985.

Friedman, M., Dietary impact of food processing, *Annu. Rev. Nutr.*, 12, 119–137, 1992.

Gibson, G.G. and R. Walker, (Eds.), *Food Toxicology — Real or imaginary problems?* London, Taylor and Francis, 1985.

Gormley, T.R., G. Downey and D. O'Beirne, (Eds.), *Food, Health and the Consumer*. Amsterdam, Elsevier Applied Sciences, 1987.

Gosting, D.C., (Ed.), *Food Safety 1990; An Annotated Bibliography of the Literature*. London, Butterworth-Heinemann, 1991.

Tannenbaum, S.R., (Ed.), *Nutritional and Safety Aspects of Food Processing*. New York, Marcel Dekker Inc., 1979.

From raw materials to consumer: aspects of dietary behavior

chapter seven

Aspects of dietary behavior

P. van Assema and G.J. Kok

7.1   Introduction

Part 1 dealt with the route of food (components) from raw material to the consumer. The relationship between the origin of food and exposure to it or its components has been discussed from chemical, microbiological, and technological points of view. Whether the consumer is ultimately exposed to food (components) or not depends on his dietary behavior. This chapter looks at dietary behavior and its determinants.

7.2   Behavior and its determinants

The concept of dietary behavior is very complex. Dietary behavior includes a multitude of behaviors and can refer to:

- food choice (e.g., buying skimmed milk instead of full cream milk in the supermarket);
- food preparation (e.g., frying an egg in margarine);
- food preservation (e.g., keeping raw meat in the refrigerator);
- (actual) food consumption.

Attempts have long been made to change certain dietary behaviors of the population for reasons of health. These nutritional interventions are aimed at groups of patients, high-risk groups, healthy people, or intermediaries such as people working in the kitchen of a restaurant.

The complexity of dietary behavior can also be illustrated by the diversity of the objectives of nutritional interventions such as:

- increasing the hygienic behavior in the catering industry;
- reducing the consumption of proteins and sodium by kidney patients to relieve the kidney(s) as much as possible;

- reducing the total energy intake to prevent or cure obesity;
- increasing the consumption of food products containing carbohydrates to improve achievements in endurance sports;
- increasing the knowledge about food preservation to prevent food poisoning, for example, resulting from microbial contamination;
- reducing alcohol consumption to decrease the number of alcohol-related traffic accidents.

In order to change dietary behavior it is important to know the factors that determine it. Why do people display a certain dietary behavior? Once the behavioral determinants have been established, specific nutritional interventions can be chosen (see further Part 3, Chapter 22). Dietary behavior, like behavior in general, is determined by many factors, in this case for example, availability of food, food policy of the government, social environment, advertising, and experience and opinions people have regarding food safety. These are all factors, either at the macrolevel (systemic) or at the microlevel (individual), which influence people's food choices, how they prepare their food, and how they preserve it. To make matters even more complex, the determinants of all specific behaviors have to be studied one by one. The reasons why a healthy person eats two oranges a day are completely different from those of an overweight person to eat low-fat cheese. Moreover, the determinants of a specific behavior can also vary from person to person.

The following sections deal with theories on factors determining dietary behavior. Also, some studies on the determinants of specific dietary behaviors will be discussed, to give an idea of the state of the art in this field. Point of departure is that a person's cultural environment is the dominating determining factor of his or her dietary behavior. Much information on someone's dietary habits is already available if the country of origin of that person is known. However, within cultures there are many differences in food choices. The key question is which factors determine the differences.

7.3 Models of behavioral determinants

Behavior in general can be explained at the macro- as well as the microlevel. Factors affecting dietary behavior at the macrolevel can be policy, advertising, availability and acceptability of products, and cultural standards and values. Theoretical models at the macrolevel which might be helpful in studying (dietary) behaviors are scarce. Moreover, they are very abstract and lack empirical support. Often, there are complex multicausal relationships between the variables in these models. On the other hand, if one wants to explain behavior at the microlevel, there are often empirical theories available with operational, quantifiable (often one-to-one) relationships. Therefore, this chapter is restricted to theories at the microlevel in which individual behavior is the dependent variable. However, factors at the macrolevel are also considered to be important prerequisites, so behavioral change interventions should also focus on them (see Part 3, Chapter 22).

Social psychologists and health education researchers have developed models for explaining individual behavior in general, i.e., health behavior, environmental (hygiene) behavior, political choices, etc. Extreme anxiety behavior or addictive behavior, however, can not be explained with these models. The majority of dietary behaviors, which are important from a health point of view, can be explained. A prerequisite for using the models is that the behavior that has to be explained is under a person's control and that the person is aware of the options. This does not mean though that people always have to be completely conscious of these behavioral choices. Many of them are made implicitly, especially those which have become habits.

For the various models, three main groups of determinants can be distinguished:

- attitude (what do people think of their behavior themselves?)
- social influence (what is the role of the social environment?)
- possibilities (either internal or external to the person) for displaying behavior.

Often, there is an overlap between the three groups.

An *attitude* towards a specific behavior reflects whether a person's general feelings are favorable or unfavorable towards that behavior, and is determined by the evaluation of all pros and cons of the behavior. It should be noted that this concerns the *perceived* positive and negative consequences of the behavior, and not necessarily the *actual* consequences. For example, a person may think that canned vegetables contain many additives. Although this does not need to be so, it still affects the person's attitude towards buying and eating the vegetables. For many, the evaluation of the health implications of a behavior is important. It is questionable, however, whether people's behavior is indeed determined by its health consequences. Some people do not eat butter, as they think it is unhealthy. On the other hand, there are also people who do not eat it because they do not like it. In general, health implications often play a role in behavioral choice, although sometimes a very marginal one.

Besides health consequences, there are other consequences which can influence someone's attitude towards a behavior. As already suggested, these are sometimes more important than the health aspects. Think of taste (raw meat may taste good), or practical considerations (it may be too much bother to remove burned parts of grilled meat). Also, considerations related to social environment may be important: eating fruit may give the image of a health fanatic. In short, many positive and negative aspects of both the desired behavior and the risk behavior play a role in behavior.

The attitude model designed by Fishbein and Ajzen (1975) is based on the assumption that people have reasons for their behavior. These reasons, however, are not necessarily rational. On the contrary, according to experts, people may have very irrational reasons for behavior, which are valid in their own conception. The model does not predict behavior effectively when people are not aware of their actual motives for behavior.

Before the other two main groups of behavioral determinants, i.e., social influence and possibilities are addressed, measurement of attitudes will be described below.

Intermezzo

Measuring attitudes. Many authors have reported on the measurement of attitudes. According to Fishbein and Ajzen (1975), a person's attitude may be measured directly by one single question, e.g., "Is this behavior good or bad?" Also, the structure of the attitude may be analyzed. In other words, the specific pros and cons the behavior is connected with, may be identified. To determine this structure, people may be asked to point out which consequences they believe the behavior is connected with (beliefs, B), using a scale ranging from extremely likely to extremely unlikely.

For example: eating two pieces of fruit a day reduces my chances of getting cancer. I think this is extremely likely/extremely unlikely, with some answers in between. Subsequently, the answers, for example, can be given marks ranging from 1 (extremely likely) to 0 (extremely unlikely).

It is also important to know whether the consequence is considered as an advantage or a disadvantage. These evaluations (E) can also be measured, for instance as follows. Reducing my chances of getting cancer is: very good/very bad, again with some intermediate answers. These can also be given a mark, for example from +3 (very good), via 0 (neutral) to –3 (very bad). Ultimately, the attitude can be expressed by multiplying the beliefs with the corresponding evaluations, followed by summation of the products thus obtained:

$$A = \sum_{i=1}^{N} B_i \times E_i$$

where A is the attitude, B the belief and E the corresponding evaluation.

The *social environment* very much affects behavior. This is especially so in dietary behaviors, as most of these are displayed in the presence of other people. The influence of the social environment is often underestimated. People are inclined to think that their behavior reflects their own attitudes, particularly if they can rationalize their behavior. This, however, has been shown to be a misunderstanding. Behavior is influenced very strongly by the social environment and the possibilities or impossibilities, sometimes so strongly that attitude has no influence at all. There are two kinds of social influence: direct and indirect. *Direct* influence refers to the clear expectations of others as to how someone should behave, e.g., an adolescent with his friends in a snackbar or drinking alcohol in company. In both situations, a certain kind of behavior is expected from the person. Not cooperating means not belonging or no longer belonging to the group of people the person belongs to or wants to belong to. *Indirect* influence is more subtle. It refers to modeling, imitating the behavior of others. Behavior is learned by observing others. This kind of social influence is called indirect, as the observed (the model) does not explicitly formulate expectations. For example, the model can be seen on film or television.

Most times, there is a relationship between the opinion of the social environment about a behavior and one's own attitude towards it. This is partly because people incorporate the opinion of the social environment into their own attitude. This is called "internalizing the opinion of the social environment." However, the relationship between attitude and social influence can also be absent. There are situations in which a person's attitude does not match the expectations of the social environment, as he or she functions in different social environments, for example at school, at a sports club, in the family. In such situations the person can behave as the environment expects, which means there is a discrepancy between the person's behavior and his or her own opinion. For example, someone may eat meat at an official dinner, while he/she actually dislikes meat.

Social psychology distinguishes two principles that explain the influence of the social environment: reward and information. The principle of *reward* is based on the basic learning process: if someone behaves in a way and expresses opinions that agree with the behavior and opinions of superiors, this is rewarded. It can yield social appreciation, status and respect. The principle of *information* implies that people want to have correct information. It does not concern actual correct information, but perceived correct information. The behavior and opinions of important others are considered to be a source of information, as long as this information agrees with the information a person already has.

Thus, the social environment affects people's behavior, as it gives rewards and is a source of information. It is clear that behavior that does not agree with one's own opinions is based on the principle of reward. On the other hand, the internalization of the opinion of the social environment is primarily based on the principle of information.

Information is supposed to influence all three groups of determinants, but not directly behavior. Information is often erroneously supposed to be the only or most important determinant of behavior. Information about nutrition and about the relationship between a behavior and a health problem is a prerequisite, but rarely a motive for behavioral change.

The third, and sometimes very important, group of influencing factors are the *possibilities* or impossibilities of displaying a behavior. There is the example of dieticians who advised women to buy skimmed milk, while there was no such milk available in the small towns where the women lived. This is an example of an impossibility external to the person. Other examples are the unavailability of money or time and non-cooperative members of the family. Impossibilities for behavior can also be internal to the person, e.g., lack of information, skills, or perseverance.

Often, influences from the social environment and possibilities and impossibilities go together. For example, studies on weight loss show that in particular family members have a negative influence on long-term changes in dietary habits (by the principle of negative rewards) and that the people involved, in spite of their intentions, are often not able to stand up against this effectively.

Apart from the three main groups of determinants, factors such as age, education, and sex determine behavior. Their effects are indirect: through attitude, social influence, and possibilities/impossibilities. Physiological variables can also be considered external variables. Physiological processes underlie attitudinal considerations such as taste. Also, hunger can reduce the importance of certain negative consequences of a behavior. Another important external variable is habit. In general, people do what they are used to doing. Especially with regard to habitual behavior, the frequency of former behavior will predict future behavior. If the behavior in question is easy and does not need much consideration, habits will predict future behavior independent from the three main groups of determinants. Dietary behavior is often habitual. The attitude model is summarized in Figure 7.1.

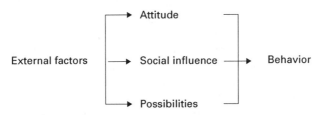

Figure 7.1 Determinants of behavior.

7.4   *Studies on determinants of dietary behavior*

Research on determinants of dietary behavior is not as developed as research on other health behaviors, such as smoking and alcohol consumption. There is no clear conceptual framework. The few studies carried out give an ad hoc impression.

One of the possible explanations for this is the earlier-mentioned complexity of the concept of dietary behavior. In addition, until recently the knowledge acquired in studies of other health behaviors had not been applied to dietary behavior: many studies on dietary behavior focus on just one determinant, such as knowledge and attitude. Also, many studies have been carried out on special subjects, such as athletes. Finally, it seems that a practical application of the study results (e.g., the study results in points of impact for an intervention) was not the leading motive among the majority of the researchers. Usually, the relationship with the problem is not clear.

The results of several studies on the relationship between one or more of the above-mentioned determinants and specific dietary behaviors will be described in the following.

In one research project, the effects of attitude, social influence, and knowledge (of product composition) on the consumption of meat, meat products, butter, and milk, were examined. The attitude towards these products was found to be a better predictor of actual consumption of the four products than social influence. The behaviors did not show a relationship with knowledge.

Studies on salt intake and consumption of skimmed, low-fat, and whole milk also showed attitude to be a stronger determinant than social influence. A positive attitude towards salt intake appeared to be primarily determined by the perceived advantage that food *tastes* better if it contains salt. In the case of milk, considerations of taste, *nutritional value* and suitability for a specific purpose, like making pudding, primarily determined the attitude. Financial considerations played no role at all. On the other hand, the opinions of family members strongly affected the choice of the milk type.

The frequency of consumption of ice cream, sweet yogurt and soda has been reported to depend more on taste considerations and opinions of others than on perceived health advantages or disadvantages of the behaviors.

The preferences for different kinds of meat were studied among elderly people (age 65 to 80). The respondents were asked to value 4 product characteristics of 11 kinds of meat. The characteristics were:

- sensoric quality (good taste, juicy, nice smell, nice appearance)
- amount of fat and unhealthiness (bad for health, bad for coronary heart diseases, overweight)
- exclusiveness (for special occasions, for the weekends, exclusive, expensive)
- convenience in preparing (short preparation time, unsuitable to prepare for more than one day).

The preference for the different kinds of meat could be explained for 16%, 14%, 13%, and 8% by the respective perceived product characteristics. There were, however, differences between preference and actual consumption.

In another study, mothers of pre-school children were interviewed. The responses of working-class mothers showed that *taste* was the most important determining factor for the choice of food in their families. In the case of middle-class mothers, however, health considerations were very important.

In a study on food choice in canteens, employees were asked prior to lunch which attitude and social considerations would play a role in their choice of lunch. The employees answered that prior to lunch they were guided by advantages such as taste, health and convenience. Behavior and opinion of the social environment were not very important. Afterwards, the first three considerations, and especially health considerations, appeared to be not so important in the actual choice of lunch. *Social influences* (friends chose the same lunch; friends said I should take this lunch) were relatively more important.

Further, the effects of experience with a food product, knowledge of nutrients, and several positive and negative aspects, such as taste, satiation, health and price on among other things, the consumption of milk, whole wheat bread, margarine and salads have been studied. *Taste* was found to be the most important determinant of all dietary behaviors. *Health* considerations were relatively more important for the elderly than for young people, as also shown in other studies. Finally, the results of a study on the determinants of dietary behavior in children showed that health considerations hardly influence children's dietary behavior at all. The same applied to financial considerations. However, *taste* considerations and the influence of the *social environment*, especially the behavior of the parents, were very strong determinants.

It has already been mentioned that, according to the theoretical model, *socio-demographic factors* have an (indirect) influence on behavior. This has been studied for some specific dietary behaviors.

Household income does not strongly affect the consumption of milk, bread, and eggs. The consumption of meat, fish, fresh vegetables, snacks, candies, whole wheat bread, and skimmed milk, however, was positively related to income. Larger families consume relatively less fresh vegetables and fresh fruit. Households with a highly educated mother consume more fruit, vegetables, milk products, meat, game, and fish than households with a less well-educated woman. Becoming older does not result in major changes in the choice of food products. However, the total energy intake decreases. No real differences between men and women in dietary behavior have been found. Women eat less, but not less varied food. In general, the interindividual differences in dietary behavior can only be partly explained by socio-demographic variables. Some studies have shown that, indeed, socio-demographic variables influence behavior indirectly through attitude or social influence, for example:

- health considerations have more influence on the dietary behavior of elderly people than on that of younger people;
- women, from higher socio-economic classes and ranging from 26 to 45 years of age have a more negative attitude towards the consumption of meat, meat products, butter, and milk than others.

7.5 Summary

A person's cultural environment is the central determinant of dietary behavior. The key question in this chapter is: what determines the differences in dietary behavior within a culture? In theory, three main groups of determinants can be distinguished at the microlevel: attitude, social influence, and possibilities. As far as the determinants of specific dietary behaviors are concerned, it can be concluded that the concept "attitude" can explain the interindividual differences in dietary behavior very well. Especially, taste considerations have quite a considerable effect on dietary behavior. For some people (mostly elderly), health implications are important; for others they are not important at all. With regard to the social aspects of dietary behavior, it is remarkable that behavior is not always related to this determinant. This might be due to the methods used to measure social influence: direct questions about the influence of the social environment are not effective, as people are not aware of the influence of the social environment or do not want to admit that they are influenced by their environment.

Information on the determinants of dietary behavior is incomplete. Research on the third main group of determinants, possibilities, is almost completely absent. Furthermore, many studies focus on just a small number of possible determinants, for example on attitudes or social influences on the behaviors only. Information on whether one determinant is more important than another cannot be acquired in this way.

More research into the determinants of specific dietary behaviors, based on a more complete theoretical framework, is necessary. The results of such studies should be aimed at points of impact for interventions.

Reference and reading list

Ajzen, I., The theory of planned behavior, in: *Organizational Behavior and Human Decision Processes* 50, 1991.

Axelson, M.L., The impact of culture on food-related behavior, in: *Ann. Rev. Nutr.* 6 , 345–363, 1986.

Bandura, A., *Social Foundations of thought and action*. Englewood Cliffs, N.J.: Prentice Hall, 1986.

Contento, I., The effectiveness of nutrition education and implications for nutrition education policy, programs and research: A review of research, *J. Nutr. Educ.*, 27, 277–418, 1995.

Dalton, S., Food choice: intention and practice, a study of intention and actual selections, in: *Hygie* 6, 9–11, 1987.

Fishbein, M. and I. Ajzen, *Belief, attitude , intention and behavior*. Reading, Mass., Addison Wesley, 1975.

Green, L.W., M.W. Kreuter, *Health promotion planning, an educational and environmental approach*. Mountain View, Mayfield, 1991.

Hayes, D., C.E. Ross, Concern with appearance, health beliefs and eating habits, in: *Journal of Health and Social Behavior* 28, 120–130, 1987.

Hochbaum, G.M., Strategies and their rationale for changing peoples eating habits. *Journal of Nutrition Education* 13 (suppl.), 59–65, 1981.

Hollis, J.F. T.P. Carmody, S.L. Connor, S.G. Fey, J.D. Matarazzo, The nutrition attitude survey: associations with dietary habits, psychological and physical well-being, and coronary risk factors, in: *Health Psychology* 5, 359–374, 1986.

Hulshof, K.F.A.M., *Assessment of Variety, clustering and adequacy of eating pattern, Dutch National food consumption survey. Thesis*. University of Limburg, Maastricht, The Netherlands.

Kok, G., H. Schaalma, H. de Vries, G. Parcel, and Th. Paulussen, Social psychology and health education, in: W. Stroebe and M. Hewstone, Eds., *Eur. Rev. Soc. Psychol. 7*, Chichester, Wiley, 1996.

Kok, G.J., H. de Vries, A.N. Mudde, V.J. Strecher, Planned health education and the role of self-efficacy: Dutch research. *Health Education Research* 6, 231–238. 225–233, 1991.

Krondle, M., M.S. Coleman, Social and biocultural determinants of food selection, in: *Progress in Food and Nutrition Science* 10, 179–203, 1986.

Lau, S.D., *Nutrition behavior analysis: food perceptions as determinants of food use*. Dissertation University of Toronto, 1985.

Michela, J.L., I.R. Contento, Cognitive, motivational, social and environmental influences on childrens food choices, in: *Health Psychology* 5, 209–230, 1986.

Murphy, B.M., *Psycho-social factors that discriminate between people who report having made desirable changes in their diets from those who have not*. Dissertation Columbia University Teachers College, 1985.

Prattala, R. and M. Keinonen, The use and the attributions of some sweet foods, in: *Appetite*, 5, 199–207, 1984.

Shepherd, R., L. Stockley, Nutrition knowledge, attitudes, and fat consumption, in: *Journal of the American Dietetic Association* 87, 615–619, 1987.

Shepherd, S., C.A. Farleigh, Preferences, attitudes and personality as determinants of salt intake, in: *Human Nutrition: Applied nutrition* 40, 195–208, 1986.

Tuorila, H., Selection of milks with varying fat contents and related overall liking, in: attitudes, norms and intentions. *Appetite* 8, 1–14, 1987.

Vries, H. de, M. Dijkstra, P. Kuhlman, Self-efficacy: the third factor besides attitude and subjective norm as a predictor of behavioral intentions, in: *Health Education Research* 3, 85–94, 1988.

Part 2

Adverse effects of food
and nutrition

chapter eight

Introduction to adverse effects of food and nutrition

V. J. Feron

8.1   Introduction

Part 2 consists of eight chapters dealing with the induction of toxic effects by food components and food products, and the mechanisms underlying these adverse effects. Disorders related to food, like obesity, will not be dealt with, as the central theme of this book is food safety, treated from a toxicological point of view.

8.2   Two major problems in food safety assessment

Food toxicologists are confronted with two major problems:

(a)   food and food products are complex chemical mixtures of variable composition;
(b)   the existing procedures for extrapolation of animal toxicity data to man are incompatible with Recommended Dietary Allowances (RDAs) (see Chapter 12, Section 12.1; and Chapter 17, Section 17.2.1) for many essential nutrients and also with the

normal use of many common foods and food products. Current guidelines for
toxicity testing of chemicals are inappropriate for (macro) food compounds, food
products, and foods. Specific approaches for the safety evaluation of foods and food
chemicals are to be pursued.

Aspects of these two problems will be discussed in several chapters of Part 2, and will
return in certain sections of Part 3 dealing with risk assessment.

8.3   Toxicity (testing) of food chemicals and foods

In order to outline the way in which adverse effects of foods and food chemicals are
currently measured and the underlying mechanisms are examined, the toxicity (testing) of
food chemicals, food products, and foods will be discussed along the following lines:

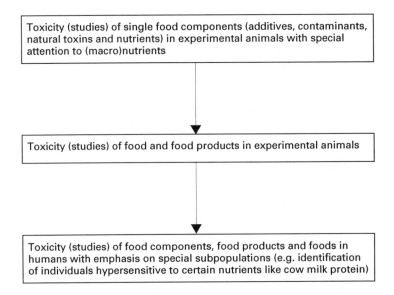

In essence, two lines are combined using this approach: one from single substances to
complex mixtures, and the other from studies in experimental animals to studies in
humans. The combination of both lines ends in the assessment of toxicological risks due
to foods and food products (complex mixtures) in humans, including high-risk groups.
The titles of the various chapters reflect this approach.

This chapter begins with a brief description of a number of general toxicological
principles (Section 8.4) followed by a discussion of the above-mentioned major problems
facing food toxicologists (Sections 8.5 and 8.6). Further, a short survey is given of the
toxicology of the various categories of food components (Section 8.7). Next, the topics
discussed in the other seven chapters of Part 2 are touched upon (Section 8.8). Finally, the
characteristics and practical aspects of toxicity testing of food chemicals and (complex)
food products are described (Section 8.9).

8.4   Toxicity of (food) chemicals

Toxicity (or hazard) is the potential of a chemical to induce an adverse effect in a living
organism e.g., man. Each chemical, and thus also each food component, whether it is an

essential amino acid, a trace element, vitamin, contaminant or additive, has its own specific toxicity. Whether a food component is of natural origin or is man-made is irrelevant for its health hazard.

Generally, information on the toxicity (hazard) of food chemicals is obtained from studies in experimental animals, *in vitro* studies, studies in volunteers, or epidemiological studies. The main goals of these studies are to determine (a) the type of adverse effects, (b) dose–effect relationships including the no-observed-adverse-effect levels, and (c) the mechanisms underlying the adverse effects.

The induction of biological effects or toxic effects largely depends on the disposition of the substances concerned. The interaction of a substance with a living organism can be divided into a kinetic phase and a dynamic phase. The kinetic phase comprises absorption, distribution, metabolism, and excretion. It concerns the fate of a substance in the body: along which routes does the substance enter the body, and in what way is it distributed, metabolized, and excreted? For example, the pathway along which a substance is absorbed may largely affect the type and intensity of its effects. An inhaled substance reaches the blood circulation through the lungs, while a food component passes the liver, which is the main organ involved in biotransformation. This means that the body has a number of defense mechanisms at various levels of the kinetic phase, metabolism, and excretion. These mechanisms are aimed at detoxication of the substances that enter the body. However, the systems involved in detoxication may be saturated with certain chemicals. But also, they may convert the parent substance into a toxic reactive intermediate (bioactivation).

8.5   Food, a complex mixture of variable composition

Part 1 has shown that the chemical composition of food can be extremely complex and variable. Food products are estimated to consist of several hundred thousands of different chemicals. Usually, the toxicity of such a complex mixture does not simply depend on the toxicities of the individual components. Interactions may occur that lead to synergism or antagonism. Moreover, the rule of additivity may apply to the induction of effects (summation of effects). This means that combined actions may occur. It is impracticable to test every single substance for toxicity. Even if the toxicity data and Acceptable Daily Intakes (ADIs) (see Chapter 17, Sections 17.3.2 and 17.3.3) of all food components were available, there would still be the problem of many possible interactions and combined actions.

Addition of new chemicals to food should meet the requirements that guarantee safety. The only adequate way to deal with the safety problems of food chemicals already in use is the development and implementation of a priority-setting system based on the amount ingested, the number of consumers, potential toxic effects of food components, or combinations of groups of food components, and possible interactions between components.

8.6   Problems in toxicity testing and extrapolation of animal data to man

For most food additives and for many contaminants, the amount allowed for human consumption is at most 1% of the highest dose shown to cause no adverse effect in an appropriate animal study.

Obviously, it is impossible to give animals 100 times the amount of a macronutrient (e.g., single-cell protein, fat substitute, chemically modified starch) anticipated to be consumed by humans. Therefore, a safety factor of 100 cannot be applied to calculate an ADI. Instead, the safety data base of such food products should be expanded beyond the traditional requirements or, in other words, the safety factor may be reduced on the basis of additional information. This information may be obtained from studies on absorption, distribution, metabolism, and excretion in humans and non-human primates, from long-term studies in appropriate primates, from studies in humans on the possible effects on vitamin and mineral state, and from very specific toxicity tests, e.g., for immunotoxicity and neurotoxicity. In brief, the safety evaluation of macronutrients requires more fundamental information on their effects on physiology and their toxicology. Such information forces toxicologists to abandon their strict safety procedures and to seek integration of their approach with that of nutritionists. This does not ask for more rules but rather requires a case-by-case approach on the basis of carefully discussed, well-reasoned safety evaluation procedures for (macro)nutrients, food products, and foods.

8.7   Categories of food components

Generally, food components are classified into four groups: nutrients, non-nutritive naturally occurring components, including antinutritives and natural toxins, and man-made contaminants and additives. For many food chemicals which are necessary for life, the margin between RDA and minimum toxic dose is often much smaller than a factor 100 or even 2, for example for fat. This is quite understandable, since nutrients play an essential role in the maintenance of homeostases. A slight overintake of nutrients may lead to exceeding the limits within which the homeostases should be kept.

As will be discussed in Section 8.8, there are large gaps in our knowledge of the toxic potential of the majority of natural food components and the consequences of their intake for human health. It is clear that this group of food components should have a high priority with regard to further toxicological research.

The requirements for testing the toxicity of man-made contaminants and their toxicological evaluation are similar to those for additives. ADIs are assessed and standards are set.

Before a chemical is admitted as a food additive, extensive toxicological research is required. The results are often the basis for assessing the recommended limit values, such as ADIs, usually applying a safety factor of 100. The levels of additives in food are usually much lower than the ADIs. Therefore, food additives are relatively safe.

Chapters 9 through 12 deal with the toxicology of the various groups of food components in general and the mechanisms underlying their toxic effects in particular. Consumers are increasingly confronted with food products and food components produced with modern technological methods.

Chapter 13 concerns the toxicology of mixtures of the chemical substances that make up our food. It looks at the different types of possible interactions between substances (antagonistic or synergistic) and of independent combined actions of substances (presence or absence of additivity or no additivity).

Not enough is known yet about the prevalence of food allergies and intolerances. Estimates vary widely and are unreliable. It is not easy to diagnose a food allergy and to identify the food component provoking the allergic reaction. The same holds for food intolerance. In Chapter 14, the different types of food allergy and food intolerance and the associated problems are discussed.

The final chapter of Part 2 is an introduction to the use of epidemiological methods in general, and to the application of epidemiology in studying associations between diet and adverse effects of food (components) in particular. The basic principles of epidemiology will be covered at an introductory level. Topics include types of study design, dietary exposure, disease outcome, causality, validity, bias, interpretation and integration of epidemiological data with animal data. Special attention will be paid to the possibilities and limitations of diet assessment methods and the use of biomarkers in studying diet–disease associations.

8.8 Rank order of hazards from food components

In general, in developed countries food safety is adequate. However, it should be noted that the information on the (chronic) toxicity of natural food components is insufficient. Further, a number of important health problems such as cardiovascular disorders, diabetes, osteoporosis, obesity, allergy, and cancer are believed to be related to nutrition. Nutritional interventions could drastically reduce the incidence of these diseases.

There is a consensus of opinion among experts (nutritionists and food toxicologists) that food hazards should decrease in the following order of importance:

8.8.1 Wrong dietary habits

These are believed to be main causes for the nutrition-related disorders mentioned above. A more balanced diet means changes in dietary habits: energy according to need, and less fat, cholesterol, salt, sugar, and alcohol, and more dietary fiber. Nutritionists and toxicologists are well aware of the fact that for nutrients the margin between physiological need and safe dose is often very small. Large safety factors cannot be applied. More basic information on the physiology and toxicology of macro- and micronutrients is required. Such information may be used for recommendations aimed at changing dietary habits (see also Part 1B, and Part 3, Chapter 22).

8.8.2 Microbial contamination

Food can serve as a vehicle or growth medium for pathogenic microorganisms. The incidence of food-borne diseases due to microorganisms is estimated at some hundred thousands of cases per year in the Netherlands and a staggering 20 million or more cases per year in the US. Worldwide, the number of cases of food-borne diseases is astronomical.

A distinction can be made between food-borne infections and microbial food intoxications. The former are caused by the pathogenic microorganisms themselves, the latter by toxins produced by microorganisms in the food. Table 8.1 lists the main microorganisms involved in either food infection or food intoxication. In view of the central theme of this book, food infections will not be dealt with any further. Food intoxications will be discussed in detail in Chapter 11 in which naturally occurring toxins are discussed.

8.8.3 Natural toxins

The number of naturally occurring non-nutritive chemicals in foods is unknown, but is probably larger than 500,000. Only a small portion has been identified chemically and only a few have been submitted to adequate toxicological examination. In contrast, synthetic pesticides, food additives, and industrial contaminants have been subjected to extensive toxicological screening. This has led the public to believe that man-made chemicals are potentially more hazardous to humans than natural chemicals.

Based on their origin, natural toxins associated with foods can be divided into four groups, as listed in Table 8.2. The toxicology of the different classes of natural toxins will be discussed in Chapter 11, using important representatives as examples.

8.8.4 Man-made contaminants

Man-made contaminants are substances unintentionally present in foodstuffs or their raw materials. They may occur as the result of production, processing, preparation, packaging,

Table 8.1 Microorganisms causing food-infections or food intoxications

Microorganism	Pathogenicity
Salmonella	infection
Shigella	infection
Escherichia coli	infection
Yersinia enterocolitica	infection
Campylobacter jejuni	infection
Listeria monocytogenes	infection
Vibrio parahaemolyticus	infection
Aeromonas hydrophila	infection
Staphylococcus aureus	enterotoxin
Clostridium botulinum	botulinum toxins
Clostridium perfringens	enterotoxin
Bacillus cereus	enterotoxin, emetic toxin
Aspergillus flavus	aflatoxins
Penicillium citrinum	citrinin
Aspergillus ochraceus	ochratoxin
Aspergillus versicolor	sterigmatocystin
Penicillium claviforme	patulin
Fusarium graminearum	zearalenone

Table 8.2 Classification of natural toxins according to their origin

Toxins	Organism	Toxic product (examples)
Bacterial toxins	Bacteria	Botulinum toxin
Mycotoxins	Fungi	Aflatoxin
Fycotoxins	Algae	Diarrhetic shellfish poison
Fytotoxins	Plants	Solanin

transport or storage of foods or their raw materials, or as a result of environmental contamination. By definition, contaminants are unintentional, but some are present as a result of intentional applications, e.g., residues of pesticides, additives to feedstuffs, or veterinary drugs.

To protect people against hazards from contaminants, governmental agencies in many countries have developed and implemented legislation in which approval and establishing of ADIs is regulated. Once a pesticide is approved, conditions leading to its safe use are imposed. For example, a safety period between the last treatment of a crop and its harvest is specified. Also, the maximum residue level must be as low as consistent with Good Agricultural Practice and always low enough to avoid exceeding the ADI.

8.8.5   *Additives*

Food additives are chemicals that are intentionally added to foods or their raw materials to preserve or improve the quality of the product. The increasing demand for food by an ever-increasing world population, as well as by changes in lifestyles in developed societies, has led to the use of additives to preserve foods or to process raw foods into nutritionally adequate ready-to-eat foods. Examples of types of additives are preservatives, antioxidants, colorings and color-preserving substances, flavorings, thickening and emulsifying agents, stabilizers, bleaching agents, moisture repellants, and defoaming agents.

In the Netherlands, for example, the admittance of additives is regulated in the Commodities Act. Regulation under this act has taken the form of a so-called positive list. This list contains all chemicals that have been approved as food additives. It also incorporates detailed specifications concerning identity and purity of the substance, the purpose of using it, and its maximum permitted concentration in foodstuffs or categories of foodstuffs designated by name.

Nearly all major additives have been subjected to a thorough toxicological evaluation on the basis of which an ADI is established. Because of this rigorous toxicological evaluation and the application of large safety factors (generally 100 or more) in calculating ADIs for food additives, this category of food components ranks at the bottom of the list of foodborne hazards, far behind nutrients, microbial toxins, food infections, natural toxins, pesticides, and environmental contaminants.

8.9 Identification of health hazards due to food chemicals and foods

The results of proper toxicological and epidemiological studies are the only scientific basis for assessing the level of exposure to a specific (food) chemical that is low enough to avoid unacceptable health risks. Toxicological data are generally obtained from various types of animal experiments, *in vitro* studies, and studies in humans. Studies in experimental animals have become the main source of toxicological data, although ideally the data should be obtained from humans because the ultimate goal is to assess the health risk from chemicals to humans. *In vitro* studies using organ and cell cultures of animal and human origin are increasingly used to study the mechanisms underlying the adverse effects.

Epidemiological studies are one type of studies in humans. The possibilities of epidemiological studies to detect and quantify adverse effects of food components and foodstuffs are discussed in Chapter 15.

The next sections introduce animal experiments, *in vitro* studies and studies in human volunteers, focusing on aspects of particular interest in testing food components, food products, and foodstuffs.

8.9.1 Animal experiments

Guidelines drawn up by the Joint FAO/WHO Expert Committee on Food Additives and the Scientific Committee for Food of the European Union provide a general outline for the toxicity testing of food components. The guidelines of the Organization for Economic Cooperation and Development (OECD) for toxicity testing of chemicals give more details of the design of studies, the way in which studies should be carried out, and the parameters to be used. However, it is evident that the end points specified in the OECD guidelines are not always appropriate for providing relevant toxicity data on a food component, particularly in the case of macroingredients such as bulk sweeteners, fat substitutes, modified starches, and novel food products. Some experienced food toxicologists believe that the traditional (guideline) approach to toxicity testing has not only impeded the development of toxicology as a science, but has priced itself out of the market as far as food chemicals are concerned. Others are somewhat more cautious with their criticism but feel that an effort should be made to relate toxicological findings more to the human situation. For instance, more attention should be paid to parameters characteristic of the cardiovascular system, the immune system, and the central nervous system.

The relevance of the major end points specified in the OECD guidelines for the hazard assessment of food chemicals is critically analyzed in the following paragraphs.

Acute toxicity. A potential food component rules itself out if it is acutely toxic to a considerable extent. Therefore, determination of LD_{50} (acute dose that is lethal to half of the exposed animals) should not be required as a major end point for a food component. Only range-finding studies (e.g., a one-week multiple dose feeding study in rats) would be necessary to ensure that the ingredient proposed for use in food has a low acute toxicity.

Subacute/subchronic repeated dose studies. These are important for examining the safety of food components. The substance is added to the feed or drinking water to imitate exposure to humans. Special attention should be paid to the composition of the diet, if the substance under investigation is a macronutrient, because in that case it usually has to be incorporated into the diet at levels as high as 20 to 60% at the expense of a comparable nutrient. Examples are alkaline treated proteins, protein concentrates from bacteria or yeasts, and chemically modified potato or maize starch. Aspects to be checked are, for example, vitamin and mineral content and their bioavailability to avoid nutrient deficiencies, which could strongly influence the results of the toxicity studies and, thus, lead to erroneous conclusions. The problems associated with toxicity testing of macronutrients, food products, and new foods have already been touched upon (Section 8.6) and will be discussed in more detail in Chapter 12.

Allergy. Testing for allergic sensitization is highly relevant. However, the commonly used animal models only detect substances that are active on the skin and/or after inhalation. Substances which are highly active in such tests are unsuitable as food components. Some food additives may cause intolerance reactions in certain individuals with symptoms similar to genuine allergic reactions. Therefore, there is a need for studying these end points in the testing of food ingredients. Currently, however, there is no animal model or *in vitro* test system available that unequivocally reveals intolerance. Testing in volunteers should be considered (see also Chapter 15).

Reproductive toxicity. Reproduction toxicity tests of food components are necessary. They should include male and female fertility and reproduction, multi-generation, and teratogenicity tests.

Long-term studies. For food components, long-term studies may not always be necessary. In the guidelines of the Scientific Committee for Food of the European Union, a decision point approach is recommended. For example, if the food ingredient is a simple ester that on hydrolysis yields products identical to substances of the normal metabolism, no testing beyond a subchronic study is needed. Similarly, chronic toxicity and carcinogenicity tests may be unnecessary for peptides, proteins, carbohydrates, and fats which by chemical analytical and metabolism studies can be shown to consist of well-known sequences of amino acids, mono- and disaccharides, and fatty acids. Nevertheless, if such substances are to be used in large amounts or will have a widespread use, long-term studies may be warranted.

Mutagenicity tests. The testing of mutagenicity as an end point is a subject of discussion concerning its relevance to food components. The present state may be summarized as follows. The significance of mutagenicity per se as an end point for food components is not clear and no regulatory agency seems willing to use positive results in mutagenicity tests alone as grounds for non-admittance of a food component. In addition, the faith in mutagenicity tests as pre-screens for carcinogenicity is declining. A positive response does not need to be proof of carcinogenicity. However, mutagenicity or genotoxicity is considered a very important end point in evaluating carcinogenicity data from animal tests. If a substance is found to be genotoxic, especially when tested *in vivo*, positive results of carcinogenicity tests make admittance as a food component very difficult if not impossible. On the other hand, quite a few non-genotoxic carcinogens are widely used as food additives, for example, butylated hydroxyanisole as an antioxidant, cyclamate, saccharin, and lactitol as artificial sweeteners, and propionic acid as a preservative. For these non-genotoxic carcinogens ADIs have been calculated in a way similar to that used for other non-genotoxic, non-carcinogenic substances (see also Chapters 19 and 21).

8.9.2 In vitro *studies*

Isolated cells, tissues, and organs are increasingly used in toxicological research. Major advantages of these *in vitro* systems are:

- toxic effects can be studied independent of other compartments in the body;
- the systems are often very sensitive, and effects can be measured or calculated directly;
- *in vitro* systems are excellent tools for screening substances for organ-directed toxicity;
- molecular studies are easier than *in vivo* studies;
- phenomena and mechanisms can be studied in human cells which allows direct comparison of effects on human cells with effects on animal cells, which possibly makes extrapolation of toxicity data from animal to man more meaningful.

On the other hand, each model system has its limitations. The major disadvantage of *in vitro* systems is that there is no integration of cells, tissues, or organs as in an intact and functioning whole animal or human physiological system, and hence, no elimination by excretion whether or not in combination with biotransformation.

Of special interest to food toxicologists are *in vitro* systems using pieces of intestine and intestinal epithelial cells, for instance to examine the mechanism of absorption of substances, and interactions at the absorption level between xenobiotics, or between micronutrients and other food components. Such *in vitro* intestinal systems are successfully used to study the mode of action of so-called antinutritive factors such as lectins.

8.9.3 *Studies in volunteers*

From an ethical point of view, studies in volunteers can only be carried out if careful evaluation of all available data leads to the conclusion (preferably drawn by an independent ethical committee) that no unacceptable risk is being run. The end points should be short-term and indicative of reversible disturbances of physiology rather than of cell, tissue, or organ damage. Of particular relevance are absorption, distribution, metabolism, and excretion studies. Such studies in man would certainly contribute to more confident interspecies extrapolation (see also Chapter 18).

8.10 Summary

This first chapter of Part 2 dealing with adverse effects of food, introduces the characteristics of the toxicology and toxicity testing of food chemicals, food products, and foods. Food is a complex mixture of chemicals, the toxicity of which also depends on possible interactions between components. The current safety evaluation procedures of food additives and contaminants are incompatible with the RDAs of many essential nutrients and also with the normal use of many common foods. Hazards posed by food decrease in the following order: wrong dietary habits (too much food, too fat, too salt, too few fresh vegetables and fruits), food infection, natural toxins including microbial toxins, man-made contaminants, and finally additives (considered among the safest food components). Major aspects of the toxicology of the various categories of food chemicals (additives, contaminants, natural toxins, and nutrients) are briefly discussed. Finally, a brief description is given of the methods for studying the toxicity of food chemicals and foods (studies in animals, *in vitro* studies and studies in volunteers) focusing on experiments in animals as the main source of toxicological data.

Reference and reading list

Aeschbacher, H.-U, Potential Carcinogens in the Diet, in: *Mut. Res.* 259, 201–410, 1991.

Concon, J.M., *Food Toxicology, Part A: Principles and Concepts.* New York, Marcel Dekker, 675, 1988.

Concon, J.M., *Food Toxicology, Part B: Contaminants and Additives.* New York, Marcel Dekker, 676–1371, 1988.

Enne, G., H.A. Kuiper and A. Valentini, *Residues of Veterinary Drugs and Mycotoxins*, Proc. of a teleconference held on Internet, April 15 - August 31, 1994. Wageningen, Wageningen Pers, 1996.

Hathcock, J.N., *Nutritional Toxicology*, Vol. I. New York, Academic Press, 515, 1982.

Hathcock, J.N., *Nutritional Toxicology*, Vol. II. New York, Academic Press, 300, 1987.

Hathcock, J.N., *Nutritional Toxicology*, Vol. III. New York, Academic Press, 159, 1989.

Kroes, R. and R.M. Hicks, Re-evaluation of Current Methodology of Toxicity Testing Including Gross Nutrients, in: *Food Chem. Toxicol.* 28, 733–790, 1990.

Miller, K., *Toxicological Aspects of Food.* London, Elsevier Applied Science, 458, 1987.

National Research Council, *Carcinogens and Anticarcinogens in the Human Diet.* Washington, D.C., National Academy Press, 1996.

Niesink, R.J.M., J. de Vries, and M.A. Hollinger, Eds., *Toxicology: Principles and Applications.* Boca Raton, FL, CRC Press, 1996.

Taylor, S.L. and R.A. Scanlan, *Food Toxicology, A Perspective on the Relative Risks.* New York, Marcel Dekker, 453, 1989.

Tu, A.T., Ed., Food Poisoning, Vol. 7 in: *Handbook of Natural Toxins.* New York, Marcel Dekker Inc., 1992.

chapter nine

Adverse effects of food additives

H. Verhagen

9.1   Introduction

The increasing demand for "ready-to-eat" foods, snacks, and a continuous assortment of foodstuffs, even if they are out of season, and the change in lifestyle that has taken place during the last centuries, have led to an increase in the use of food additives.

Food additives are substances that man adds to food intentionally to provide protection against contamination with microorganisms, to prevent oxidative deterioration of oils, fats, and shortenings, to keep food appealing and tasteful, and to improve its texture, etc. By using additives, foodstuffs can be kept for long periods of time, and restricting oneself to seasonal foods is no longer necessary. Food additives thus fulfill valuable functions in our daily food and as such, cannot be left out. Currently, there may be as many as 2800 substances in use as food additives, of which the majority (about 2500) are naturally occurring flavoring substances. However, the most important additives are only a handful, and most of the additives of natural origin are used in trace amounts. In the 1970s, the US Food and Drug Administration estimated the use of sucrose, corn syrup, dextrose, and salt at 93% (w/w) of the total use of food additives. If black pepper, caramel, carbon dioxide, citric acid, modified starch, sodium bicarbonate, yeasts, and yellow mustards are included, the percentage comes to about 95%. In fact, food additives can be considered food-oriented substances.

Food additives can be divided into two broad groups: synthetic and natural substances. The latter group includes substances of plant and in some cases, of animal origin. Synthetic food additives are extensively tested for toxicity before they are allowed for use in food. Several additives of natural origin have also been tested.

Obviously, only substances that have been shown to pose no serious toxicological risks at levels anticipated for consumption are admitted for use as food additive. This also applies to substances for which there is conclusive evidence for nongenotoxic carcinogenicity from lifetime bioassays in rodents, for example, the antioxidant butylated hydroxyanisole and the sweetener saccharin. For synthetic substances, no-observed-adverse-effect levels (NOAELs) (see Sections 16.3.2.1, 17.3.2, and 19.2.2) are established. Subsequently, acceptable daily intakes (ADI) are calculated by applying safety factors (SF): ADI = NOAEL/ SF. Also for the nongenotoxic food additives mentioned above NOAELs are used. Synthetic food additives are thus the safest components of the diet. In practice, the only adverse reactions to food additives are intolerances. It should be noted, however, that intolerances are not restricted to synthetic substances only.

Admitted synthetic food additives are put on a so-called positive list, indicating that only substances on that list are allowed to be used in food.

In view of the fact that the toxicological risks associated with the intake of synthetic food additives are minimal, the next sections will only deal with the most relevant toxicity data, as obtained (at the very high dose levels above the NOAEL) in experimental animals. Of each type of synthetic food additive, a few examples will be given. Intolerance reactions are discussed in Chapter 14.

9.2   Food colorings

Color is a property of foodstuffs that makes them visually attractive. The use of artificial colorings started at the beginning of the 19th century. At that time, there were no restrictions, and several cases of abuse have been recorded. For example, many people died from eating sweets and puddings colored with arsenic derivatives, and cheese dyed with red lead and vermilion (HgS). Foods were frequently colored to mask that they had been diluted with cheap ingredients. For example, at the turn of the century, milk was colored yellow to hide skimming and dilution with water. This practice was so widespread that people refused uncolored milk for fear of adulteration. Nowadays, food adulteration is prohibited by law, and foods are colored either by naturally occurring pigments (e.g., dried algae meal, paprika, beet powder, grape skin extract, caramel, carrot oil, ferrous gluconate, iron oxide) or by artificial food colorings (e.g., tartrazine and erytrosine). Most food colorings used are synthetic. They are cheaper, more intense, and more stable than their natural counterparts. Concerning artificial food colorings, there is a controversy whether there is an association between the intake of the colorings and intolerance reactions in children. Here, the various aspects of tartrazine, as an example of such a food coloring, will be briefly described.

9.2.1   Tartrazine

Tartrazine is a yellow synthetic azo dye. Several clinical symptoms have been attributed to tartrazine, including asthma, hyperactivity of children, and urticaria (hives).

Much attention has been paid to the induction of effects in asthma patients after the intake of tartrazine. A number of studies reported a high incidence of intolerance of tartrazine among aspirin(acetylsalicylic acid)-intolerant asthmatics. On the whole, however, little evidence has been found against the use of tartrazine in cases of asthma, even among those who are intolerant to aspirin.

With regard to hyperactivity of children, there is a controversy regarding the association between tartrazine and the hyperactivity. So far, studies on this potential problem have not provided conclusive evidence for such an association. A similar controversy links a possible association between tartrazine and urticaria. In this case too, no relationship has been found.

9.3 Preservatives

Preservatives keep food edible for long periods of time by preventing the growth of microorganisms such as bacteria and fungi. Although the public perceives preservatives in particular as hazardous, they are not only harmless at the levels ingested but in fact beneficial in that they reduce or prevent the risks due to bacterial and fungal contamination (see Part 1, Chapter 2).

9.3.1 Nitrate and nitrite

Nitrates and nitrites are used to preserve meats. For example, they contribute to the prevention of growth of *Clostridium botulinum*, the bacterium that produces the well-known highly potent botulinum toxin. The adverse effects after intake of nitrates and nitrites are methemoglobinemia and carcinogenesis, the latter resulting from the formation of nitrosamines.

Bacteria in the oral cavity can reduce nitrate to nitrite. Nitrite oxidizes (ferrous) hemoglobin to methemoglobin, which cannot bind oxygen. This may lead to a state of anoxia. The consumption of meat with high levels of nitrate and nitrite as well as of other dietary nitrate sources, such as drinking water and spinach, has resulted in life-threatening methemoglobinemia, especially in young children. Newborns are (transiently) deficient in NADH-reductase, the major system responsible for methemoglobin reduction.

Nitrite (either ingested directly or indirectly via the reduction of nitrate) also reacts with secondary amines under the formation of a variety of nitrosamines, e.g., dimethylnitrosamine, diethylnitrosamine, and N-nitrosopyrrolidine.

$$R_1{\diagdown}\atop{R_2}{\diagup} N — NO$$

$R_1 = R_2 = CH_3$ Dimethylnitrosamine
$R_1 = R_2 = C_2H_5$ Diethylnitrosamine

N — nitrosopyrrolidine

Nitrosamine formation can take place in food and *in vivo*. The acidic conditions in the stomach favor nitrosamine formation. Nitrosamines are mutagens as well as carcinogens. They induce cancer in a variety of organs, including the liver, respiratory tract, kidney, urinary bladder, esophagus, stomach, lower gastrointestinal tract, and pancreas. Nitrosamines need biotransformation for their activation. The bioactivation of nitrosamines is mediated by cytochrome P-450. It involves oxidative N-dealkylation, followed by a sequence of rearrangements to yield the alkylating alkylcarbonium ions (see Figure 9.1).

It should be noted that a decrease in the incidence of botulism may be accompanied by an increase in the formation of carcinogenic nitrosamines, as a result of an increase in the nitrite level of the meat (products).

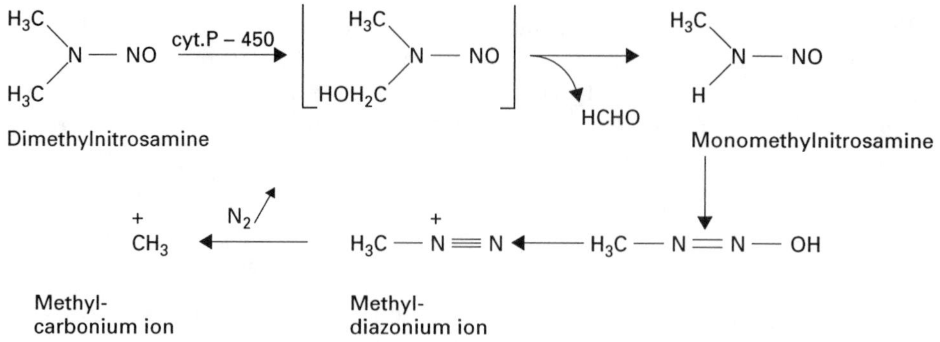

Figure 9.1   Metabolic activation of dimethylnitrosamine.

9.4 *Antioxidants*

Antioxidants are used to protect oils, fats, and shortening against oxidative rancidity and to prevent the formation of toxic degradation products and polymers.

Many foods may undergo oxidation, but particularly those containing fats are susceptible to changes in color, odor, taste, and nutritional value. Unsaturated fatty acids are readily peroxidized in the presence of molecular oxygen. The peroxidation products may induce toxic effects. Also, in biological systems peroxidation of lipids may have severe adverse consequences. Peroxidation of polyunsaturated fatty acids is believed to be involved in disturbing the integrity of cellular membranes, the pathogenesis of hemolytic anemia, and pulmonary and hepatic injury. Secondary peroxidation products, e.g., hydroxynonenal, can form adducts with DNA.

The peroxidation of lipids consists of the following steps (LH = lipid):

$$\begin{array}{lll}
\textit{initiation:} & LH \xrightarrow{\text{catalyst}} & L^{\cdot} + H^{\cdot} \\
& LH + O_2 \xrightarrow{\text{catalyst}} & L^{\cdot} + {}^{\cdot}OOH \\
\textit{propagation:} & L^{\cdot} + O_2 \longrightarrow & LOO^{\cdot} \\
& LOO^{\cdot} + LH \longrightarrow & LOOH + L^{\cdot} \\
\textit{termination:} & LOO^{\cdot} + LOO^{\cdot} \longrightarrow & LOOL + O_2 \\
& LOO^{\cdot} + L^{\cdot} \longrightarrow & LOOL \\
& L^{\cdot} + L^{\cdot} \longrightarrow & LL
\end{array}$$

In the *initiation* step, the unsaturated lipid LH undergoes hydrogen abstraction under the formation of a lipid radical $L^{\cdot}$. This process can be catalyzed by light, heat, traces of transition metals, and enzymes. The carbon-centered radical tends to be stabilized by intramolecular rearrangement to form a conjugated diene, which readily reacts with molecular oxygen (O_2) to yield a lipid peroxide radical, $LOO^{\cdot}$. This, in turn, is capable of inducing the initiation of lipid peroxidation by abstracting a hydrogen atom from another lipid molecule, also leading to the *propagation* of the oxygenation reaction. The *termination* step is characterized by the combination of two radicals. Lipid radicals can combine to form dimers, polymers, alcohols, and peroxides. Under normal oxygen tension, the rearrangement of two lipid peroxide radicals ($LOO^{\cdot}$) is most likely to yield LOOL and O_2.

Lipid hydroperoxides undergo degradation, leading to the formation of secondary peroxidation products, such as alkanes (e.g., ethane and pentane), aldehydes (e.g., malondialdehyde and hydroxynonenal), ketones, alcohols, and esters.

The purpose of using food antioxidants is to protect food from organoleptic deterioration, decrease in nutritional value, and formation of toxic products by removing radicals. Two types of antioxidants can be distinguished: radical scavengers and synergists.

Radical scavengers, like the phenolic substances butylated hydroxyanisole (BHA) and butylated hydroxytoluene (BHT), interfere with the propagation step, thereby terminating the lipid peroxidation:

$$AH + L^{\cdot} \longrightarrow A^{\cdot} + LH$$
$$AH + LO^{\cdot} \longrightarrow A^{\cdot} + LOH$$
$$AH + LOO^{\cdot} \longrightarrow A^{\cdot} + LOOH$$

AH = phenolic antioxidant   LH = lipid

The antioxidants themselves are converted to resonance-stabilized intermediate radicals $A^{\cdot}$, which is illustrated for BHA in Figure 9.2. The resulting phenoxy radical $A^{\cdot}$ may either be regenerated to the parent antioxidant AH by reducing agents or further oxidized to a stable quinone, or combine with other phenoxy or lipid peroxy radicals.

Synergists may either regenerate the parent radical scavenging antioxidants from phenoxy radicals ($A^{\cdot}$) formed in the interference with the propagation step, or act as a sequestering agent for transition metals, active catalysts in the initiation, and propagation steps of lipid peroxidation.

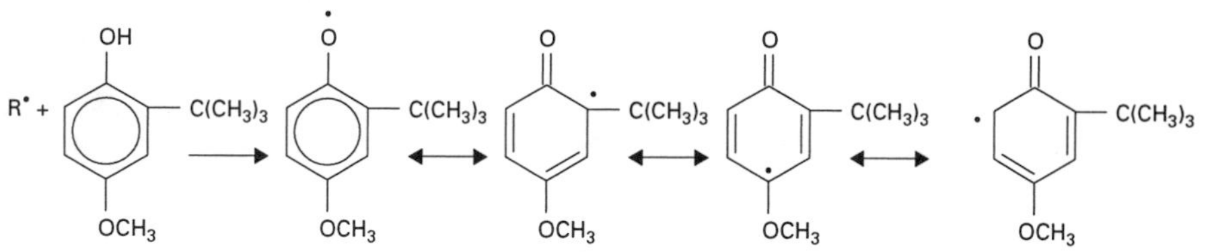

Figure 9.2   Radical scavenging by BHA; $R^{\cdot}$ = lipid free radical.

Well-known radical scavengers are α-tocopherol (vitamin E), BHA, BHT, ascorbic acid (vitamin C) and gallate esters (propyl-, octyl- and dodecylgallate). Synergistically-acting antioxidants include ascorbic acid, citric acid, and ethylenediaminetetraacetic acid (EDTA).

9.4.1   Butylated hydroxyanisole

The acute toxicity of BHA is relatively low. Its oral LD_{50} in rats is 2.5 to 5 mg/kg body weight. Owing to many years of use without adverse effects (except for a few cases of allergic reactions), BHA was given the Generally Recognized As Safe (GRAS) status by the US Food and Drug Administration. In the early 1980s, experimental data became available on the induction of tumors by BHA in rodents (rats, hamsters, and mice). Changes such as hyperplasia, papillomas, and carcinomas were observed in the forestomach (an organ that is absent in man). These changes were time- and dose-dependent (Figure 9.3).

The International Agency for Research on Cancer (IARC) evaluated that there was sufficient evidence for carcinogenicity of BHA in experimental animals to classify BHA as a (IARC) class 2B carcinogen (i.e., possibly carcinogenic for humans). However, no conclusive evidence for genotoxicity has been found. This means that BHA does not directly interact with DNA. Such carcinogens are known as nongenotoxic carcinogens and they are assumed to have threshold doses. Induction of hyperplasia is believed to play an essential role in the mechanism underlying the tumorigenicity of BHA. Recently, however, a

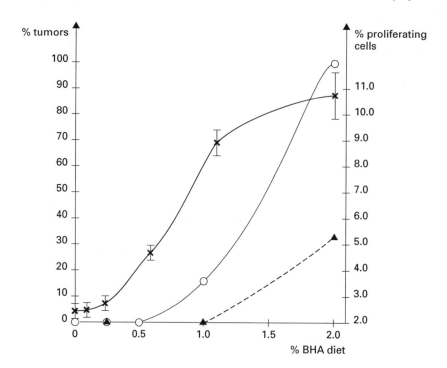

Figure 9.3 Dose-response curves for dietary BHA-induced cell proliferation (at 9 days x), carcinomas (at 2 years ▲) and papillomas and carcinomas (at 2 years o). Source: Clayson et al., 1991.

proposal has been made for the mechanism underlying the tumorigenicity of BHA, in which its biotransformation plays an essential role.

The main metabolic pathways of BHA in all species studied, including man, are glucuronidation and sulfation (Figure 9.4). Both conjugation reactions lead to detoxication and elimination of the ingested BHA. A minor metabolic pathway in several species, including man, is oxidative O-demethylation to tertiary butylhydroquinone (TBHQ, Figure 9.4). O-demethylation is relatively more important at lower dose levels. TBHQ also undergoes glucuronidation and sulfation.

In the proposed mechanism, BHA-induced tumor formation is believed to result from a sequence of reactions. This includes O-demethylation of BHA, oxidation of TBHQ to tertiary butylsemiquinone (TBSQ), and tertiary butylquinone (TBQ), conjugation of TBQ with glutathione (GSH) and ultimately formation of reactive oxygen species in the redox cycling of the TBQ-glutathione conjugate. Redox cycling leads to the formation of superoxide anion radicals ($O_2^{\cdot-}$). These radicals can spontaneously dismutate to hydrogen peroxide (H_2O_2). In the so-called Haber-Weiss reaction $O_2^{\cdot-}$ and H_2O_2 can react to form hydroxyl radicals ($OH^\cdot$):

$$O_2^{\cdot-} + H_2O_2 \xrightarrow{Fe^{2+}} OH^\cdot + OH^- + O_2$$

Hydroxyl radicals readily react with biomacromolecules, such as proteins, DNA and RNA. The formation of 8-hydroxy-2'-deoxyguanosine in DNA has been reported to lead to the induction of mutations and to tumor development.

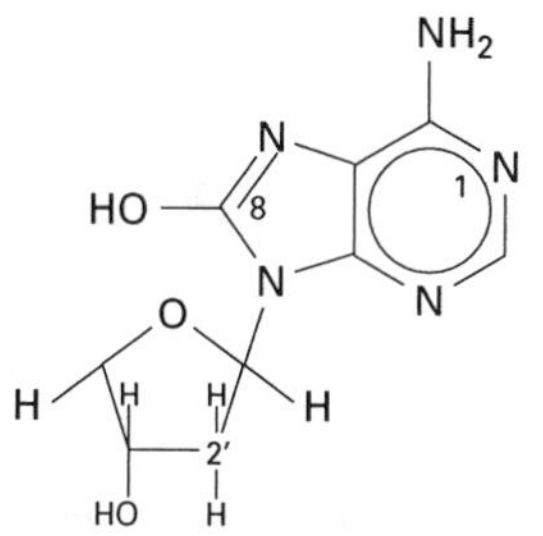

8 – OH – 2' – Deoxyguanosine

The consequences for the assessment of the risk of BHA-induced genotoxicity are not yet clear. The above findings contribute to the elucidation of the mechanism underlying the carcinogenicity of BHA (hazard identification). To assess the risk of genotoxicity more knowledge of the kinetics of the sequence of steps is needed.

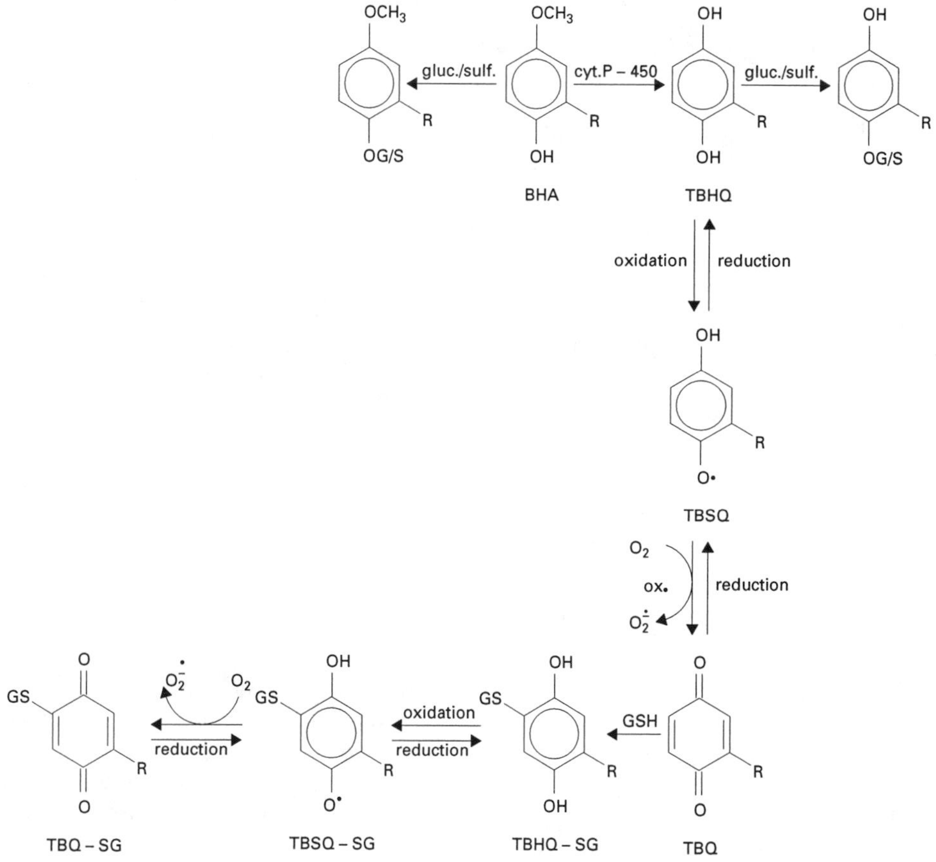

R = tertiary butyl – C(CH$_3$)$_3$
G/S = glucuronate/sulfate
BHA = butylated hydroxyanisole
TBHQ = tertiary butylhydroquinone
TBSQ = tertiary butylsemiquinone
TBQ = tertiary butylquinone
SG = glutathione conjugate

Figure 9.4 Metabolic inactivation and activation of BHA.

9.5   Emulsifiers

This group of additives includes thickening, gelatizing and stabilizing agents. They are used to improve the texture of food. Examples are agar-agar, tragacanth, sorbitol, mannitol, glycerol, gelatin and cellulose.

9.5.1   Sorbitol

Sorbitol is found in high levels in rowan berries (*Sorbus aucuparia*, Rosaceae), and also in cherries, prunes, apples, pears, peaches, apricots, and algae. It is also a synthetic substance. Sorbitol is a stabilizer as well as a sweetener.

Sorbitol acts as a diuretic and as a laxative. Large amounts of sorbitol may cause formation of gas, swelling of the belly, and diarrhea, accompanied by pain. It is metabolized for 70% to CO_2. No ADI has been estimated for sorbitol. A daily intake of 40 g is considered to be acceptable.

9.6   Flavoring agents

The most widely used flavor enhancer is salt (sodium chloride, NaCl). It is also a preservative and a nutrient. Generally, it is primarily regarded as a food additive. A well-known toxic effect of NaCl is high blood pressure.

9.6.1   Monosodium glutamate (ve-tsin)

Monosodium glutamate (MSG) is found in seaweed (*Laminaria japonica*). It is also a synthetic product. MSG is an excitatory neurotransmitter. It has been shown to cause permanent lesions of the hypothalamus in newborn rats and mice. Presumably, this is attributable to immaturity of the blood-brain barrier. Further, in young mice and rats, lesions of the retina have been reported after large doses of glutamate.

Humans have also been found to be sensitive to food to which MSG has been added as a flavor enhancer. The symptoms, known as "Chinese restaurant syndrome," include loss of feeling, general weakness, and heart palpitations.

9.6.2   Safrole

This substance is a typical member of a series of propenylbenzenes. These also include methyleugenol and estragole. The propenylbenzenes are natural and synthetic flavoring agents. Sassafras, containing high levels of safrole, used to be added to sarsaparilla root beer. Nowadays, safrole is still present in the diet as a (minor) component of various herbs and spices, e.g., cloves.

Safrole and related substances have been shown to be carcinogenic. Possible metabolic activation routes are 1'-hydroxylation, followed by sulfation, and epoxidation of the double bond in the propenyl group (Figure 9.5).

In the case of safrole, biotransformation data suggest that the 1'-hydroxy sulfate ester is the ultimate carcinogenic species capable of binding to DNA. Administration of 1'-hydroxysafrole to sulfation-deficient mice resulted in a lower tumor incidence than administration of the metabolite to normal mice.

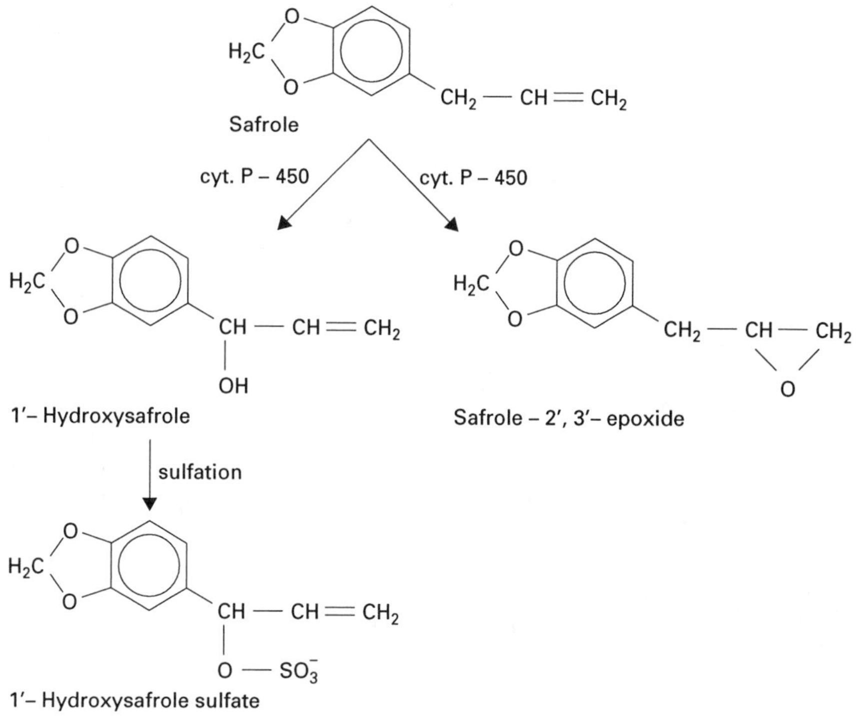

Figure 9.5 Possible routes of metabolic activation of safrole.

9.6.3 Saccharin

Saccharin (1,2-benzisothiazole-3(2H)-one 1,1-dioxide) is a well-known non-nutritive artificial sweetener. It was discovered by accident. In 1879, Constantin Fahlberg, a graduate student at John Hopkins University in Baltimore, was working on the synthesis of toluene derivatives. One day, while having lunch with unwashed hands, he discovered his bread to taste extraordinarily sweet. Soon after, the sweet taste could be attributed to one of the derivatives he synthesized: saccharin.

Saccharin has a low toxicity. Still, it has been extensively examined as a carcinogen. Early studies in experimental animals reported an increase in incidence of bladder tumors in the offspring of mother animals fed saccharin throughout pregnancy. Interpretation of these results was complicated by the presence of contaminants in the saccharin. Interpretation of later findings concerning an increase in the number of bladder tumors was complicated by the fact that in the meantime saccharin had been shown to be a promoter of other bladder carcinogens, like methylnitrosurea.

Saccharin is not metabolized, and has not been found to be genotoxic. On the basis of additional data from animal experiments and *in vitro* genotoxicity studies as well as from epidemiological studies, saccharin can be considered as a nongenotoxic carcinogen for which a threshold dose can be set. Its ADI is 2.5 mg/kg body weight.

9.6.4 Aspartame

Aspartame is another artificial sweetener. It is a dipeptide, consisting of L-aspartic acid and the methyl ester of L-phenylalanine.

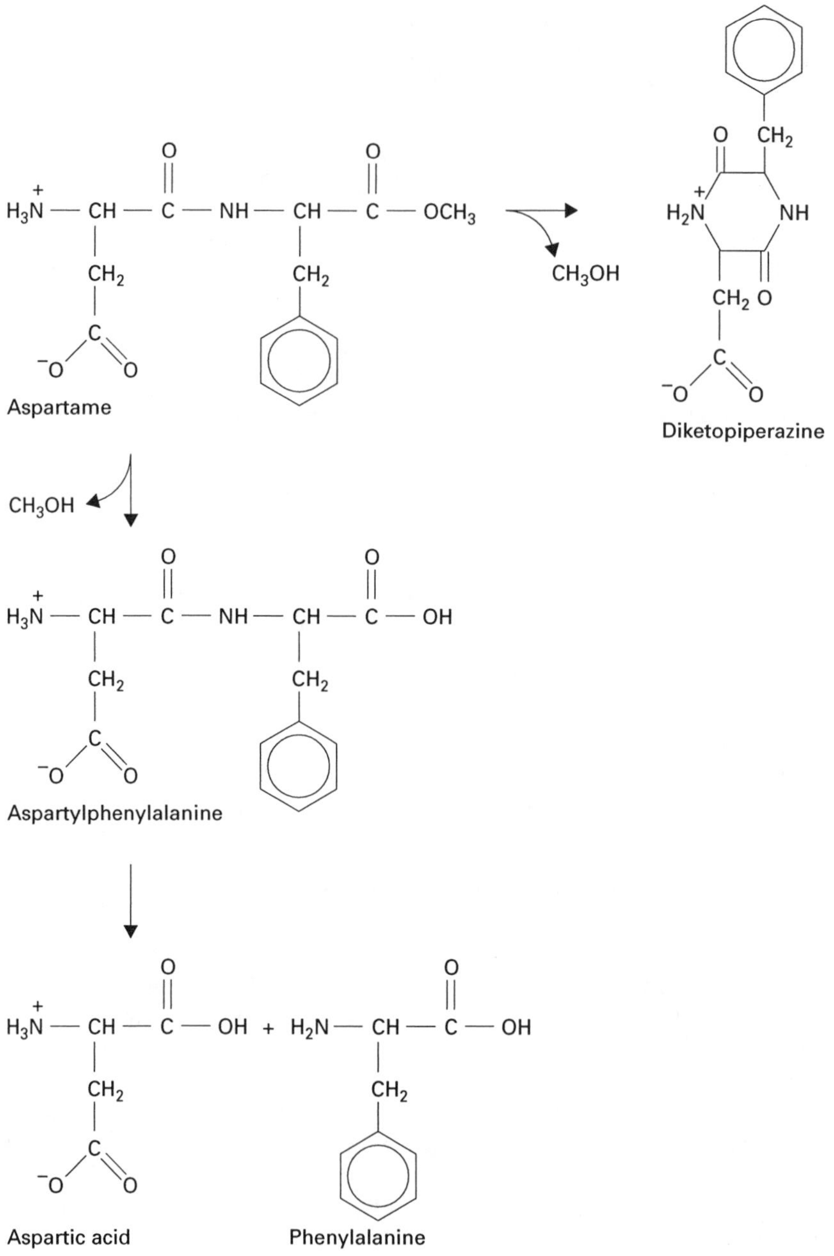

Figure 9.6 Hydrolysis of aspartame.

In the gastrointestinal tract, aspartame undergoes complete hydrolysis into its three components aspartic acid, phenylalanine, and methanol (Figure 9.6).

Although aspartame has been approved (the ADI is set at 50 mg/kg body weight per day) as a sweetener in many countries, there are still some toxicological aspects under consideration. A small number of urticarial reactions have been demonstrated. In general, it is agreed that the methanol and aspartic acid formed from aspartame by hydrolysis are safe. As far as the third component, phenylalanine, is concerned, there is the possibility of combined action. Interactions between phenylalanine and other amino acids are suggested at the level of amino acid transport, leading to nervous disturbances as a result of decreased neurotransmitter levels. Formation of phenylalanine from aspartame can actually pose a risk for so-called homozygous phenylketonurics. These people lack the ability to hydroxylate phenylalanine, the first step in its metabolism.

9.7 Summary

The synthetic food additives, and some of the naturally occurring food additives, have extensively been screened for toxicity. Limit values have been assessed for dietary intake (by humans) on the basis of extrapolation of data obtained in experimental animals. The probability that food additives cause adverse effects in humans at the levels recommended for dietary intake is at least minimal and probably negligible. An exception should be made for those food (component)s that cause hypersensitivity or allergy.

In this chapter, therefore, the attention is focused on the examination of the type of toxic effect and the underlying mechanism after administration of food additives (at high doses) to experimental animals. A number of illustrative examples are given.

The preservatives nitrate and nitrite are known to induce acute as well as long-term adverse effects: methemoglobinemia and cancer. Bacteria in the oral cavity can reduce nitrate to nitrite. Nitrite oxidizes hemoglobin to methemoglobin, which fails to bind oxygen. The consumption of dietary sources with high levels of nitrate (e.g., drinking water) and nitrite (e.g., meat and meat products) has resulted in life-threatening methemoglobinemia, especially in children. Nitrite (either ingested directly or indirectly via the reduction of nitrate) reacts with secondary amines under the formation of a variety of nitrosamines, e.g., dimethylnitrosamine, diethylnitrosamine and N-nitrosopyrrolidine. Nitrosamines are mutagens as well as carcinogens. They induce cancer in a variety of organs, including the liver, kidney, urinary bladder, stomach, and pancreas. Nitrosamines need bioactivation, mediated by cytochrome P-450, followed by a sequence of rearrangements to yield the alkylating alkylcarbonium ions.

The synthetic antioxidant butylated hydroxyanisole (BHA) has been shown to be carcinogenic in rodents. It is a nongenotoxic carcinogen. In tests such as the Ames test, it proved to be nonmutagenic. It has been suggested that biotransformation plays an essential role in the carcinogenicity of BHA. It may undergo metabolic activation via a sequence of steps. The first step is oxidative O-demethylation to tertiary butylhydroquinone (TBHQ). This is oxidized to tertiary butylquinone (TBQ) via two one-electron steps. TBQ is nonenzymatically conjugated with glutathione to form ultimately TBQ-SG. Redox cycling of TBQ and TBQ-SG produces reactive oxygen species, the most reactive being the hydroxyl radical. The formation of hydroxyl radical–DNA adducts has been reported to lead to the induction of mutations and to tumor development.

The flavoring agents safrole, methyleugenol, and estragole are hepatocarcinogens. Data on the biotransformation of safrole suggest that the 1'-hydroxy sulfate ester is the ultimate carcinogen. Further, the adverse effects of the food color tartrazine, the emulsifier sorbitol, the flavoring agent MSG, and the artificial sweeteners saccharin and aspartame are briefly discussed.

Reference and reading list

Branen, A.L., P.M. Davidson and S. Salimen, *Food Additives*. New York, Marcel Dekker Inc., 1990.

Chenault, A.A. (Ed.), Is it fit to eat?, in: *Nutrition and Health*. New York, CBS College Publishing, 524–575, 1984.

Clayson, D.B., F. Iverson, E.A. Nera and E. Lok, Early indicators of potential neoplasia produced in the rat forestomach by non-genotoxic agents: the importance of induced cellular proliferation, in: *Mut. Res.* 248, 321–332, 1991.

Enomoto, M., Naturally occuring carcinogens of plant origin: Safrole, in: *Bioactive Mol.* 2, 139–159, 1987.

Gangolli, S.D., P.A. van den Brandt, V.J. Feron, C. Janzowski, J.H. Koeman, G.J. Speijers, B. Spiegelhalder, R. Walker, J.S. Wisnok, Nitrate, Nitrite and N-nitroso compounds. *Eur. J. Pharmacol.* 292, 1–38, 1994.

Kleinjans, J.C.S., Food toxicity: the toxicological history of aspartame, in: Niesink, R.J.M., J. de Vries, M. Hollinger, Eds., *Toxicology: Principles and Applications*. Boca Raton, CRC Press Inc., 1996.

Ministry of Agriculture, Fisheries and Food, *Nitrate, nitrite and N-nitroso compounds in food*. The 20th report of the Steering Group on Food Surveillance. The Working Party on Nitrate and Related Compounds in Food. Food Surveillance Paper no. 20, London, Her Majesty's Stationery Office.

Ommen, B. van, A. Koster, H. Verhagen and P.J. van Bladeren, The glutathione conjugates of tert-butyl hydroquinone as potent redox cycling agents and possible reactive agents underlying the toxicity of butylated hydroyanisole, in: *Biochem. Biophys. Res. Commun.* 189, 309–314, 1992.

Parke, D.V., D.F.V. Lewis, Safety aspects of food preservatives. *Food Add. Contam.* 9, 561–577, 1992.

Reddy, C.S. and A.W. Hayes, Food-borne toxicants, in: A.W. Hayes (Ed.), *Principles and Methods of Toxicology*, 2nd edition. New York, Raven Press, 67–110, 1989.

Smith, J., *Food Additive User's Handbook*. Glasgow and London, Blackie, 1991.

Verhagen, H., P.A.E.L. Schilderman and J.C.S. Kleinjans, Butylated hydroxyanisole in perspective, in: *Chem.-Biol. Interact.* 80, 109–134, 1991.

chapter ten

Adverse effects of food contaminants

J.P. Groten

10.1 Introduction

From Part 1 of this textbook it is already clear that food contaminants originate from many sources, including human activities. This and other factors make it difficult to give an equivocal definition of the term food contaminant. For example, the substance involved may be either synthetic or natural. Also, its presence in the food may be accidental or deliberate. So, the definition varies from country to country and from book to book, but in any case, unlike food additives, food contaminants are unwanted in foodstuffs.

The naturally occurring contaminants (toxins) are the subject of Chapter 11. This chapter looks at the technical contaminants of which the main source is the large-scale application of fertilizers, pesticides, growth stimulants, and antibiotics in farming. Further, contaminants can originate from production and use of other synthetic chemicals, e.g., packaging and canning materials (see Part 1). Technological food contaminants can be divided into two subcategories: *metals* (inorganic as well as organic derivatives) and *organic chemicals*. The majority of the human population is chronically exposed to low levels of these contaminants. In a number of cases, e.g., Hg, Cd, Pb, and polyhalogenated aromatic hydrocarbons, the substances may accumulate in tissues and organs. Following chemical contamination, acute toxic effects seldom occur, and if so, mostly in occupational settings. At present, acute toxicity of contaminants is not a matter of great concern as far as food safety is concerned. An overall evaluation of the health risks due to dietary intake of contaminants, however, also includes chronic toxicity.

Table 10.1 Relative importance of actual food hazards

1	Microbiological contamination	100,000
2	Nutritional imbalance	100,000
3	Environmental contaminants, pollutants	100
4	Natural toxicants	100
5	Pesticide residues	1
6	Food additives	1

Source: Ashwell, 1990.

The toxicology of contaminants focuses primarily on long-term effects, such as mutagenicity, carcinogenicity, and teratogenicity. This is based on the assumption that low doses of contaminants may cause long-term effects in humans because of (a) the long human lifespan, (b) interactions between contaminating food components, and (c) effects of other (dietary) factors.

Only small amounts of contaminants are ingested. Therefore, the manifestation of toxic effects may be delayed. Further, there is the possibility of combined actions. In that case, it is difficult to find a relationship between the presence of a certain contaminant in food and a toxic effect.

It shoud be noted that in general food contaminants *do not give rise to concern*, as they usually do not exceed limit values. In Table 10.1, six main categories of hazard are listed, in order of relative importance. The ranking, based on criteria such as severity, incidence, and onset of biological symptoms was originally proposed by Wodicka in 1971. It gives a good idea of the proportion of risks associated with the intake of food contaminants to the risks from the intake of additives (lower risk), nutrients, and contaminants of microbial origin (higher risk). There is a growing awareness regarding the dramatically increasing number of man-made enviromental pollutants. This may give rise to concern about other man-made compounds like food additives, whereas in general, additives do not pose health hazards.

The cases presented in this chapter are not intended to give a complete survey of all categories of food contaminants. Mainly, the toxic effects of food contaminants and the underlying mechanisms of the major groups are discussed. For extensive reviews on the toxicology of food contaminants, see the literature references at the end of this chapter.

10.2 Metals

Of the approximately 100 elements in the Earth's crust, 20 to 30 are known to be necessary to the human body. This section discusses some nonessential metals, including organometallic complexes. The toxicology of essential minerals is the subject of Chapter 12.

Knowledge of the mechanisms underlying toxic effects is needed for an adequate setting of limit values such as NOAEL and ADI. In order to understand this procedure, the following terms should be understood: *critical organ, critical effect*, and *critical concentration*. Prior to the discussion of the toxicology of the food contaminating nonessential metals, these terms need to be defined:

Intermezzo

- *Critical concentration*: target cell/organ concentration at which adverse (reversible/ irreversible) functional changes occur. These changes are called *critical effects*.
- *Critical organ*: organ in which the critical concentration is reached first under specified conditions for a given population.

Table 10.2 Calculated hypothetical total daily intake of cadmium and contributing sources

Individual	Source of cadmium	Intake in µg
Non-smoker living in rural area	air	0.0005
	food	4
	water	2
	total	6
Smoker living near cadmium source and eating contaminated food	air	25
	food	84
	water	2
	tobacco	4
	total	115

Source: Hallenbeck, 1985.

- PPC_{10} is the concentration at which in 10% of the population the critical organ is affected. Further, the terms *subcritical concentration* and *subcritical effect* are used. They relate to the conditions under which the first disturbances can be expected. They warn that critical concentrations (causing critical effects) may soon be reached.

10.2.1 Cadmium

For people not occupationally exposed to cadmium (Cd), dietary intake is the main route of exposure to cadmium. This is shown by Table 10.2, listing data on the routes of exposure and the daily intake by adults. Smoking of 20 cigarettes per day causes an intake of 4 µg of cadmium. In comparison with dietary intake, this does not seem to be too much. However, one should take into account that the absorption of Cd is much lower after oral intake (4 to 8%) than on inhalatory exposure (15 to 40%).

Cd is present in nearly all foodstuffs. In noncontaminated regions, food usually contains less than 0.1 mg Cd/kg. High concentrations (1 to 10 mg/kg) can be found in the organs of cattle, in seafood, and in some mushroom species.

Like most inorganic contaminants, Cd is only absorbed to a small extent from the gastrointestinal tract. Its ability to accumulate in the body may be accounted for by its long biological half-life.

After uptake from the gastro-intestinal tract, Cd is bound to the low molecular weight protein metallothionein (Mt) in the cells of the intestinal wall and in the liver. Mt plays a key role in the homeostases of trace elements such as Zn and Cu in the organism, and in the detoxication of nonessential metals such as Cd, Hg, and Pt. Cd–Mt complexes are gradually released from the intestinal wall and the liver into systemic circulation. After renal excretion by glomerular filtration, the complexes are reabsorbed by the renal proximal tubule cell. It is believed that free ionic Cd resulting from lysosomal degradation of Cd–Mt causes damage to the kidneys.

Several factors are known to affect the absorption of Cd and its distribution over the body. The latter, for example, is determined by the Cd content of the diet. At low doses, Cd accumulates mainly in the kidneys. After high doses, the intestinal Mt-pool is saturated and free Cd will reach the liver. As a result, acute oral toxicity has been observed mainly in the liver and the erythropoietic system, while long-term exposure to low Cd levels (orally and inhalatory) has been found to result in toxic effects in the lungs, kidneys, and bones. It is important to note that in most long-term animal studies where renal and other effects were examined, the renal effects preceded or occurred simultaneously with the other effects.

Also, many dietary factors are known to influence the absorption and distribution of Cd in humans. Various metals may interfere very efficiently with the uptake of Cd.

High-fiber and low-fat diets with adequate mineral levels of calcium, zinc, iron, and phosphorus are known to lead to a lower total body retention of Cd than diets high in fat with a marginal mineral status.

Cadmium has been shown to interfere with the metabolism of vitamin D, calcium, and collagen. These effects manifest themselves as osteoporosis and osteomalacia in humans as well as in animals. An illustrative example is the Itai Itai bone disease ("ouch ouch" disease). This disease occurred as an epidemic among the inhabitants of the Fuchu area in Japan, who for a long time ingested rice that was highly contaminated with Cd (300 to 2000 µg Cd per day). The etiology of this disease points to a combination of factors. Not only the exposure to Cd, but also a deficient diet (low in protein, calcium, and vitamin D) were found to be responsible for the development of this disease in that particular area.

In occupational settings and in studies in rodents, it has appeared that long-term inhalatory exposure to Cd is associated with an increase in prostate cancer and lung cancer. However, the potential carcinogenicity of cadmium has not been clearly shown in oral studies.

In addition to the dose and dietary factors, the *speciation* of Cd in the diet appears to be an important factor in determining its uptake from food. There is a clear need for data on the form in which Cd is present in food. In animal tissues, Cd occurs mainly as metallothionein complexes. Foods originating from plants are an even more important source of dietary Cd than food of animal origin. In plants, Cd is bound to phytochelatins, proteins that have several properties in common with metallothioneins. Information on the toxicological risks due to the oral intake of Cd bound to metallothionein is limited. Cd and Cd–Mt differ in intensity of toxicity. After parenteral administration, Cd–Mt is more nephrotoxic than inorganic Cd. This seems not to be the case after oral administration. It has been suggested that after a low intake, the metabolic routes of both Cd forms are similar. After uptake from the gastro-intestinal tract, Cd may be released from Cd–Mt. After a high intake, there seems to be a difference in metabolic fate, leading to a higher availability of Cd after intake of inorganic Cd. A difference in Cd availability between foods will certainly have important consequences for the evaluation of risks due to Cd intake and the estimation of tolerable Cd levels in different types of food. The kidneys are the critical organs for long-term oral exposure to Cd, and renal effects always precede or occur simultaneously with other effects. Epidemiological studies in occupational settings have shown that 10% of a population of industrial workers at the age of 45 shows symptoms of renal dysfunction once the renal Cd concentration has reached a level of 200 mg/kg kidney cortex. For the general population, it has been calculated that this level will be reached in 45-year old individuals after a daily dietary Cd intake of ±400 µg. (Table 10.3).

Table 10.3 Calculated average cadmium kidney cortex concentration at age 45 for non-smokers with cadmium intake via food only

		100	200	300	400	500	600	700
Average daily cadmium intake at age 50		100	200	300	400	500	600	700
Geometric mean cadmium concentration in kidney cortex* (mg/kg)		61	102	143	183	224	265	305
Estimated proportion (%) with kidney cortex cadmium above their individual-critical concentration	A	2.7	11	22	34	44	53	60
	B	1.8	7.8	17	26	35	44	51

Note: Body weight = 70 kg.

A: PCC_{50} = 250 mg Cd/kg and PCC_{10} = 180 mg Cd/kg, log-normal distribution of critical concentrations.

B: PCC_{50} = 300 mg Cd/kg and PCC_{10} = 200 mg Cd/kg, log-normal distribution of critical concentrations.

* Assumed to have a log-normal distribution with geometric SD of 2.

Source: Friberg et al., 1985, 1986.

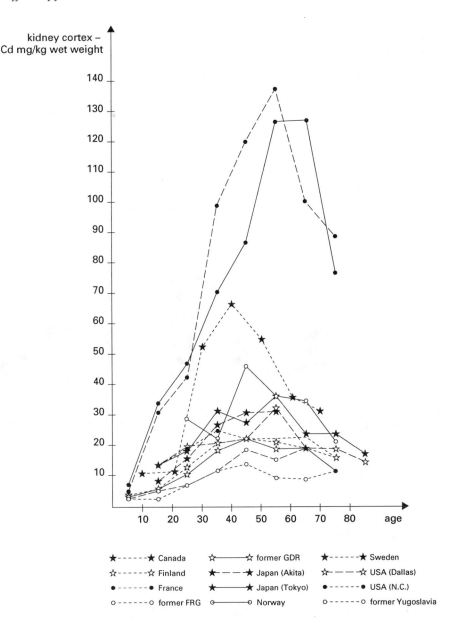

Figure 10.1 Average concentration of cadmium in kidney cortex in relation to age. Results from 12 different studies in 10 countries are summarized. The averages are based on data from smokers, non-smokers, females, and males combined. Source: Friberg et al., 1985, 1986.

The FAO/WHO provisional tolerable daily intake (see Chapter 16, Section 16.3.2.1) of Cd, 70 µg, leads to kidney cortex concentrations of 40 to 60 mg/kg (compare these figures with the actual renal Cd concentrations in 10 countries, given in Figure 10.1).

Differences in Cd availability resulting from the intake of different forms of Cd or the effects of other dietary factors are not taken into account in the estimation of tolerances. Another factor contributing to the Cd retention in the body is the smoking behavior of the population. A cigarette contains 0.8 to 2 µg Cd, from which 25 to 40% is absorbed on inhalation. This means that smokers may have higher Cd tissue levels than non-smokers (Figure 10.2).

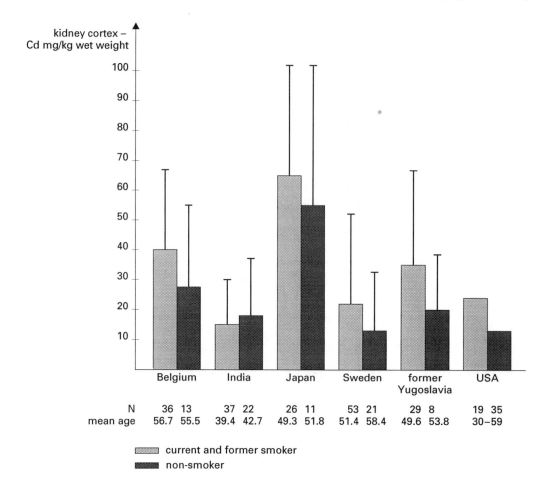

Figure 10.2 Cadmium kidney cortex concentration (geometric mean values) in relation to smoking habits among subjects (30 to 69 years of age) studied in Belgium, India, Japan, and (former) Yugoslavia. Also included in the figure are data from Sweden and the US (subjects aged 30 to 59). Source: Friberg et al., 1985, 1986.

10.2.2 Mercury

The widespread use of mercury (Hg) in industry and agriculture (e.g., as fungicidal derivatives) has led to serious environmental pollution and, as a result, to contamination of food, particularly of fish and meat. Tissues and organs of fish have such a high affinity for mercury (Hg) that it can accumulate by a factor 9000 compared to the environment. In fish, mercury is present as methylmercury. Methylmercury is the most toxic form and is formed by intestinal bacteria and bacteria in the slimes of the skin. Bacterial methylation of mercury has also been found in the organic fraction (sediment) of aquatic systems. Methylmercury is more extensively absorbed from the gastrointestinal tract than inorganic mercury, namely 85 to 95% and 8 to 12%, respectively.

The toxicity of Hg is related to the absorbability from the gastrointestinal tract. Insoluble mercurous chloride or elemental mercury does not cause toxic effects in man up to 500 g, while organic Hg complexes are far more toxic than inorganic Hg. After absorption, Hg^{2+} ions are bound to metallothionein in the blood. The metabolism of Hg–Mt is similar to that of Cd–Mt. Nephrotoxicity may occur after long-term accumulation.

Hg in the organic form is almost completely absorbed from the gastrointestinal tract. A large part of the body burden of organic mercurials is found in the red blood cells. Methylmercury has been shown to pass the blood-brain barrier more readily than other Hg forms; inorganic Hg cannot pass the barrier. This makes the nervous tissue, especially the brain (atrophy of cerebral cortex), one of the target organs. The fetal brain is more sensitive to methylmercury than the adult brain (teratogenicity!). Dysfunction of the nervous system is believed to be preceded by biochemical disturbances such as inhibition of protein (enzyme) synthesis. The first symptoms of methylmercury intoxication — convulsion, anorexia, weight loss, and fatigue — are difficult to recognize. They are very unspecific and show high interindividual variation. Characteristic effects are paresthesia (tingling of extremities), loss of coordination, reflex changes, and mental deterioration. Methylmercury intoxication is mostly seen in occupational settings. Besides the speciation of Hg, interactions of Hg with other trace elements such as copper, zinc, and selenium are important. For instance, Korean fishermen who consumed Hg-containing tuna fish appeared to be extremely tolerant of Hg. They showed no neurotoxic effects, although their mercury blood levels rose to above 10 mg/l. The critical blood concentration of mercury is 20 µg/l (neurotoxic effect!). It has been suggested that the fishermen's resistance is related to the high selenium content of tuna fish.

The provisional tolerable weekly intake (PTWI) (see Chapter 16, Section 16.3.2.1) of mercury is 0.21 mg (WHO, 1973). The daily dietary Hg intake mainly depends on fish consumption and methylmercury levels in fish. In general, Hg intake is less than 10 µg per day.

10.2.3 Lead

Contamination of food with lead (Pb) appears to be inevitable. Lead originates from natural sources as well as from human activities (see Part 1). The majority of organic lead in the environment is accounted for by the anti-knock gasoline additive tetraethyllead. Since the introduction of lead-free gasoline, the concentration of air-borne Pb and the lead content of food are decreasing. Contamination of food of vegetable origin with lead is rather high. In food of animal origin, the lead content is very low, if not nil. For example, the milk from cows grazing on grass with 100 times the normal lead level contained only 4 times more lead than the milk from cows grazing on uncontaminated grass. Increases in the lead level of foods are mostly due to indirect contamination from packaging material and handling.

Lead is absorbed more easily by children (about 40%) than by adults (about 10%). The distribution of lead can be described by a three-compartment model, including bone tissue (95%), blood (2%), and soft tissues (3%). Lead blood levels (PbB) parallel the concentrations in soft tissues. Therefore, the effects of Pb are usually related to the PbB level. A clinical manifestation of lead poisoning is anemia due to a decreased lifespan of the erythrocytes and interaction with several enzyme systems in heme synthesis (Figure 10.3).

The most sensitive indicator of hematological changes after exposure to lead is inhibition of the enzyme δ-aminolevulinic acid (δ-ALA) dehydratase. ALA blood levels increase at PbB levels of 40 to 80µg/l. However, the toxicological significance of the inhibition of this enzyme (i.e., change in hemoglobin levels) is not yet fully understood. Anemia clinically manifests itself at higher PbB levels. The WHO (1987) has set the lowest-observed-adverse-effect level (LOAEL) (see Chapter 21, Section 21.4.4.3) at 200 µg/l blood. Recent findings have shown that exposure to low lead levels may lead to neurological disorders which used to go unnoticed, in particular in the developing brains of children. Prenatal exposure to lead is also of great concern, as Pb passes the placenta, and the blood-brain barrier of the fetus.

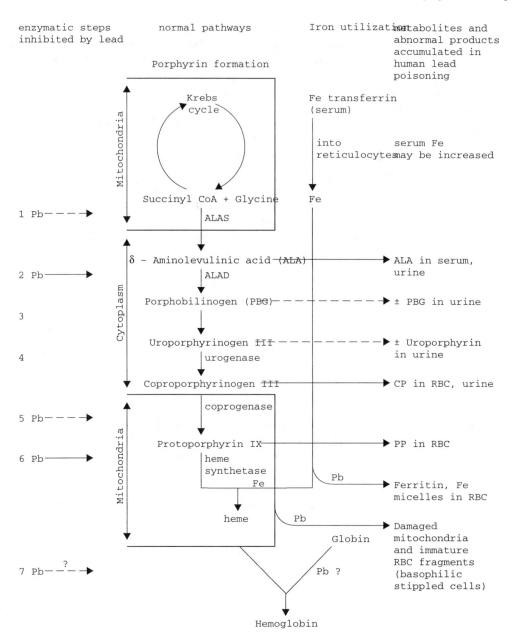

Figure 10.3 Lead interferes with the biosynthesis of heme at several enzymatic steps (Source: Goodman et al., 1990).

The absorption of Pb is higher in children than in adults. Generally, children are exposed to higher lead levels from the environment (dust, pica). Furthermore, children are particularly sensitive. Today's concern about lead intake has drawn the attention to neurotoxicity at prenatal lead blood levels and to the effect of lead on the development of the child after birth. For babies, the lead intake from milk powder and dust is estimated at ±35 μg and ±40 μg per week respectively. The lead intake by babies from drinking water with a lead content of 50 μg/l is estimated at ±25 μg per day. However, the lead intake by sucklings should not exceed 25 μg/kg per week according the WHO standardization. This means that the average lead level in drinking water is probably too high for babies.

10.3 Organic chemicals

Well-known types of organic substances occurring as food contaminants are pesticides, drugs, antibiotics, and industrial chemicals. Food-contaminating industrial chemicals include mainly polyhalogenated aromatic hydrocarbons, carbamates, and plasticizers. Chemicals such as organic solvents (chloroform, benzene, methylene chloride) and plastics (styrene and acrylo polymers) cause concern in occupational settings (skin, lungs) rather than in food consumption.

The following subsections deal with pesticides, notorious environmental pollutants such as polychlorinated biphenyls, dibenzodioxins and dibenzofurans, and feed additives.

10.3.1 Pesticides

Pesticides are hazardous compounds which are used to control or eliminate unwanted species of insects (insecticides), acarides (acaridicides), fungi (fungicides), higher plants (herbicides), rodents (rodenticides), or nematods (nematodicides).

The biocidal action of pesticides includes a variety of disturbances of physiological processes, such as inhibition of acetylcholinesterase by insecticidal organophosphates, blockade of neurotransmission by chlorinated hydrocarbons, and inhibition of oxidative phosphorylation by herbicidal dinitrophenols.

The insecticidal organophosphates and carbamates inhibit acetylcholinesterase, the enzyme that regulates neurotransmission by hydrolyzing acetylcholine. Another group of insecticides, the chlorinated hydrocarbons, cause blockade of neurotransmission by interaction with the sodium/potassium channels, resulting in inhibition of nerve membrane depolarization. A third group of pesticides, the herbicidal dinitrophenols, are uncouplers of the oxidative phosphorylation. Most of these biological targets also exist in man. Therefore, it is not surprising that accidental massive poisoning following inaccurate use of pesticides regularly occurs all over the world.

Parathion, an insecticidal organophosphate

Carbaryl, an insecticidal carbamate

Aldrin, an insecticidal
chlorinated hydrocarbon

Dinitro – o – cresol,
a herbicidal dinitrophenol;

Organochlorine and carbamate pesticides can also induce long-term effects: cancer and malformations. The former pesticides are highly lipid-soluble and are only slowly broken down. Therefore, they persist in the environment for a long time and accumulate in food chains.

Table 10.4 NOAELs and ADIs of some pesticides

Pesticide	NOAEL (μg/kg B.W.) (species)	ADI (μg/kg B.W.)	Safety factor	Max. Residue* in meat, fish (μg/kg)
Mutagenic and carcinogenic				
aldrin/dieldrin	25 (rat/dog)	0.1	250	1–86
DDT	50 (rat)	5	10**	3–10
methyl parathion	100 (human)	1	100	—
captan	1.25×10^4 (monkey) 1×10^5 (rat)	100	125	26–40
Mutagenic and non-carcinogenic				
dichlorvos	33 (human)	4	8	
malathion	5000 (rat) 200/day (human)	20	10	4–96

* High residues are found particularly in meat, fish, and poultry.

** In spite, of its carcinogenic potential for humans, the applicable safety factor for DDT is only tenfold. The ADI is conditional; only permission for application when no available subtitutes can be used.

Source: Concon, 1988.

The carcinogenic organochlorine pesticides include aldrin and dieldrin. They need metabolic activation to become carcinogenic. The main targets in rats and mice are liver and lung. An example of the carcinogenic insecticidal carbamates is carbaryl. The members of this group become carcinogenic on conversion to nitroso compounds in the reaction with nitrite.

Further, pesticides have been reported to induce malformations when given to mammals during pregnancy, e.g., aldrin, dieldrin, and carbaryl. However, carcinogenicity and teratogenicity have not yet been confirmed in valid epidemiological studies. As a result, the ADIs of most pesticides are based on animal data. This calls for continuous attention to potential hazards to humans due to the presence of pesticides in food. Therefore, safety factors of 100 or higher are applied (Table 10.4). This means that the ADI is usually 1% of the no-adverse-effect level observed in the most sensitive species. In the few cases where toxicity data in humans were available, a safety factor of 10 has been applied, accounting for intraspecies variation in man.

The organochlorine pesticides have been largely replaced by carbamates. The latter are less persistent in the environment. Further, their carcinogenic potential is lower than that of the organochlorine pesticides.

10.3.2 Halogenated Aromatic Hydrocarbons

10.3.2.1 Polychlorinated biphenyls

The polychlorinated biphenyl (PCB) content of animal food is decreasing in recent years. This is due to the ban on the use of PCBs. Notwithstanding, the levels of PCBs can still be high because of their low biodegradability.

PCBs are known to cause:

- chloracne
- induction of phase I (i.e., oxidases, reductases) as well as of phase II (i.e., conjugases) xenobiotic-metabolizing enzymes
- cancer
- teratogenic effects
- neurotoxic effects

The most prominent effect in humans is persistent chloracne on the skin of the head and chest. This skin disease is believed to result from acanthosis and hyperkeratosis of the skin. Hair follicles are ultimately plugged by keratinaceous material and the glands around the follicles become cystic. The majority of studies on enzyme induction by PCBs concerned the cytochrome P-450-dependent monooxygenase. Evidence has been obtained that the mechanism underlying the induction of cytochrome P-450 isoenzymes consists of a sequence of events, the first of which is binding to a receptor protein, the so-called Ah (aromatic hydrocarbon) receptor. The ligand–Ah receptor complex is transferred to the nucleus. Interactions of the complex with structural genes result in stimulation of the transcription of those genes. This leads to an increasing synthesis of the enzymes coded for by the genes.

The mechanism of the carcinogenicity of PCBs involves promotion rather than initiation. They stimulate the growth of tumors (induced) in liver, skin and lungs. Although PCBs have been reported to be carcinogenic in animals, there are only a few reports suggesting that these compounds are also carcinogenic in man. Tumors were found in 8 of 22 people involved in a rice oil accident in Japan, and in 7 of 92 industrial workers exposed to arochlor.

Several of the toxic effects are similar to those of pesticides like dieldrin and aldrin: teratogenicity and neurotoxic effects. In both cases, the underlying mechanisms are not yet exactly known.

10.3.2.2 Polychlorinated dibenzodioxins and dibenzofurans

Polychlorinated dibenzodioxins (PCDDs) and polychlorinated dibenzofurans (PCDFs) originate from several sources.

PCDD/PCDF emission can result from the incineration of domestic waste containing low-molecular chlorinated hydrocarbons and PCBs. PCDDs and PCDFs are also formed during the production of organochlorine compounds such as polychlorobenzenes, polychlorophenols, and PCBs. The most toxic polyhalogenated aromatic hydrocarbon is 2,3,7,8-tetrachlorodibenzo-*p*-dioxin (TCDD), a well-known contaminant of the herbicide 2,4,5-trichlorophenoxyacetic acid(2,4,5-T). TCDD has an oral LD_{50} of 22 to 45 µg/kg in rats.

Many TCDD-induced effects are similar to effects caused by PCBs and other structur-ally-related compounds. Hepatic monooxygenase activity is elevated. PCDDs and PCDFs are also highly teratogenic (0.25 µg/kg). Further, immunosuppression and thymic atrophy have been reported in experimental animals (after 10 µg/kg).

Enzyme induction, immunosuppression, and thymic atrophy by TCDD and structur-ally-related compounds are believed to be mediated via stereospecific and irreversible binding to the Ah receptor (see Section 10.3.2.1). An essential structural requirement is coplanarity. The structures of the halogenated aromatic hydrocarbons involved should be as planar as that of TCDD.

10.3.3 Antibiotics in use as feed additive

An important concern of veterinary toxicology is the possible transmission of harmful substances from meat, milk, and other foodstuffs to the human population. This concerns primarily antibiotics in use as feed additives. They include tetracyclines, nitrofurans, and sulfonamides. Recently, the detection of metabolites of the nitrofuran furazolidone in meat products revived the discussion on the acceptability of this veterinary drug. Oxytetracycline and also furazolidone are suspected of being carcinogens, and oxytetracycline has been reported to react with nitrite to yield (carcinogenic) nitrosamines.

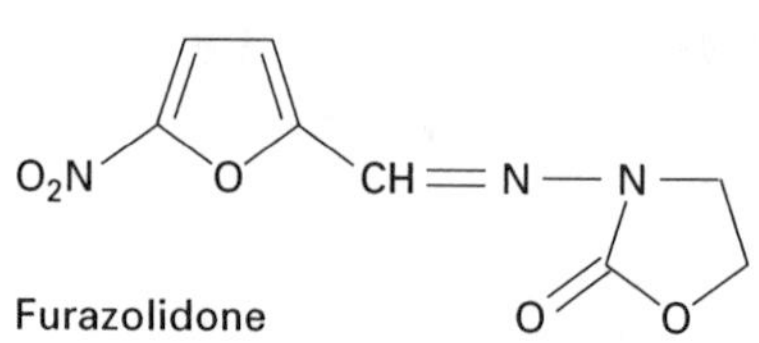

Furazolidone

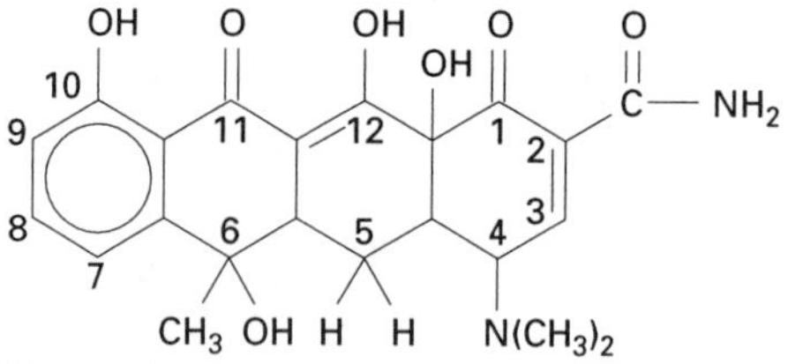

Tetracycline

congener	substituent(s)	position(s)
Chlortetracycline	— Cl	(7)
Oxytetracycline	— OH, — H	(5)
Demeclocycline	— OH, — H; — Cl	(6; 7)
Methacycline	— OH, — H; $=$ CH$_2$	(5; 6)
Doxycycline	— OH, — H; — CH$_3$, — H	(5; 6)
Minocycline	— H, — H; — N(CH$_3$)$_2$	(6; 7)

In addition, the majority of the antibiotics in use as feed additives pose a serious (indirect) health hazard to humans. Ingestion of these antibiotics may lead to an increased resistance of bacteria. This may imply:

- transfer of antibiotic-resistant bacteria to humans via food intake, originating from animals treated with antibiotics or infected by resistant bacteria;
- transfer of the resistance factor (R-factor) from resistant non-pathogenic bacteria to other bacteria which will lead to widespread resistance.

10.4 Summary

Food contaminants are substances that are unintentionally present in food. This chapter deals with nonnatural contaminants originating from production and technological applications. This category of food contaminants can be divided into two subcategories: metals and organic chemicals.

Exposure levels are quite low, and the appearance of toxic effects is usually delayed. Therefore, causal relationships are not easily to establish. The toxicology of food contaminants primarily focuses on long-term effects such as carcinogenicity, teratogenicity, and neurotoxicity. Absorption, disposition, and toxicity of food contaminants are determined by a large number of factors, including dietary habits, age, sex, speciation, and dietary factors, such as fat, proteins, and minerals. For heavy metals, extensive research has been carried out to establish dose–response relationships and to elucidate the underlying toxicity mechanisms. For pesticides and complex mixtures of halogenated aromatic hydrocarbons, on the other hand, the dose–response relationships are still unknown. As a result, the safety factor, accounting for the differences between the acceptable intakes and the actual intakes, is higher for mixtures of organic chemicals than for heavy metals. This does not mean that heavy metals are of less concern. The testing of mixtures of substances for toxicity at relevant dose levels must be emphasized.

Reference and reading list

Ashwell, M., How safe is our food? A report of the British Nutrition Foundation's eleventh annual conference, in: *J. Royal College Phys.* 24 (3), 233–237, 1990.

Concon, J.M., *Food Toxicology* (in two parts). New York, Marcel Dekker Inc., 1988.

Fiedler, H., H. Frank, O. Hutzinger, W. Partzefall, A. Riss, and S. Safe Dioxin '93, in: *Organohalogen Compounds*. Vol. 13 and 14, 1993.

Friberg, L., C.G. Elinder, T. Kjellström and G.F. Nordberg, *Cadmium and Health: a Toxicological and Epidemiological Appraisal*. Boca Raton, Florida, CRC Press, 1985/1986.

Friberg, L., G. Nordberg and V.B. Vouk, *Handbook of the Toxicology of Metals*. Amsterdam, Elsevier, 1986.

Goodman, L.S., A. Gilman, and A. Goodman Gilman, (Eds.), *Pharmacological basis of therapeutics*. New York, MacMillan P.C., 1990.

Graham, H.D., *The Safety of Foods*. Connecticut, The AVI Publishing Company Inc., 1982.

Groten, J., E. Sinkeldam, J. Luten and P. van Bladeren, Cadmium accumulation and metallothionein concentrations after 4-week dietary exposure to cadmium chloride or cadmium-metallothionein in rats, in: *Toxicol. Appl. Pharmacol.* 111, 504–513, 1991.

Hallenbeck, W.H., Human health effects of exposure to cadmium, in: *Experientia* 40, 136–142, 1985.

Lu, F.C., Acceptable daily intake: inception, evolution and application, in: *Reg. Toxicol. Pharmacol.* 8, 45–60, 1988.

Miller, K., *Toxicological Aspects of Food*. London, Elseviers Applied Science, 1987.

Poland, A. and J.C. Knutson, 2,3,7,8-tetrachlorodibenzo-*p*-dioxin and related halogenated aromatic hydrocarbons. Examination of the mechanism of toxicity, in: *Ann. Rev. Pharmacol. Toxicol.* 22, 517–554, 1982.

Safe, S., Polychlorinated biphenyls (PCBs), Dibenzo-p-dioxins (PCDDs), dibenzofurans (PCDFs), and related compounds: environmental and mechanistic considerations which support the development of toxic equivalency factors (TEFs). *Crit. Rev. Toxicol.* 21, 51–88, 1990.

Ter Haar, Sources and pathways of lead in the environment, in: *Proc. Int. Symp. Environ.* Health aspects lead, 1973.

Vroomen, L., M. Berghmans, P. van Bladeren, J. Groten, C. Wissink and H. Kuiper, In vivo and in vitro metabolic studies of furazolidone: a risk evaluation, in: *Drug Metab. Rev.* 22, 663–676, 1990.

chapter eleven

Adverse effects of naturally occurring nonnutritive substances

H. van Genderen

11.1 Introduction

By far the majority of the non-nutritive components are harmless. However, a number of naturally occurring substances have been identified that induce adverse effects. These originate mainly from plants and microorganisms.

In this chapter, such substances occurring in common foods are discussed. Cases of adverse effects following the intake of unusual foods, particularly in tropical regions, are not included. Their occurrence is too incidental. The following categories are discussed:

– (low-molecular) endogenous toxins of plant origin;

– toxic contaminants of microbial origin;
– plant proteins that interfere with the digestion of the absorption of nutrients.

11.2 Endogenous toxins of plant origin

Low-molecular endogenous toxins of plant origin are products from the so-called secondary metabolism in plants. In phytochemistry, a distinction is made between primary and secondary metabolism. Primary metabolism includes processes involved in energy metabolism such as photosynthesis, growth, and reproduction. Macro- and micronutrients are products of primary metabolism. Secondary metabolism is more or less species-, genus- and family-dependent. Each plant contains a large variety of secondary metabolites that function as pigments, flavors, protecting agents, or otherwise. The number of identified secondary metabolites involved in plant–animal interactions is estimated at 18,000.

Relatively few secondary metabolites in food plants have been shown to be toxic. They may induce a wide variety of effects, including growth inhibition and neurotoxicity, but also mutagenicity, carcinogenicity, and teratogenicity. For the majority of these substances, the information on their toxicity is limited, and often completely lacking. The studies involved were usually concerned with cases, and not with underlying mechanisms.

Testing of plant substances for toxicity is not provided for by official food safety regulation. Isolation and purification of amounts needed for toxicity testing are expensive. In general, industry has no interest in giving such studies financial support. Flavors of plant origin are an exception. Many of these are used in the production of food additives. As such, they come under the regulation of additives. The toxicity of some plant flavors has been examined by modern methods. Important additional information comes from experiences acquired with farm animals. These animals very often ingest one edible plant species or relatively simple and homogeneous feed mixtures over long periods of time, which may be comparable to chronic toxicity testing in experimental animals.

From a food safety point of view, two groups of non-nutritive natural food components can be distinguished:

– those that have given or still give rise to concern, but at present do not pose actual hazards;
– those that are of important toxicological relevance. The dietary intake of some of these substances has led to mass poisoning.

Table 11.1 lists a number of examples of the first group.

The second group is dealt with in the next subsection. It includes α-aminopropionic acid derivatives (N-oxalyl-diaminopropionic acid and β-cyano-L-alanine), agaritine, biogenic amines (serotonin, tryptamine and tyramine), cyanogenic glycosides (amygdalin, primasin, dhurrin, linamarin and lotaustralin), glucosinolates (sinigrin, progoitrin and glucobrassicin), glycoalkaloids (solanine, chaconine and tomatidine), and pyrimidine glycosides (vicine and convicine).

11.2.1 Nonnutritive natural food components of important toxicological relevance

11.2.1.1 α-Aminopropionic acid derivatives

α-Aminopropionic acid derivatives occur in peas of certain *Lathyrus* species. These substances are known to cause skeletal malformations (osteolathyrism) and neurotoxic effects

Table 11.1 Nonnutritive natural food components that have given or still give rise to concern

Substance/origin	Main toxic effect(s)	Comments
Erucic acid (fatty acid)/rapeseed	fibrotic myocardial lesions (in the rat)	varieties free from erucic acid have been bred
Cyclopropane and cyclopropene fatty acids/cottonseed oil	promotion of aflatoxin-induced carcinogenesis (in the trout)	the acids are largely removed during food processing; the main problem is their presence in feed
Carotatoxin (poly-acetylene)/carrots	neurotoxicity	low levels in edible carrots
Thujone (α– and β–) (monoterpene)/spices (component of absinthe liqueur)	neurotoxicity	absinthe is prohibited; for use as food additive, not more than 10 ppm is allowed.
D-limonene (monoterpene)/citrus oil	nephrotoxicity (in male rats, not in female rats and other animals)	intake is considered to pose no toxicological risks to man
Cucurbitacin E (triterpene)/squash or zuchini (as glucoside)	irritation gastrointestinal tract; vomiting, diarrhea	rarely present in bitter summer squash; may originate from cross-breeding with a wild species
Safrole (phenol derivative, occurring in plants)/spices (mainly sassafras oil)	liver cancer	allowed for use as food additive in the EU, prohibited in the US.
Coumarin/various plants (e.g., woodruff) and spices	hepatotoxicity (in rats)	see above under safrole
Quercetin/many plants (free and as glycosides)	mutagenicity (in Ames'test; in mammalian test systems mainly negative)	evaluation needs further research
β-aminopropionitrile/seed of *Lathyrus odoratus*	skeletal malformations (osteolathyrism) and neurotoxicity (neurolathyrism)	

(neurolathyrism). The peas are easily grown on poor soil and are often used as feed. Both diseases have occurred as epidemic in Northern India in years with a poor harvest. At present, osteolathyrism has largely disappeared. Neurolathyrism still poses a serious health problem.

Neurolathyrism is associated with the long-term intake of the peas of *L. sativus*. The disease is characterized by muscular weakness, degeneration of spinal motor nerves, and paralysis. The peas have been found to contain a neurotoxin: N-oxalyl-diaminopropionic acid (ODAP). In addition, the peas may be contaminated with a vetch species (*Vicia sativa*), also containing a neurotoxic α-aminopropionic acid derivative, β-cyano-L-alanine.

The neurotoxicity of the amino acids is attributed to their structural relationship with the neurotransmitter glutamic acid. ODAP and β-cyano-L-alanine are believed to bind irreversibly to the glutamate receptors on specific nerve cells. Long occupation of the glutaminergic receptors has been reported to result in neurodamage.

HOOC — C — NH — CH$_2$ — CH — COOH

with O double bond on C and NH$_2$ on CH

N – oxalyl – diamino –
propionic acid

NC — CH$_2$ — CH — COOH

with NH$_2$ on CH

β – Cyano – L – alanine

HOOC — CH$_2$— CH$_2$—CH — COOH

with NH$_2$ on CH

Glutamic acid

11.2.1.2 *Agaritine*

Agaritine is a member of a series of hydrazine derivatives, occurring in mushrooms, including the common edible mushroom *Agaricus bisporus*. It is the most important derivative.

Agaritine undergoes degradation on cooking. It is partly left intact when heated in oil. In the body, it is hydrolyzed by glutamyltransferase into glutamic acid and 4-(hydroxymethyl)phenylhydrazine (Figure 11.1). Agaritine has proved to be mutagenic in the *Salmonella*/mammalian microsome assay. If glutamyltransferase is added, the mutagenicity increases, suggesting that 4-(hydroxymethyl)phenylhydrazine is a more potent mutagen. Since the 4-(hydroxymethyl) phenyldiazonium ion is highly mutagenic, it is assumed to be the ultimate mutagen. 4-(Hydroxymethyl)phenylhydrazine induces tumors in soft mouse tissues at the injection site. Recently, also in mice, tumors have been found after mushroom feeding. Further studies are needed to confirm this.

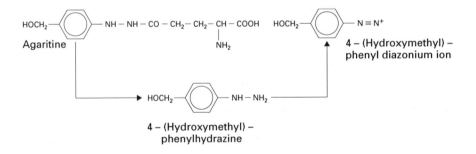

Figure 11.1 Hydrolysis of agaritine, followed by its activation.

11.2.1.3 *Biogenic amines*

Biogenic amines are formed by decarboxylation of amino acids. The term biogenic amines usually refers to the catecholamine neurotransmitters dopamine, norepinephrine, and epinephrine, the indoleamine neurotransmitter serotonin, and the mediator of inflammation histamine.

Examples of biogenic amines as food components are serotonin in bananas and pineapple, tryptamine in tomatoes, and tyramine in certain kinds of fully mature cheese. The precursor of serotonin is 5-hydroxytryptophan, that of tryptamine, tryptophan, and that of tyramine, tyrosine. Biogenic amines in food can also originate from fermentation (beer, wine, cheese) or bacterial contamination (meat). Tyramine in fermentation products results from the bacterial decarboxylation of tyrosine.

Well-known toxic effects include hypertension, palpitations and severe headache. Under normal conditions, tyramine is detoxicated by monoamine oxidase (MAO). Patients taking MAO-inhibitors as antidepressants may suffer from headache, and attacks of palpitation and hypertension, if they consume foods containing considerable amounts of tyramine.

11.2.1.4 Cyanogenic glycosides

Cyanogenic glycosides are monosaccharide or disaccharide conjugates of cyanohydrins. There is evidence that the cyanohydrins are derived from amino acids. Cyanogenic glycosides are widely present in plants where they are the principal precursors of hydrocyanic acid. Their presence is believed to provide protection against herbivores.

Representatives of importance identified in edible plants are:

- amygdalin, the gentiobiose conjugate of mandelonitrile. It is present in bitter almonds, apple pips, and kernels of cherries, apricots, and peaches;
- primasin, the D-glucose conjugate of mandelonitrile. It is also found in bitter almonds and other fruit kernels;
- dhurrin, the D-glucose conjugate of p-hydroxybenzaldehyde cyanohydrin. It occurs in sorghum and related grasses;
- linamarin, the D-glucose conjugate of acetone cyanohydrin. It occurs in pulses, linseed, and cassava;
- lotaustralin, the D-glucose conjugate of 2-butanone cyanohydrin. See for its occurrence under linamarin.

Many cases of cyanide poisoning in man after dietary intake have been reported.

The formation of hydrogen cyanide from cyanogenic glycosides in plants takes place via a sequence of enzymic hydrolyses. In a first step, the glycosides are hydrolyzed by β-glycosidases to the cyanohydrins and mono- or disaccharides (see Figure 11.2). The cyanohydrins undergo further hydrolysis by lyases to hydrogen cyanide and the carbonyl compounds involved.

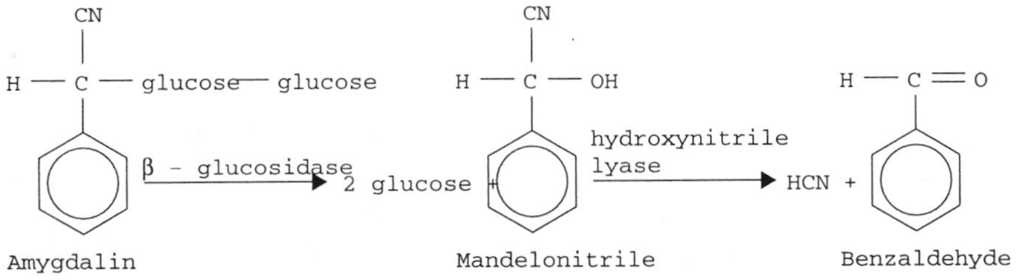

Figure 11.2 Hydrolysis of amygdalin.

Hydrolysis of the cyanogens requires tissue disruption, such as crushing of the wet, unheated tissues. The destruction of the compartmental organization of the cells brings the glycosides in contact with the hydrolytic enzymes.

The occasional intake of small amounts of cyanogenic glycosides does not involve danger. The cyanide formed is generally detoxicated by conversion to thiocyanate. This reaction is catalyzed by the sulfurtransferase rhodanase.

Overloading of the detoxication route by taking in large amounts of cyanogenic glycosides can lead to cyanide intoxication. Fatal poisonings of children have been reported as a result of eating 7 to 10 bitter almonds.

In addition, there are the toxicological risks due to chronic consumption of improperly prepared cassava. Damage to the nervous system after chronic intake of cassava in a number of African countries is believed to be a long-term effect of cyanide or, perhaps, of thiocyanate, resulting from insufficient removal of the cyanogen.

11.2.1.5 Glucosinolates

Glucosinolates are thioglucosides. They have a sulfur atom between the glucosyl group and the aglycon. Glucosinolates also derive from amino acids.

$$R-C\underset{N-O-SO_3^-}{\overset{S-C_6H_{11}O_5}{\Big\backslash}}$$

Glucosinolate

R = a variety of alkane and aromatic groupings

Glucosinolates are thyroid agents. Their main effects are hypothyroidism and thyroid enlargement.

The glucosinolates themselves are not the active agents. They need activation by hydrolysis. All thioglucosides of natural origin are associated with enzymes that can hydrolyze them to an aglycone, glucose and bisulfate. The aglycone can undergo intramolecular rearrangements to yield isothiocyanate, nitrile, or thiocyanate (Figure 11.3).

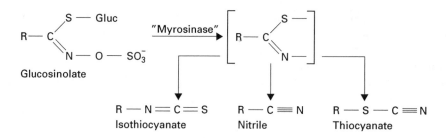

Glucosinolate

R — N = C = S
Isothiocyanate

R — C ≡ N
Nitrile

R — S — C ≡ N
Thiocyanate

Figure 11.3 Hydrolysis of glucosinolates.

Thiocyanates contribute to the antithyroid activity, isothiocyanates are alkylating agents, and the nitriles have also been found to be toxic. Glucosinolates occur in plants belonging to the Cruciferae. The main food sources are cabbage, broccoli, turnips, rutabaga, and mustard greens. Each cruciferous plant may contain up to 10 different glucosinolates. Major representavies are sinigrin (in the above general structure of glucosinolates, R = allyl), progoitrin (R = 2-hydroxy-3-butenyl) and glucobrassicin (R = 3-indolylmethyl).

Sinigrin occurs in cabbage species and black mustard. Its hydrolysis product, allylisothiocyanate, has been shown to be a mutagen in the Ames' test. Swelling of the throat in rats fed on a diet containing Ethiopian rapeseed has also been attributed to the formation of the reactive metabolite. At high concentrations, allylisothiocyanate acts as lachrymator and vesicant. There are no indications that the present consumption of mustards can lead to the induction of adverse effects.

Progoitrin is a major component of rutabaga and a minor one of cabbage, kale, Brussels sprouts, and cauliflower. As its name indicates, progoitrin is a goitrogen or antithyroid. Two types of reactive metabolites are believed to be responsible for the goitrogenic activity: (2-hydroxy-3-butenyl) isothiocyanate and 5-vinyloxazolidine-2-thione (goitrin). Goitrin is formed from the isothiocyanate in a cyclization reaction (Figure 11.4).

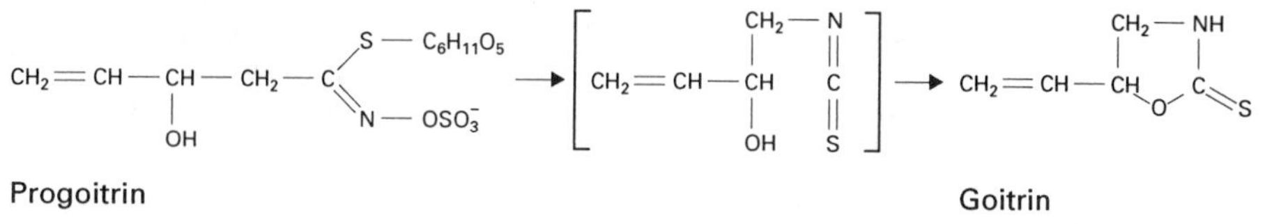

Figure 11.4   Formation of 5-vinyloxazolidine-2-thione.

Oxazolidine-2-thiones inhibit the production of thyroid hormones by preventing the incorporation of iodine in tyrosine.

Glucobrassicin occurs in a variety of cabbage species. Hydrolysis of this glucosinolate results in the formation of a number of products: indole-3-acetonitrile, indole-3-carbinol (I3C) and indole. I3C can cause sedation, ataxia, and sleep. Further, given orally, it is a potent inducer of hepatic as well as intestinal phase I and phase II drug-metabolizing enzymes. On parenteral administration and in isolated hepatocytes, however, it does not induce enzymes. Under the acidic conditions of the stomach, I3C undergoes oligomerization to yield products such as diindolylmethane. Diindolylmethane is also an enzyme inducer.

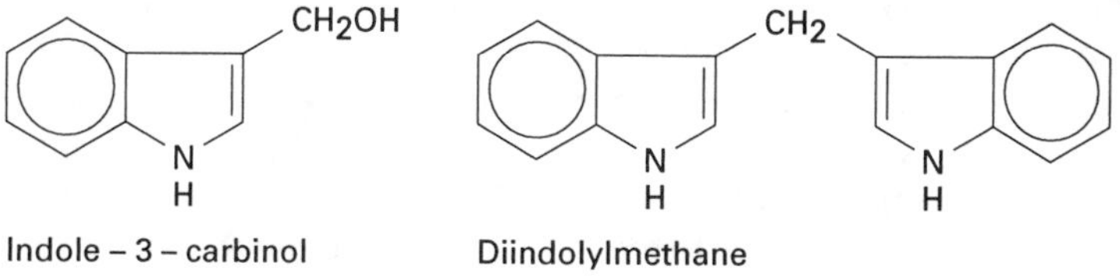

Indole – 3 – carbinol   Diindolylmethane

In some cases, the prevention of cancer has been related to the intake of glucosinolates. The formation of I3C is believed to decrease tumor induction by a variety of carcinogens. Further, feeding of cabbage to experimental animals prior or during the treatment with carcinogens was found to result in inhibition of tumor induction. Cabbage-feeding after administration of the carcinogens led to promotion of carcinogenesis. Probably, the protection against carcinogens is related to a more effective detoxication resulting from enzyme induction.

11.2.1.6 Glycoalkaloids

Steroidal alkaloids are mainly present as glycosides in the family of the Solanaceae, including the potato and the tomato. The major glycoalkaloids in potatoes are α-solanine and α-chaconine, both glycosides of solanidine.

Solanine and chaconine are potent irritants of the intestinal mucosa and cholinesterase inhibitors, the first being the most active. Poisoning with either substance results in gastrointestinal and neurological symptoms. The gastrointestinal symptoms can include vomiting and diarrhea, and the neurological symptoms include irritability, confusion, delirium, and respiratory failure, which may ultimately result in death. Further, poisoning is often accompanied by high fever.

In general, the glycoalkaloid contents of potato tubers do not pose adverse effects in humans. Serious poisonings have been reported following the consumption of potatoes with high glycoalkaloid contents ($\geq$200 mg/kg). Potatoes that have been exposed to light, and those that are diseased by fungal infection or have been mechanically bruised may contain toxic levels of glycoalkaloids.

Results of epidemiological studies on birth defects in regions with fungous potato disease suggested a relationship between the severity of the fungal infection and the occurrence of spina bifida and anencephaly. In animal studies, teratogenic effects and fetal mortality have been observed at dose levels that caused maternal mortality, either on administration of the pure alkaloids or on feeding with diseased potatoes. As far as testing for teratogenicity is concerned, the WHO Expert Task Group on Updating the Principles for the Safety Assessment of Food Additives and Contaminants in Food stated: "If the test substance injures reproduction or development at levels comparable with levels that cause toxicity in adults, then no special concern should be attached to the results of the reproduction/development toxicity studies." Recently, a Dutch expert, however, added to this statement: "It is advisable that for the selection of new varieties the guideline of about 60–70 mg glycoalkaloids/kg is followed in potato breeding, until an appropriate acceptable level has been set." The major glycoalkoloid in tomatoes is α-tomatidine, with tomatidenol as the aglycone. It is present in all parts of the plant. In the fruit, the concentration decreases during ripening. Poisonings in humans due to the consumption of tomatoes have not been reported.

11.2.1.7 Pyrimidine glycosides

For more than a century, a disease caused by the ingestion of fava beans has attracted the attention of toxicologists. The disease, called favism, is characterized by acute hemolysis, in serious cases accompanied by jaundice and hemoglobinuria. It is mainly found in Mediterranean populations with a congenital deficiency of NADPH-dependent glucose-6-phosphate dehydrogenase (G6PD). Fava beans contain two pyrimidine glycosides that have been shown to induce hemolysis: vicine and convicine. The aglycons are divicine and isouramil, respectively.

Divicine and isouramil are powerful reducing agents. In red cells, they are readily oxidized by oxyhemoglobin under the formation of methemoglobin, H_2O_2, and Heinz bodies (thought to consist of denaturated hemoglobin). The oxidation products undergo reduction by glutathione, and H_2O_2 is reduced by glutathione peroxidase. The oxidized glutathione produced by these reactions is reduced by NADPH, generated from glucose-6-phosphate and G6PD.

The defect leading to hemolysis lies in the red cells which have insufficient G6PD, i.e., diminished levels of reduced glutathione, to protect them against oxidative attack.

11.3 Toxic contaminants of microbial origin

In addition to the naturally occurring food components discussed in the preceding section, several important groups of toxic contaminants of microbial origin may enter the food production chain. These may be produced by fungi (mycotoxins), marine algae, or bacteria.

11.3.1 Mycotoxins

Mycotoxins are secondary fungal metabolites. They induce toxic effects upon inhalation or consumption by humans or animals. Mass-poisonings by mycotoxins are unusual in humans.

The history of human intoxications by mycotoxins (mycotoxicoses) dates back to the Middle Ages, when epidemics of hallucinations, delirium, convulsions, and gangrene were not uncommon. By the 1850s, the *ergot alkaloids* (products of the fungus *Claviceps purpurea*) were identified as the causative agents of the disease. Renewed interest in mycotoxin-caused diseases resulted from the death of thousands of turkeys and ducks (Turkey X disease) in England in the early 1960s. The animals were fed diets containing peanut meal contaminated by so-called *aflatoxins*, products of the fungus *Aspergillus flavus*. In the last three decades, more than 100 mycotoxins have been identified throughout the world. Two more classes of mycotoxins posing health hazards are the *ochratoxins* and *trichothecenes*.

11.3.1.1 Ergot alkaloids

Recurrent poisonings by ergot alkaloids (ergotism) in the past resulted from the consumption of *Claviceps purpurea*-infected rye as bread. All ergot alkaloids are derivatives of lysergic acid, with ergonovine (ergometrine) and ergotamine as the most important ones.

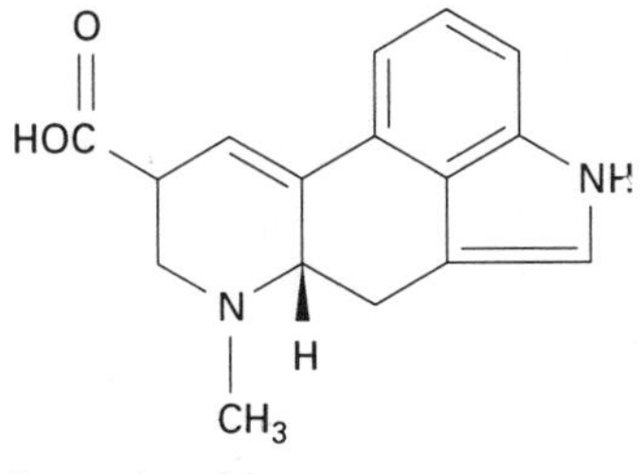

Lysergic acid

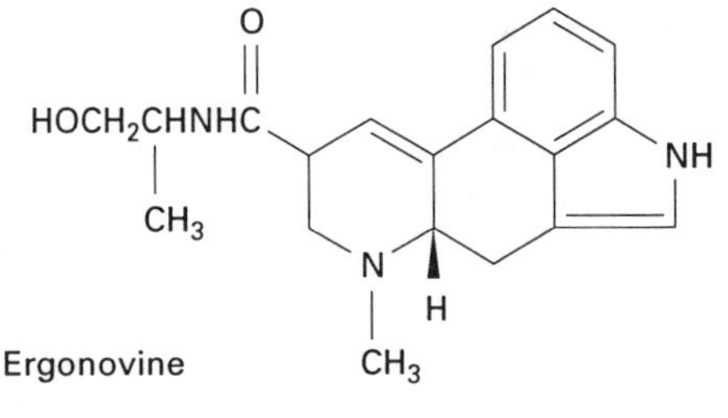

Ergonovine

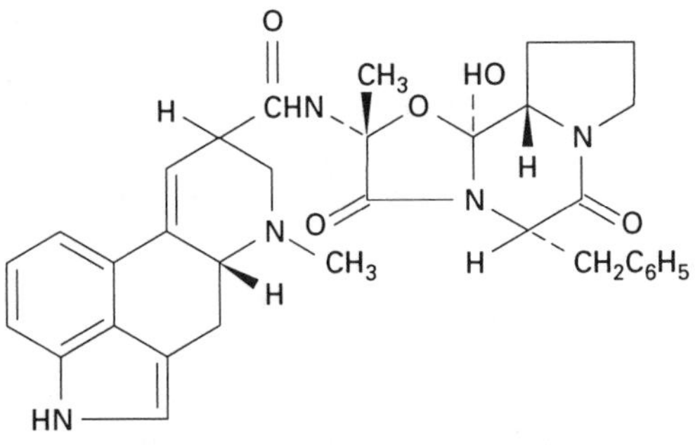

Ergotamine

Ergotism can manifest itself in two ways: a gangrenous type and a convulsive type. The first syndrome is characterized by intense tingling of and hot and cold sensations in the limbs, followed by progression to gangrene and mummification of the extremities. Gangrenous ergotism is largely due to long and intense peripheral vasoconstriction. Ergot alkaloids are partial α-adrenergic agonists. They promote vasoconstriction.

The convulsive syndrome includes central nervous system symptoms such as vomiting, headache, numbness, muscle spasm, and convulsions. Ergonovine has been reported to increase uterine motility, which may cause abortion.

Epidemic ergotism has almost been eliminated. In 1977 and 1978, cases were reported in Ethiopia. In rye, low levels of contamination with ergot alkaloids may still occur.

11.3.1.2 *Aflatoxins*

Aflatoxins are highly substituted coumarin derivatives that contain a fused dihydrofuran moiety. They are divided into two major groups: the B-group (with a cyclopentanone ring) and the G-group (with a lactone ring), based on blue and green fluorescence (Figure 11.5).

Foodstuffs most likely to become contaminated by aflatoxins are peanuts, various other nuts, cottonseed, corn, and figs. Human exposure can also occur from intake of aflatoxins from tissues and milk (in particular aflatoxin M1, a metabolite of aflatoxin B1) from animals that have eaten contaminated feeds.

The group of aflatoxins includes hepatotoxicants and carcinogens. Hepatotoxicity seen in experimental animals is characterized by bile duct epithelium proliferation, fatty infiltration, and centrilobular necrosis. Aflatoxin B1 is highly hepatotoxic and one of the most potent hepatocarcinogens in rats. In many cell systems, it has also been demonstrated to be a mutagen. The hepatotoxicity as well as the mutagenicity and carcinogenicity are believed to depend on its activation by cytochrome P-450 to the 2,3-epoxide. This potent electrophile can covalently bind to proteins and form adducts with DNA.

Epidemiological data indicate that there is a difference in risk of liver cancer between populations in Asia and Africa on the one hand and in North America on the other. Recent studies on the occurrence of hepatitis B virus suggest that chronic infection may contribute to a higher incidence of liver cancer in aflatoxin-exposed populations. In experimental animals, aflatoxin B1 has been shown to suppress cell-mediated immunity.

11.3.1.3 *Ochratoxins*

Ochratoxins are a group of seven dihydroisocoumarin derivatives. The isocoumarin moiety is linked to phenylalanine by an amide bond. Further, some of the ochratoxins distinguish themselves from other mycotoxins by possessing a chlorine atom (Figure 11.6). Ochratoxins have been identified in grains, soybeans, peanuts, and cheese.

Epidemiological studies on the cause of nephropathy in several areas of Yugoslavia, Rumania, and Bulgaria in the late 1950s presented evidence implicating ochratoxin A, present in foodstuffs infected by *Aspergillus ochraceus* and a number of other *Aspergillus* and *Penicillium* species. A similar nephropathy was observed in swine in Denmark and the US. Symptoms include necrosis, fibrosis, and decreased glomerular filtration. In cattle, the ochratoxins undergo degradation by ruminal microorganisms. In addition to nephropathy, ochratoxins have been reported to induce teratogenic effects, renal adenomas, and hepatomas in mice. Tests for mutagenicity, however, gave negative results.

11.3.1.4 *Trichothecenes*

The trichothecene mycotoxins constitute a group of more than 80 sesquiterpenes, derivatives of 12,13-epoxytrichothecene. In the first half of this century, outbreaks of a mycotoxicosis associated with the consumption of contaminated food were reported in Russia. The disease, called alimentary toxic aleukia (ATA), caused atrophy of bone marrow, agranulocytosis, necrotic angina, sepsis, and death. Later, it was related to the infection of

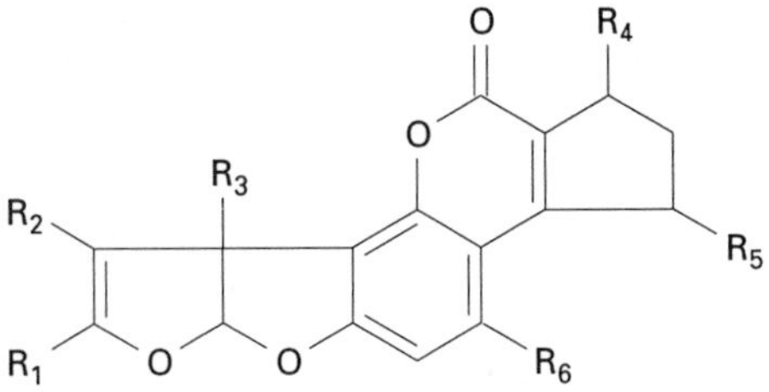

Aflatoxin B1
and its derivatives

Aflatoxin	R_1	R_2	R_3	R_4	R_5	R_6
B1	H	H	H	$=O$	H	OCH_3
B2	H_2	H_2	H	$=O$	H	OCH_3
B2a	HOH	H_2	H	$=O$	H	OCH_3
M1	H	H	OH	$=O$	H	OCH_3
M2	H_2	H_2	OH	$=O$	H	OCH_3
P1	H	H	H	$=O$	H	OH
Q1	H	H	H	$=O$	OH	OCH_3
R_0	H	H	H	OH	H	OCH_3

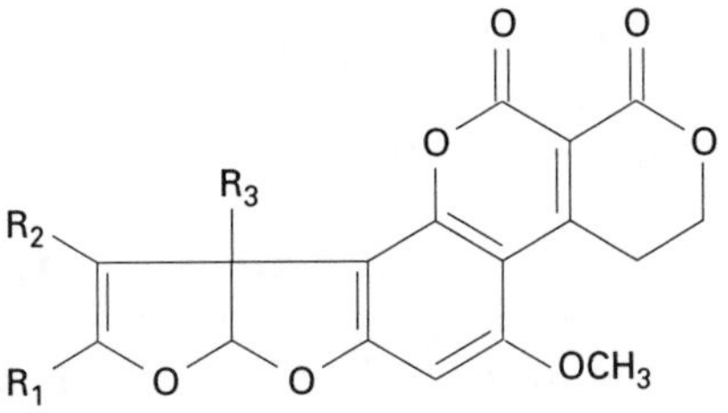

Aflatoxin G1
and its derivatives

Aflatoxin	R_1	R_2	R_3
G1	H	H	H
G2	H_2	H_2	H
G2a	OH	H_2	H
GM1	H	H	OH

Figure 11.5 Structures of aflatoxins.

grains with *Fusarium* species. The most important source of trichothecenes is the fungus genus *Fusarium*. The main contaminants of grains are T-2 toxin and 4-deoxynivalenol (vomitoxin).

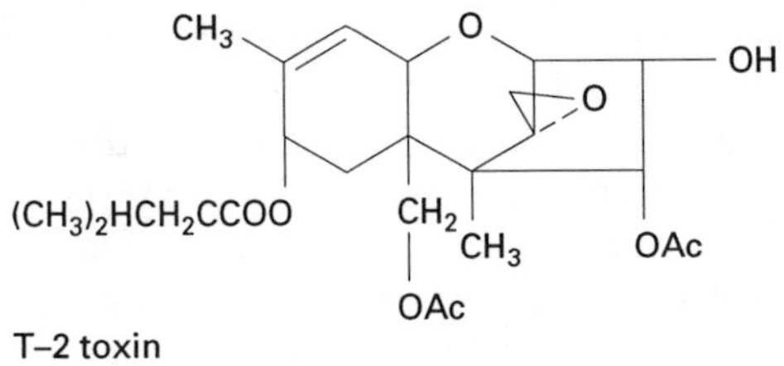

T–2 toxin

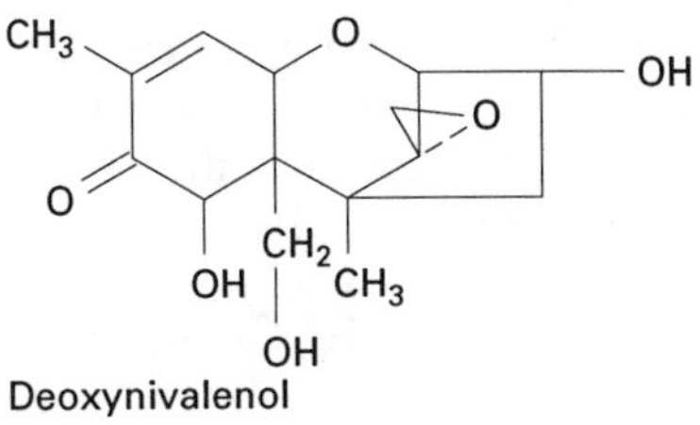

Deoxynivalenol

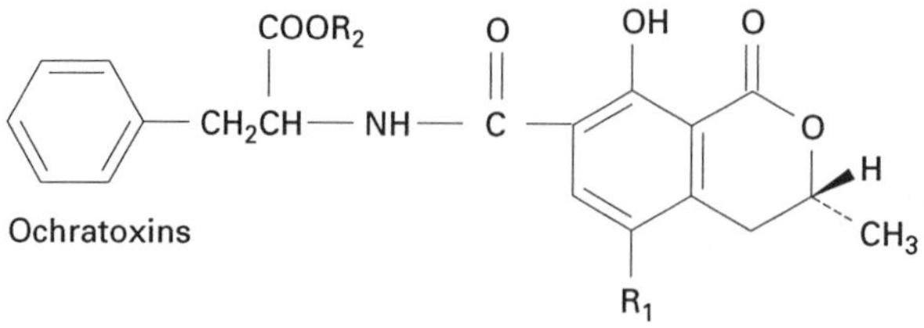

Ochratoxins

Ochratoxin	R_1	R_2
A	Cl	H
B	H	H
C	Cl	C_2H_5
methylester A	Cl	CH_3
methyl or ethylester B	H	CH_3 or C_2H_5

Figure 11.6 Structures of ochratoxins.

There are two forms of trichothecene-caused toxicosis: an acute form, characterized by neurological signs, and a chronic form, characterized by signs of dermanecrosis, leukopenia, and gastrointestinal inflammation, and hemorrhages.

Many toxic effects of trichothecenes are believed to originate from inhibition of protein synthesis. Trichothecenes are generally recognized as the most potent inhibitors of protein synthesis in eukaryotic cells. The inhibition can take place at the initiation, elongation as well as termination phases.

In animals, the trichothecenes are rapidly metabolized to nontoxic compounds. They undergo deacetylation, hydroxylation, and glucuronidation in the liver and kidneys. This detoxication mechanism may contribute to the reduction of the risks in humans from dietary intake of trichothecenes.

11.3.2 *Toxins originating from marine algae or plankton*

Only a few of the large number of marine organisms capable of producing toxins are involved in food poisoning. Poisonings following the ingestion of toxins produced by algae or plankton form significant public health problems in seafood consumption.

Here, two types of marine illnesses will be discussed: shellfish poisoning, a disease resulting from the consumption of shellfish that have ingested toxic algae, and ciguatera poisoning. The latter is caused by ingestion of contaminated fish, in which the toxin has accumulated via a food chain. The alga involved is consumed by a small herbivorous fish. Larger fish feeding on the smaller fish, concentrate the toxin further in the chain. *Shellfish poisoning* manifests itself in two forms: paralytic shellfish poisoning and diarrheic shellfish poisoning.

Paralytic shellfish poisoning is a neurological syndrome. It is characterized by a sequence of events. Within a few minutes after consumption, signs such as numbness of the lips, tongue, and fingers manifest themselves. After extension of the numbness to the limbs, this is followed by muscular incoordination, paralysis, and death.

Paralytic shellfish poisoning is caused by a mixture of several toxins variously termed *paralytic shellfish poison* (PSP) and saxitoxin (in fact, a component of the mixture).

PSP is produced by toxic species of the dinoflagellate genus *Gonyaulax*. Bivalve shellfish (clams and mussels) concentrate the toxin ingested with these organisms. The shellfish are toxic during seasons of algae bloom (so-called "red tide"), i.e., when the concentration of algae is high. Mussels pose the greatest hazard. Paralytic shellfish poisoning is believed to be due to interference of the toxins with ion transport. Saxitoxin is known to block sodium conductance.

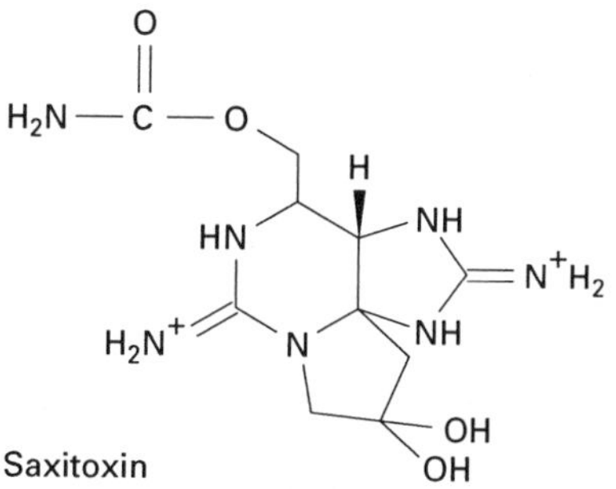

Saxitoxin

Diarrheic shellfish poisoning is characterized by gastrointestinal complaints, including diarrhea, vomiting, nausea, and abdominal spasms. Recently, toxins involved in this poisoning have been chemically identified. They constitute a group of derivatives of a C_{38} fatty acid, okada acid.

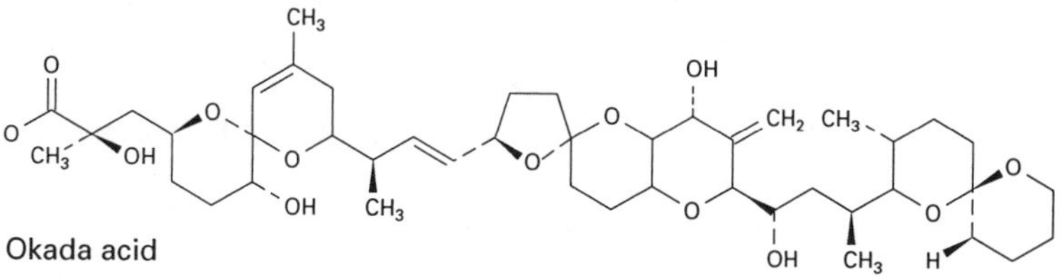

Okada acid

These diarrheic shellfish poisons (DSP) are produced by the dinoflagellate species *Dinophysis* and *Prorocentrum*.

Ciguatera fish poisoning results from the consumption of nondirect plankton feeders. Fish species constituting a food chain concentrate the ciguatera toxins. The fish acquire the toxins by ingestion of the photosynthetic dinoflagellate *Gambierdiscus toxicus*. This kind of intoxication is found in the South Pacific and the Caribbean.

The various ciguatera toxins do not all contribute to the poisoning to the same extent. The main cause is ciguatoxin. Symptoms of ciguatera poisoning are paresthesia in lips, fingers and toes, vomiting, nausea, abdominal pain, diarrhea, bradycardia, muscular weakness, and joint pain. The mechanism underlying these symptoms may be based on the neuroactivity of ciguatoxin. It increases sodium permeability, leading to depolarization of nerves.

11.3.3 Bacterial toxins

Food contaminated by bacteria forms a major source of human disease.

A distinction can be made between food-borne infections, caused by the pathogenic bacteria themselves, and food intoxications, resulting from toxin production by the bacteria. Here, two examples of the latter type of bacterial disease will be discussed.

A major concern of the food industry is contamination of food with *Clostridium botulinum*, not primarily because of a high incidence, but because of the extreme toxicity of the enterotoxin produced by this bacterium. Botulinum toxin is generally regarded as the most acutely toxic chemical known. The toxic syndrome it causes is known as botulism. Cases of botulism have mainly been associated with the consumption of inadequately processed home-canned meat and vegetables.

The bacterium in question has received much attention with respect to food-borne bacterial illnesses. However, there is increasing evidence that other bacteria are also important. For a long time a well-known type of fish poisoning, tetrodotoxin intoxication, was attributed to the production of the toxin by the fish itself, the pufferfish. Recently, however, it has been found that tetrodotoxin is formed by bacteria (*Schewanella putrefaciens*) in the intestines of the pufferfish.

Botulinum toxins consist of at least seven immunologically distinct types labeled A through G. All toxins are similar heat-labile proteins, varying in molecular mass from 128,000 (type F) to 170,000 (type B). Types A, B, and E are commonly associated with human botulism. *Botulinum* toxin is highly potent, with a mouse-LD_{50} of 2 ng/kg i.p. Signs and symptoms usually appear 12 to 36 hr after ingestion of the toxin. Initial symptoms include nausea, vomiting, and diarrhea. These are followed within 3 days by predominantly neurologic symptoms such as headache, dizziness, double vision, weakness of facial muscles, and difficulty with speech and swallowing. Progression of the toxicity leads to paralysis of the respiratory muscles and diaphragm, resulting in failure of respiration and death, usually in 3 to 10 days.

Botulinum toxins produce their toxic effects by blocking the release of acetylcholine at the endings of cholinergic nerves. They bind irreversibly to the neuromuscular junction and impair presynaptic release of the neurotransmitter. The toxic part of all types of botulium toxins consists of two subunits: one smaller (molecular mass 50,000, L) and the other larger (molecular mass 100,000, H). The subunits are linked by a disulfide bridge. The neurotoxicity results from the complementary action of the two subunits. The H subunit binds to receptors at the nerve endings, enabling the L subunit to interfere with the release of acetylcholine.

Treatment involves gut decontamination as well as the use of antitoxins. Gastric lavage, emesis, and activated charcoal may be used if consumption is recent. Neutralization of the toxins and prevention of progression of the toxicity may be achieved by administration of antitoxins.

Pufferfish poisoning is characterized by tingling of the lips and vomiting, followed by paralysis of the chest muscles and death. The toxin, *tetrodotoxin*, is extremely toxic. Toxic consumptions result in a mortality of about 60%.

Tetrodotoxin

The neuroactivity of tetrodotoxin is comparable to that of saxitoxin (see Section 11.3.2). However, it is more prolonged. Tetrodotoxin also disrupts sodium conductance. At the pH of the extracellular fluid, the guanidine group of tetrodotoxin is ionized. It is known that free guanidinim ions can compete with sodium ions for common receptors on the sodium channels of the nerve membranes.

11.4 Antinutrional plant proteins

Main groups of these proteins are protease inhibitors and lectins. The first group is constituted by substances that interfere with the digestion of proteins. Lectins may disturb the absorption of nutrients.

Protease inhibitors are proteins that inhibit proteolytic enzymes. They occur mainly in plants. Well-known are the trypsin inhibitors in legumes (e.g., soybeans), vegetables (e.g., alfalfa), cereals, and potatoes. Protease inhibitors have been reported to inhibit the growth of rats, chickens, and other monogastric animals. Apart from that, an important finding is enlargement of the pancreas in rats, chickens, mice, and young guinea pigs following the administration of trypsin inhibitors in feeding experiments. This effect is attributed to an increase in trypsin synthesis by the pancreas in response to enzyme inhibition. In the pancreas of rats fed on diets rich in trypsin inhibitors during long periods of time, nodular hyperplasia and adenomas have been observed. These symptoms did not show themselves in pigs, dogs, calves, and Cebus monkeys. Mechanistic studies revealed that the so-called Kunitz trypsin inhibitor forms stable one-to-one complexes with the protease. The inhibitor binds to the acitve site of the protein substrate, followed by hydrolysis of a peptide bond between two amino acids, viz. arginine and isoleucine. A disulfide bridge prevents dissociation of the inhibitor. As a result, it remains bound to the enzyme.

Lectins are proteins of plant origin. They especially occur in legumes such as peanuts, soybeans, kidney beans, and peas. Initially, they were called hemagglutinins, as they can agglutinate the red blood cells. After the finding that lectins can bind to specific receptors on a variety of cells including epithelial cells lining the small intestine and lymphocytes, the name lectins has found general acceptance.

Bean lectins have been shown to affect the morphology of the small intestine (proliferation of the intestinal epithelium, Figure 11.7), to inhibit absorption of nutrients and to disturb the immune function of the gut. These effects are attributed to a sequence of events: binding to receptors on epithelial cells of the small intestine, uptake by the cells through endocytosis, and stimulation of the protein synthesis. Some members of this group of proteins have been structurally identified. The lectin concanavalin A, occurring in jack beans, was found to be a lipoprotein, the protein part being composed of 4 identical polypeptides of 237 amino acids. Each subunit appeared to contain binding sites for Ca, Mn, and sugar moieties.

Figure 11.7 Section through jejunum from rats fed on a control (A) or a soybean lectin-containing diet (B). Source: Huisman et al., 1989.

Apart from inhibition of absorption, lectins can also be toxic. A highly toxic example is ricin, present in castor beans. Many poisonings leading to death have been reported, such as children dying after ingestion of raw castor beans. Feeding insufficiently heated raw material proved to be fatal for animals. Ricin consists of two polypeptide chains. One chain is used for binding to the cell, the other inactivates the ribosomal subunits involved in protein synthesis (after uptake through endocytosis). Inhibition of the protein synthesis results in the disappearance of essential enzymes, and ultimately in death.

11.5 Summary

This chapter described a selection of toxic food components of natural origin and their main effects in man and animals. With the exception of the toxic plant proteins, lectins, and protease inhibitors, all are secondary metabolites produced by higher plants, fungi, algae, or bacteria. As a threat to health, the toxins produced by marine algae are most important, followed by the mycotoxins. There is a need for more research in these fields.

Reference and reading list

Cheeke, P.R. and L.R. Shull, *Natural Toxicants in Feeds and Poisonous Plants.* Westport Conn., AVI Publ. Cy., 1985.

Cheeke, P.R., (Ed.), *Toxicants of Plant Origin,* 4 volumes. Boca Raton, Florida, CRC Press Inc., 1989.

Harborne, J.B., *Introduction to Ecological Biochemistry,* 3rd ed. Academic Press, 1988.

Hardegree, M.C. and A.T. Tu, (Eds.), Bacterial toxins, Vol. 4 in: Handbook of natural toxins. New York, Marcel Dekker Inc., 1988.

Hauschild, A.H.W., *Clostridium Botulinum,* in: *Foodborne Bacterial Pathogens,* M.P. Doyle (Ed.). New York, Marcel Dekker, 1989.

Huisman, J., T.F.B. van der Poel and I.E., Liener, *Recent Advances of Research in Antinutritional Factors in Legume Seeds.* Wageningen, Pudoc, 1989.

Keeler, R.F. and A.T. Tu, (Eds.), Plant and fungal toxins, Vol. 1 in: Handbook of natural *toxins.* New York, Marcel Dekker Inc., 1983.

Keeler, R.F. and A.T. Tu, (Eds.), Toxicology of plant and fungal compounds, Vol. 6 in: *Handbook of natural toxins.* New York, Marcel Dekker Inc., 1991.

Moss, J. and A.T. Tu, (Eds.), Bacterial toxins and virulence factors in disease, Vol. 8 in: *Handbook of natural toxins.* New York, Marcel Dekker Inc., 1995.

Rechcigl, M., (Ed.), *Handbook of Naturally Occurring Food Toxicants.* Boca Raton, Florida, CRC Press Inc., 1983.

Smith, L.D.S. and H. Sugiyama, *Botulism: The Organism, its Toxins, the Disease,* 2nd ed. Springfield, IL, Charles C. Thomas, 1988.

Sugiyama, H. and J.N. Sofos, Botulism, in: *Developments in Food Microbiology* 4, R.K. Robinson (Ed.). London, Elsevier Applied Science, 1988.

Tu, A.T., (Ed.), Marine toxins and venoms, Vol. 3 in: *Handbook of natural toxins.* New York, Marcel Dekker Inc., 1988.

Tu, A.T., (Ed.), Food poisoning, Vol. 7 in: Handbook of natural *toxins.* New York, Marcel Dekker Inc., 1992.

chapter twelve

Adverse effects of nutrients

A.A.J.J.L. Rutten

12.1 Introduction to the toxicological aspects of nutrient intake

This chapter focuses on the toxicological aspects of a special group of substances, the nutrients. With respect to nutrient intake two points are of high toxicological importance.

First, attention should be paid to the margin between physiological need and toxic intake, i.e., dose. On the one hand, nutrients are necessary for life and good health, on the other, they may pose life threatening risks. When the intake of nutrients is very low, this may lead to lethal deficiencies, whereas a very high intake may cause toxic effects. The requirements for optimal nutrient intake are based on both deficiency and toxicity data. The optimal intake of a nutrient may be defined as the intake that meets the minimal physiological needs of an organism for that nutrient, and does not cause adverse effects. An example of the implications of overintake is the acute vitamin A toxicity in Arctic and Antarctic explorers on the consumption of polar bear liver containing about 600 mg retinol per 100 g liver. The explorers were informed by the Eskimos that eating polar bear liver may cause drowsiness, headache, vomiting, and extensive peeling of the skin.

A second point that deserves attention with respect to nutrients is the possible interaction between components of a diet. If there is an interaction, there is no adequate procedure to evaluate the toxicological safety, since the traditional procedure for the evaluation of toxicological safety is inappropriate. For example, if a meal consists of protein-rich fish or fish products, and green leafy vegetables, like spinach, interaction may occur leading to the formation of nitrosamines (e.g., dimethylnitrosamine) in the stomach. Dimethylnitrosamine has been shown to induce tumors in experimental animals.

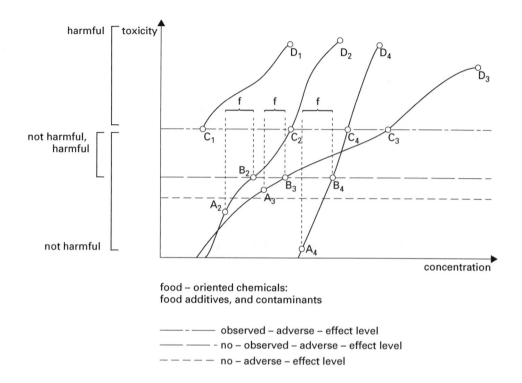

food – oriented chemicals:
food additives, and contaminants

— ·· — observed – adverse – effect level
— — · — no – observed – adverse – effect level
— — — — no – adverse – effect level

Figure 12.1 Impact of concentration on health in the case of food-oriented substances, such as food additives and contaminants. A, no-observed-adverse-effect concentration (acceptable daily intake, ADI); B, no-observed-adverse-effect level (NOAEL); C, minimum-observed-adverse-effect level; D, lethally high concentration; C_1–D_1, genotoxic substances such as nitrosamines; A_2–D_2, food contaminants such as nitrite; A_3–D_3, food additives such as benzoic acid; A_4–D_4, toxins of microbial origin such as botulinum toxin; f, safety margin.

The actual toxicological risks associated with the intake of excessive amounts of nutrients differ from nutrient to nutrient. For instance, induction of toxic effects is hard to imagine after vitamin C intake, while vitamin A poisoning following the consumption of livers of animals high in the food chain, as in the example described above, is well-known. If common nutrients pose health hazards, they must be either highly active or accumulate to a high degree in tissues. In order to gain more insight into the toxicological aspects of nutrient intake, it is useful to divide food chemicals into two groups: food-oriented and body-oriented chemicals.

Food-oriented chemicals have no nutritional value and are primarily associated with food. The group of food-oriented chemicals includes food additives (preservatives such as benzoic acid), antioxidants (butylated hydroxyanisole, BHA), (sweeteners, such as sorbitol), food contaminants (nitrate and nitrite, lead and cadmium, polycyclic aromatic hydrocarbons), and natural toxins (aflatoxins). Assessment of the toxicological risks from the intake of food-oriented chemicals is based on the results of extensive, carefully regulated toxicological screening. Therefore, food-oriented chemicals are considered to be relatively safe (see Figure 12.1). The toxicology of these substances is discussed in Chapters 9, 10, and 11 in more detail.

Body-oriented food chemicals are the nutrients. Nutrients are necessary for growth, maintenance, and reproduction of living organisms (body-oriented). They are divided into two groups: macronutrients (fats, carbohydrates, and proteins) and micronutrients (vitamins and minerals, including trace elements).

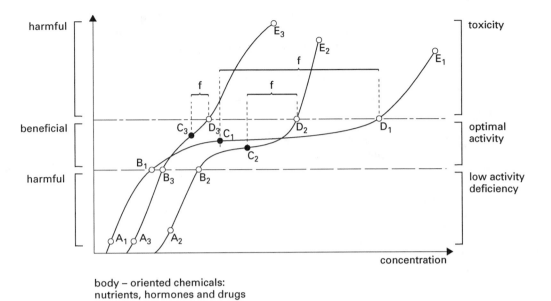

body – oriented chemicals:
nutrients, hormones and drugs

Figure 12.2 Impact of concentration on health in the case of body-oriented substances, such as nutrients, hormones and drugs. A, lethally low concentration; B, minimum concentration compatible with good health; C, concentration for optimal health (nutrients; Recommended Dietary Allowance, RDA); D, maximum concentration compatible with good health; E, lethally high concentration. A_1–E_1, hydrophilic vitamins such as vitamin C and B_1; A_2–E_2, lipophilic vitamins such as vitamin A and E, or selenium; A_3–E_3, macronutrients such as fat; f, safety margin.

For the intake of nutrients recommended dietary allowances (RDAs) are set by official committees. RDAs are defined as the intake levels of essential nutrients that (on the basis of present scientific knowledge) meet the needs of practically all healthy persons. Generally, these levels are considered to be safe (see Figure 12.2 and Table 12.1).

In the next sections, the present knowledge of the toxicity and safety of a number of selected nutrients are discussed. A more extensive study of nutritional toxicology is beyond the scope of this textbook. (For information about nutrients that are not discussed here see the reference list).

12.2 Macronutrients

For a good understanding of the toxicological aspects of the intake of macronutrients, it is important to know that the dietary levels of the three categories of macronutritients are closely related to each other. All three are sources of energy. If the energy intake percentage of one of the macronutrients changes, this will inevitably affect the intake percentage of another. This is shown in Figure 12.3. Official committees recommend a dietary protein intake of 7% of the total energy intake (E), a fat intake of about 30 to 35 energy% and a carbohydrate intake of about 60 energy% (see Figure 12.3a). In the Western countries, the estimated average intake of protein is 15 energy%, of fat, about 40%, and of carbohydrate, about 45 energy% (see Figure 12.3b).

12.2.1 Fats

A twofold increase in dietary fat intake will result in the consumption of about 80 energy% of lipid without a significant intake of carbohydrate (see Figure 12.3c). A large number of

Table 12.1 Recommended dietary allowances (RDAs),
minimum-observed-adverse-effect levels and safety margins for various
macronutrients and micronutrients

Nutrient	Unit	Recommended dietary allowance[1]	Minimum-observed-adverse-effect level	Safety margin
Macronutrients				
Fat	energy%	35	50	1.4
Carbohydrate	energy%	60	90	1.5
Protein	energy%	7	30	4.3
Micronutrients				
Vitamins				
Lipophilic				
Vitamin A	μg	1000	15,000	15
Vitamin D	μg	10	50	5
Vitamin E	mg	10	>900	>80
Hydrophilic				
Vitamin C	mg	60	>12,000	>200
Nicotinic acid	mg	20	1000	50
Thiamin	mg	1.5	>500	>300
Vitamin B_6	mg	2	2000	1000
Vitamin B_{12}	μg	2	>100	>50
Biotin	μg	100	>10,000	>100
Minerals				
Iron	mg	10	180	18
Sodium	mg	500	2500	5
Potassium	mg	2000	18,000	9
Calcium	mg	800	>2500	>3
Trace elements				
Iodine	μg	150	>2000	>130
Zinc	mg	15	150	10
Selenium	μg	150	5000	33
Fluorine	mg	1	10	10
Copper	mg	3	>35	>13
Manganese	mg	5	10	2
Molybdenum	μg	250	10,000	40

[1] The allowances are expressed in terms of average daily intake over time for adults.

studies showed that dietary intake levels of fats, ranging from 40 to 50% of the total energy intake, already lead to a variety of adverse if not toxic effects.

The higher incidence of cancer (of epithelial origin, e.g., breast cancer, Figure 12.4) is well-known. When the total intake of fat is low, polyunsaturated fats appear to be more effective than saturated fats in carcinogenesis. The role of fat in carcinogenesis may be ascribed to tumor promotion. Lipid peroxidation (see Part 1, Chapter 6 and Part 2, Chapter 9) products have been reported to cause cell proliferation. Lipid peroxidation may also be involved in the induction of toxic effects in a number of other ways.

The formation of products such as hydroperoxides and unsaturated aldehydes (e.g., hydroxynonenal) may cause toxic interactions at various levels. Not only membranes and enzymes appear to be primary targets, but also DNA. Hydroxynonenal as well as hydroxyl radicals — formed by metal ion-catalyzed reduction of hydroperoxides — may form DNA-adducts (see Part 2, Chapter 9).

If lipid peroxidation results in excessive consumption of reducing equivalents, oxidative stress may occur. Oxidative stress may lead to the disturbance of homeostases, viz.

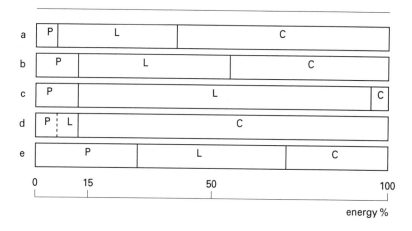

Figure 12.3 Effect of changes in the lipid (L), carbohydrate (C) and protein (P) content of the diet on nutritional balance expressed in terms of energy percentage (energy%) of the total intake. a) Intake of macronutrients recommended by official committees (protein 7 energy%, lipid 30-35 energy%, carbohydrate 58–63 energy%); b) Estimated intake of macronutrients (protein 15 energy%, lipid 40 energy%, carbohydrate 45 energy%); c) Twofold increase in fat intake (protein 15 energy%, lipid 80 energy%, carbohydrate 5 energy%); d) Twofold increase in carbohydrate intake (protein 7 energy%, lipid 7 energy%, carbohydrate 86 energy%). e) Twofold increase in protein intake (protein 30 energy%, lipid 40 energy%, carbohydrate 30 energy%).

thiol homeostasis and Ca^{2+} homeostasis. Thiol groups are then oxidized, and as a result, thiol group-dependent enzymes, like enzymes mediating Ca^{2+} transport, are inactivated. Also, indirect effects may be associated with dietary fat intake. This may concern interactions between food components. Fat intake is known to affect both bioactivation and detoxication of substances. High-fat diets appeared to enhance tumor induction in rats treated with aflatoxin B1 and diethylnitrosamine. Activation of cytochrome P-450 isoenzymes is believed to be involved in the increase in tumor incidence by these chemical carcinogens. An example of an interaction on the detoxication level is the depletion of the antioxidant vitamin E after high intake of polyunsaturated fatty acids. Vitamin E provides protection against peroxidation in general, including that of fatty acids.

12.2.2 Carbohydrates

A twofold increase in carbohydrate intake will result in a considerable decrease in lipid and protein intake (see Figure 12.3d). The adverse effects after excessively high carbohydrate intake are attributed to decreased intakes of the other macronutrients, rather than to the toxicity of carbohydrates.

A high dietary intake of specific carbohydrates has been reported to affect the health of small groups of the population. The absence of disaccharidases in the brush border of the intestinal mucosa connected with genetic as well as contracted disorders gives rise to absorption disturbances and chronic diarrhea. Deficiencies of the disaccharidases sucrase and maltase are rare. On the other hand, lactase deficiency occurs rather frequently. Symptoms of lactose intolerance are usually mild or absent unless large quantities are taken, e.g., a liter of milk, which contains 50 g of lactose. The cause of lactase deficiency may be of three types. First, there is the rare congenital lactase deficiency, with symptoms showing shortly after birth. Secondly, there is a very common ethnic form which affects a large part of the human population. In Asians and many Africans, the enzyme activity disappears at varying ages between infancy and adulthood. Lactase cannot be induced in

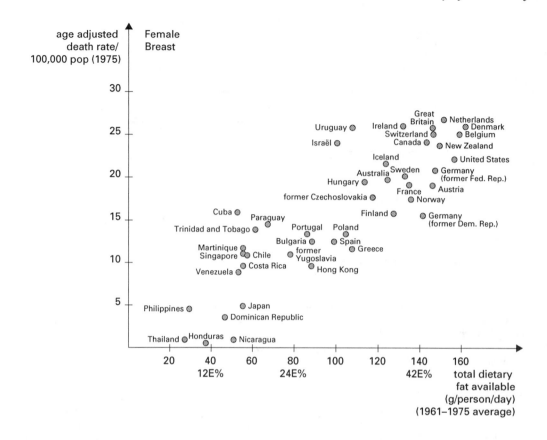

Figure 12.4 Relationship between age-adjusted mortality from breast cancer and total dietary fat available for consumption in different countries. Source: Carroll, 1980.

adults who have lost it. The frequency distribution of lactase deficiency in adults is believed to be the result of natural selection. The third cause of lactase deficiency is disease of the small intestine.

Lactase deficiency is an illustrative example of the importance of information on and education in toxicology. Many people cannot tolerate large quantities of milk or dairy products and hence, suffer from digestion disorders.

12.2.3 Proteins

A twofold increase in protein intake (30 energy%) (see Figure 12.3e) resulted in acceleration of the processes that lead to renal glomerular sclerosis. Further, it has been suggested that habitual high protein intake contributes to osteoporosis. However, protein intakes slightly higher than the physiological need are generally believed to be safe, because excess nitrogen is efficiently eliminated. This occurs mainly in the liver, where amino acids are metabolized to urea. Based on these findings, official committees recommend an upper limit of twice the RDA for protein. Oxidation of sulfur-containing amino acids has both nutritional and safety implications.

The nutritional value of proteins is determined by their amino acid composition, digestibility, and the utilization of absorbed amino acids originating from the proteins. The sulfur-containing amino acid content of many vegetables is low, particularly that of legumes. During processing and/or storage of food proteins, the sulfur-containing amino acids may be oxidized, resulting in a lower availability, i.e., in a reduction of the nutritional value.

The oxidized amino acids (e.g., lysinoalanine) have been shown to be toxic when large quantities of their free forms are consumed. Little is known about the mechanisms underlying the toxic effects of the oxidized sulfur-containing amino acids.

Protein intake may also indirectly lead to the induction of adverse effects. A well-known example of interactions between food components resulting in the formation of toxic products is nitrosamine formation. Secondary amines from fish protein may react with nitrite, originating from vegetable intake resulting in the formation of nitrosamines (for nitrosamine formation, see Part 2, Chapter 9). If vitamin C is also a component of the diet, the formation of nitrosamines can be prevented. Vitamin C inhibits the nitrosation reaction.

12.3 Micronutrients

12.3.1 Vitamins

For vitamins, health risks are traditionally associated with deficiencies. If a large intake range is considered, there is the risk of toxicity (see Figure 12.2). As far as the margin between physiological need and toxic dose is concerned, two groups of vitamins are distinguished: the lipophilic vitamins (vitamins A, D, E, and K) and the water-soluble vitamins (vitamin C, biotin, niacin, pantothenic acid, and folate, and the vitamins B). For the first group the margin may be relatively narrow, for the latter very wide.

Excessive vitamin intake can lead to a variety of toxic effects. In a number of cases, vitamin-induced toxicity is well-known. In other cases, vitamins are only slightly toxic or rather harmless. Although the vitamin content of the diet usually does not lead to toxic effects, it will be of increasing importance to take care of the standards set for vitamin intake in view of the recent trend of vitamin supplementation and fortification. In addition, vitamins are more and more used in processed food as naturally occurring antioxidants instead of synthetic antioxidants.

Also, long-term consumption of high doses of vitamins may be hazardous, even though they are rapidly eliminated. The lipophilic vitamins A and D pose the highest risk, as they can accumulate in the body.

12.3.1.1 Lipophilic vitamins

Vitamins are illustrative examples of body-oriented substances with their specific functions in organisms. It is mainly through the mechanisms underlying these functions that at high intake levels vitamins may be toxic to the organism (hypervitaminosis). Therefore, this section studies the toxicity of the lipophilic vitamins (A, D, E, and K) in relation to their intake, preceded by brief descriptions of their physiological functions.

Vitamin A represents a group of substances necessary for reproduction, cellular differentiation, and proliferation of epithelia, growth, integrity of the immune sytem, and normal eye sight. Retinal, formed from retinol, is involved in the so-called visual cycle. In this cycle, the retinal pigment rhodopsin (visual purple) is bleached on exposure to light. Next, a stimulus is sent to the rods in the retina. The bleaching of visual purple enables the human eye to see at night. In case of vitamin A deficiency, one of the symptoms is night blindness.

The large group of retinoids comprises naturally occurring substances with some vitamin A activity, such as retinol, retinaldehyde, and retinoic acid, and a large number of synthetic, structurally-related substances with or without vitamin A activity. In foods of animal origin, vitamin A is present as retinyl ester. The sources richest in vitamin A are fish liver oils. Further, considerable amounts are also present in fortified whole milk and eggs.

Food consumption data showed that the average daily dietary intake by adult men is about 1500 µg retinol equivalents (RE). The RDA for vitamin A is 1000 µg RE. If consumed

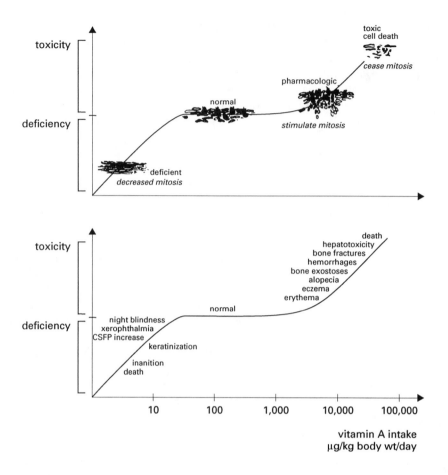

Figure 12.5 Response of a typical mucous epithelium to vitamin A intake (scheme at the top), and clinical symptoms of altered cell function as a result of vitamin deficiency as well as of vitamin A toxicity (bottom curve). Source: Int. Vit. A Consultative Group, Nutrition Foundation, 1980.

in very high doses, vitamin A causes, either acutely or chronically, a large number of adverse effects, including headache, vomiting, diplopia, alopecia, dryness of the mucous membranes, desquamation, bone abnormalities, and liver damage (see Figure 12.5).

Any toxic effects usually result from continuous high daily intake. High intakes of 15 times the RDA can be reached by consuming large amounts of liver or fish liver oils, and food with vitamin A supplements. A high incidence (>20%) of spontaneous abortions and birth defects, including malformations of the cranium, face, heart, thymus, and central nervous system, has been observed in fetuses of women who ingested therapeutic doses of 500 to 1500 µg/kg body weight of 13-cis retinoic acid during the first 3 month of pregnancy. High daily doses of retinyl esters or retinol may cause similar abnormalities.

The mechanisms underlying vitamin A toxicity are very complex, as a sequence of events is triggered. The main toxic effects are related to its function in differentiation and proliferation of cells. The kinetic behavior of vitamin A is largely determined by binding to blood proteins and receptor proteins, and by cellular transport. High intake of vitamin A (hypervitaminosis A) may result in saturation of protein binding which may lead to membrane damage (by free vitamin A).

Vitamin D is necessary for normal bone growth and mineral homeostasis. Exposure of the skin to ultraviolet light catalyzes the synthesis of vitamin D_3 (cholecalciferol) from

7-dehydrocholesterol. Another major form, vitamin D_2 (ergocalciferol), is formed from ergosterol in plants on exposure to ultraviolet light.

Fortified foods (e.g., margarine), milk, eggs and butter are the major sources of vitamin D. The daily vitamin D intake is estimated at 2 µg of cholecalciferol for adults. Presumably, vitamin D pools in the body are replenished in most people by regular exposure to sunlight. The RDA for vitamin D is 5 µg of vitamin D_3 for adults and 10 µg for young adults. Vitamin D is potentially toxic, especially for young children. The effects of excessive vitamin D intake include hypercalcemia and hypercalciuria, leading to deposition of calcium in soft tissues, and irreversible renal damage (nephrocalcinosis) and cardiovascular damage. Although the toxic dose has not been established for all ages, hypervitaminosis D in young children has been related to the consumption of as little as 45 µg vitamin D_3 per day.

Vitamin E is the collective name for an important group of natural products: the tocopherols. There are four members: α, β, γ, and δ, differing from each other in the number and position of the methyl groups attached to the chroman ring, or the saturated carbon side chain. The major and most potent form of vitamin E is α-tocopherol.

The tocopherol content of food (vegetable oils, wheat germ, nuts, green leafy vegetables) varies greatly. During storage and processing large amounts may be lost. The intake of vitamin E is estimated at about 10 mg per day, which is also the RDA. Compared to the other lipophilic vitamins, vitamin E is relatively nontoxic when taken orally. High intake may result in symptoms associated with the pro-oxidant action of vitamin E. Most adults appear to tolerate oral doses of 100 to 800 mg per day.

Vitamin K is necessary for the maintenance of normal blood coagulation. The vitamin is found in green leafy vegetables. Even if large amounts of vitamin K were ingested over a long period of time, no toxic effects were reported. However, administration of a substance structurally related to vitamin K, menadione, may result in hemolytic anemia, hyperbilirubinemia, and kernicterus in the newborn. The underlying mechanism is believed to involve interaction with sulfhydryl groups.

12.3.1.2 Hydrophilic vitamins

To this group belong vitamin C, biotin, niacin, pantothenic acid, folate and the vitamins B: thiamin (B_1), riboflavin (B_2), vitamin B_6 and vitamin B_{12}. Relatively large amounts of these vitamins can be ingested without adverse consequences. They are rapidly excreted from the body, as they are water soluble.

Daily intakes of the antioxidant *vitamin C* (L-ascorbic acid) up to 1 g did not lead to toxic effects. The harmless use of vitamin C is also shown by Figure 12.2. There is a relatively wide margin between the RDA and the toxic dose. If high doses were taken over a long period of time, however, vitamin C appeared to induce toxic effects. Well-known adverse effects after doses as high as 1 g or more are gastrointestinal disturbances such as diarrhea, nausea, and abdominal cramps. Increased peristalsis, resulting from a direct osmotic effect on the intestine, is believed to be the cause. Occasionally, these effects are attributed to sensitization associated with urticaria, edema, and skin rashes. Toxic effects following high doses of vitamin C usually disappear within one or two weeks. They can be prevented by using buffered solutions of vitamin C or by intake after meals.

Thiamin, as thiamin pyrophosphate, is a co-enzyme required for oxidative decarboxylation of α-keto acids and for transketolase in the pentose phosphate pathway. It occurs in considerable amounts in germs of grains, peas, and nuts, and in yeast. Even at very high doses, oral intake of thiamin does not lead to toxic effects; it is rapidly excreted into the urine. Following parenteral or intravenous administration of thiamin, a variety of toxic effects have been reported, but usually only at doses several hundred times higher than the RDA.

Niacin, another hydrophilic vitamin is nicotinic acid. A derivative of this acid, nicotinamide, functions in the body as a component of two cofactors: nicotinamide adenine dinucleotide (NAD) and nicotinamide adenine dinucleotide phosphate (NADP). Niacin is found in the liver, kidneys, meat and fish, and wheat bran, in the germs of grains, and in yeast. High doses of nicotinic acid, but not of nicotinamide, may lead to vascular dilatation, or flushing.

Vitamin B_6 occurs in two forms: pyridoxal phosphate and pyridoxine (or pyridoxol). Pyridoxal phosphate is the metabolically active form of the vitamin. It serves primarily as a cofactor in transamination reactions. Vitamin B_6 is found in kidneys, liver and eggs. The acute toxicity of vitamin B_6 is low. Toxic effects have not been observed in man following an intravenous dose of 200 mg or oral doses of more than 200 or 300 mg. If taken in gram quantities for months or years, however, vitamin B_6 can cause ataxia and severe sensory neuropathy.

12.3.2 Minerals

A fifth group of substances of vital importance to the body is of mineral origin. Mineral salts comprise a large number of elements necessary for growth and maintenance of the cellular and metabolic systems. In food, either of plant or of animal origin, minerals are present as complex salts. An important factor in the toxicity of minerals is their solubility in an aqueous environment, e.g., the contents of the digestive tract. Sodium and potassium salts are readily soluble in water and thus available for uptake from the intestine. Several other elements, such as iron, calcium, and phosphorus, are present in complex salts which are relatively insoluble. These elements are not easily absorbed from the gut. After intake, the major part of the insoluble salts appears in the feces. In the next paragraphs, dietary intake, RDAs (see also Figure 12.2) and toxic effects are described for several minerals.

Iron is an essential element involved in the transport of oxygen in the body. The iron is bound to porphyrin in hemoglobin (red blood cells) and in myoglobin (muscle cells). Iron is a well-known component of raw liver, beef, millet, and wheat. In vegetables and cereals, it is present as phytate or phosphate. From these relatively insoluble salts, iron is almost unavailable for uptake in the intestinal epithelial cell. Intestinal absorption is possible after the iron has been released, and reduced to the ferrous form.

Poisoning by iron has been reported after the intake of ferrous sulfate. It usually occurs incidentally, and then particularly in children. Acute symptoms are nausea and vomiting within 1 hour, followed by diarrhea and gastrointestinal bleeding and, ultimately, circulatory collapse and death.

The body of an adult contains approximately 1200 g of *calcium,* which is mainly present in the bones (body-oriented). Calcium is present in dairy products, including milk and cheese. The daily intake of calcium varies greatly with age and sex, ranging from 530 to 1200 mg. Intestinal absorption of calcium is variably influenced by a large number of factors, such as vitamin D, protein, lactose, phytic acid and dietary fiber, fat, and phosphate. From relatively high intakes, above 800 mg/day by normal adults, approximately 15% was absorbed. The RDA for calcium is 800 mg for adults and 1200 mg for young adults. In general, calcium toxicity is rarely observed. No adverse effects have been observed in many healthy adults consuming up to 2500 mg of calcium per day. However, high intakes may interfere with the intestinal absorption of essential elements such as iron and zinc. Ingestion of very large amounts may result in hypercalcemia, and decreased renal function in both sexes.

Sodium is the major cation in blood and extracellular fluids. The physiological function of sodium is primarily the regulation of osmolarity and membrane potentials of cells. The estimated need for sodium depends on the degree of physical activity and ambient temperature, and ranges from 300 to 500 mg/day for healthy adults.

In the majority of foodstuffs, the sodium originates from the addition of sodium chloride. Single excessive intake of sodium chloride leads to water transfer from the cells to the extracellular space, ultimately resulting in edema and hypertension. As long as the water need can be met, the kidney can excrete the excess of sodium and this effect is reversible. Continuous overconsumption of sodium (2500 mg/day), particularly in the form of sodium chloride, has been found to cause hypertension. Based on these data obtained in human studies, it is obvious that the safety margin for sodium is relatively small.

Potassium is the major intracellular cation. Its concentration in the cell is 30 times higher than in blood and the interstitial fluid. The low extracellular potassium concentration is of high physiological importance in the transmission of nerve impulses (in order to control the skeletal muscle contractility) and the maintenance of normal blood pressure. Dietary sources of potassium are potatoes, soya flour, and fresh fruits. The estimated daily need for potassium is 2000 mg. The safety margin for potassium is relatively small: an RDA of 2000 mg and a toxic dose of 18,000 mg/day.

Acute intoxication of adults has been reported to result from sudden enteral or parenteral potassium intakes up to about 18 g (hyperkalemia). Acute hyperkalemia may lead to cardiac arrest. High potassium blood concentrations may also result from renal failure, adrenalin insufficiency and shock after injury.

Chloride, the major inorganic anion in the extracellular fluid, is necessary for fluid and electrolyte balances. Further, it is an essential component of gastric juice. The only known "dietary" cause of hyperchloremia is dehydration.

Magnesium is primarily present in muscles, soft tissues, extracellular fluid, and bones. About 70% of the daily magnesium intake is covered by the consumption of vegetables and grains. There is little or no evidence that large oral intakes of magnesium are harmful to people with normal renal function. Impaired renal function is often associated with hypermagnesemia resulting from magnesium retention. Early symptoms include nausea, vomiting and hypertension. Hypermagnesemia occurs most frequently following the therapeutic use of magnesium-containing drugs, and not on dietary ingestion of magnesium.

Phosphorus is an essential mineral component of bone tissue, where it occurs in the mass ratio of 1 phosphorus to 2 calcium. Phosphorus is present in nearly all foods. The mean daily intake is estimated at about 1500 mg, while the RDA is 1200 mg. In a number of species, excess phosphorus, i.e., a calcium–phosphorus ratio of 0.5, led to a decrease in the calcium blood level and secondary hyperparathyroidism with loss of bone. The phosphorus levels in normal diets are not likely to be harmful.

12.3.3 Trace elements

Examples of the large group of trace elements are: zinc, iodine, selenium, copper, manganese, fluorine, chromium, and molybdenum. Trace elements are often co-factors of enzymes, and are therefore essential nutrients. The range between the dose necessary for good health and the toxic dose is relatively small for a number of trace elements (see Table 12.5.1). The trace elements that will be discussed here have an intermediate safety margin (10-33): zinc, copper, selenium, and fluorine.

Zinc is a co-factor of a variety of enzymes mediating metabolic pathways, such as alcohol dehydrogenation, lactic dehydrogenation, superoxide dismutation, and alkaline phosphorylation. It occurs especially in meat, (whole) grains and legumes. The RDA for zinc is 12 to 15 mg, depending on the age, while the zinc intake is about 10 mg/day.

Acute toxicity, including gastro-intestinal irritation and vomiting, has been observed following the ingestion of 2 g or more of zinc in the form of sulfate. Effects of relatively low intakes are of greater concern. After dietary intakes of 18.5 or 25 mg by volunteers, impairment of the copper state has been observed. Further, daily intake of 80 to 150 mg during several weeks caused a decrease in the high-density lipoprotein serum level.

Zinc intakes of 20 times the RDA for 6 weeks led to impairment of the immune system, and intakes of 10 to 30 times the RDA for several months led to hypercupremia, microcytosis, and neutropenia. For these reasons, chronic ingestion of zinc exceeding 15 mg/day is not recommended.

Copper is also incorporated in a number of enzymes including cytochrome oxidase and dopamine hydroxylase. It is found in green vegetables, fish, and liver. The copper intake varies from 1.5 to 3.0 mg/day for adults. This is also the RDA.

In general, toxicity from dietary sources is extremely rare. Liver cirrhosis and disturbances of brain functions (e.g., coarse tremor and personality change) have been reported. No adverse effects are to be expected from intakes of up to 35 mg/day for adults. Storing or processing acidic foods or beverages in copper vessels can add to the daily intake and cause toxicity from time to time.

Selenium can be of plant as well as of animal origin. It occurs in seafoods, kidneys, liver, and various types of seeds, e.g., grains. The level in plants depends on the selenium content of the soil in which the plants are growing. Selenium plays an important role in (lipid) peroxide detoxication. The detoxication is catalyzed by a selenium-containing enzyme, glutathione peroxidase.

The daily intake of selenium varies from 80 to 130 µg. The RDA is set at 150 µg. The toxic dose is about 30 times the RDA. Acute intoxication has been reported after ingestion of about 30 mg. Symptoms were nausea, abdominal pain, diarrhea, nail and hair changes, peripheral neuropathy, fatigue, and irritability. Chronic dietary intake of approximately 5 mg/day has been found to result in fingernail changes and hair loss (selenosis). In the seleniferous zone of China, a daily dietary intake of 1 mg of sodium selenite for more than 2 years resulted in thickened but fragile nails and garlic-like odor of dermal excretions.

Fluoride is present in low but varying concentrations in drinking water (1 mg/l), plants (e.g., tea), and animals (fish, 50 to 100 mg per 100 g). It accumulates in human bone tissue and dental enamel. Its beneficial effects on dental health have clearly been demonstrated.

Fluoride is toxic, if consumed in excessive amounts. The normal daily intake is 1 to 2 mg. Daily ingestion of 20 to 80 mg of fluoride leads to fluorosis. This is characterized by calcification resulting in effects on kidney function, and possibly muscle and nerve function. A single intake of 5 to 10 g of sodium fluoride by a 70 kg adult has been reported to cause death. Fluoride intakes above the level of 10 mg per day are not recommended for adults.

12.4 Summary

Two points are of high toxicological importance with respect to nutrient intake: the margin between physiological need and toxic intake and the possible interaction between food components. The nutrients are commonly divided into two groups: the macronutrients (fats, carbohydrates, proteins) and the micronutrients (vitamins and minerals, including trace elements). For the intake of nutrients, recommended dietary allowances (RDAs) are set. These are defined as the intake levels that meet the needs of practically all healthy persons. RDAs are considered to be safe.

As far as the intake of macronutrients is concerned, it is important to know from a toxicological point of view that the dietary levels of the three categories of macronutrients are closely related to each other; they are all sources of energy. An increase in dietary intake of one category will result in decreases in intake of the others.

A variety of toxic effects may result from fat intake. The higher incidence of cancer is well-known. The role of fat in carcinogenesis may be ascribed to tumor promotion. Further, effects that result from lipid peroxidation may be induced. Not only membranes and enzymes have been shown to be primary targets of peroxidation products, but also DNA. Indirect effects may also be associated with dietary fat intake. The increase of tumor incidence in rats treated with aflatoxin B1 and diethylnitrosamine following high-fat intake is believed to be caused by activation of cytochrome P-450 enzymes. Interactions between fat and other food components have also been reported at the level of detoxication. High intake of polyunsaturated fatty acids may lead to depletion of the antioxidant vitamin E.

Adverse effects after high carbohydrate intake are attributed to a decreased intake of the other macronutrients, rather than to actual toxicity of the carbohydrates. Special attention should be paid to the high-risk groups that are congenitally intolerant to particular food components, e.g., lactose.

Protein intakes slightly higher than the physiological needs are generally believed to be safe. High protein intake has been reported to result in acceleration of the processes that lead to renal (glomerular) sclerosis. Oxidation of sulfur-containing amino acids has been shown to form toxic products. Protein intake may also indirectly lead to the induction of toxic effects. Secondary amines from fish protein may react with nitrites originating from vegetable intake, and this may lead to the formation of nitrosamines. Dimethylnitrosamine has been shown to induce tumors in experimental animals.

An important determining factor in the induction of toxic effects by micronutrients is their solubility. Intake of lipophilic vitamins such as vitamins A and D poses the highest toxicological risk, as they can accumulate in the body. Relatively large amounts of the hydrophilic vitamins can be ingested without adverse consequences. They are rapidly eliminated, as they dissolve well in water. Minerals occur in food as complex salts. Several elements are not easily absorbed from the gut, as they occur as relatively insoluble salts.

If consumed at very high doses, the lipophilic vitamin A causes a large number of toxic effects, either acutely or on the long term, including liver damage and developmental disturbances. The main toxic effects are related to its function in differentiation and proliferation of cells. The kinetic behavior of vitamin A is largely determined by its binding to blood proteins and receptor proteins. Well-known effects after excessive intake of another lipophilic vitamin, vitamin D, are hypercalcemia and hypercalciuria, leading to deposition of calcium in soft tissues, and irreversible renal and cardiovascular damage.

A third lipophilic vitamin, the antioxidant vitamin E, is relatively non-toxic when taken orally. High intake may result in symptoms associated with the prooxidant action of the vitamin.

As far as the toxic effects of the hydrophilic vitamins are concerned, the gastrointestinal disturbances after high intake of vitamin C are well known. The toxic effects of vitamin C usually disappear within 1 or 2 weeks.

In general, the availability of the mineral elements iron, calcium, and phosphorus after oral intake is too low to induce toxic effects. These elements are present as relatively insoluble salts, and thus almost unavailable for absorption from the intestinal content.

Intake of the trace element selenium is known to lead to the induction of toxic effects by the element itself as well as adverse effects on the detoxication of other substances. Acute poisoning symptoms following high doses has been reported to include nausea, and nail and hair changes. In areas where the selenium content of the soil is low, disturbances of the detoxication of substances that cause lipid peroxidation may be expected. The detoxication is catalyzed by the selenium-containing enzyme glutathione peroxidase.

Reference and reading list

This chapter is largely based on:

Basu, T.K., (Ed.), *Drug–Nutrient Interactions*. Worcester, Billings & Sons, Ltd., 1988.

Comporti, M., Lipid peroxidation. Biopathological significance. Mol. Aspects *Med.* 14, 199–207, 1993.

Davidson, S., R. Passmore, J.F. Brock and A.S. Truswell, (Eds.), *Human Nutrition and Dietetics*, 7th edition. Edinburgh, Churchill Livingstone, 1979.

Guengerich, F.P., Influence of nutrients and other dietary materials on cytochrome P-450 enzymes. *Am. J. Clin. Nutr.* 61 (3 Suppl.), S 651-S 658, 1995.

Hathcock, J.N., (Ed.), *Nutritional Toxicology*, Volumes I, II and III. New York, Academic Press, Inc., 1982, 1987 and 1989, respectively.

Recommended Dietary Allowances (RDAs), Subcommittee on the 10th edition of the RDAs, Food and Nutrition Board, Commission on Life Sciences, National Research Council. Washington, D.C., National Academy Press, 1989.

chapter thirteen

Toxicology of mixtures in the light of food safety

H. van Genderen

13.1 Introduction

This chapter deals with health effects resulting from the combined actions of food components. These actions include synergism leading to increases in toxicity (of nonnutritive components in particular), as well as interactions between nonnutritive components and nutrients resulting in deficiencies.

In the medical field, there are many examples of both increases and decreases in toxicity after taking combinations of drugs. An interesting example is the case of the anticoagulant dicoumarol and certain other drugs. To maintain the desired prolongation of the prothrombin time, the dose of dicoumarol is critical. Drugs such as aspirin, phenylbutazone, and sulfonamides that displace dicoumarol from its binding sites on plasma proteins, enhance the effect of the anticoagulant. The administration of such drugs during treatment with anticoagulants has resulted in serious cases of bleeding and even fatalities. The anticoagulant effect of dicoumarol may also be reduced if it is administered in combination with drugs that induce the enzyme, mediating the metabolic elimination of dicoumarol. The enzyme involved is mixed-function oxidase (MFO). Examples of MFO inducers are the barbiturates. Simultaneous treatment with barbiturates requires a higher dose of the anticoagulant. On the other hand, cessation of barbiturate administration has been reported to result in an unexpected enhancement of the anticoagulant effect.

As far as nutrition is concerned, combined action of food components has only occasionally been reported. The more frequent occurrence of observable effects of combined actions of drugs is largely due to the combined use of a relatively small number of drugs. Further, drugs are prescribed at effective dose levels, and usually there is medical surveillance. In food, on the contrary, large numbers of components are present at dose levels intended or expected to be far below the effect level. In the case of food additives, the acceptable daily intake is often a hundredth part of the no-observed-adverse-effect level (NOAEL) in experimental animals. Acute toxic effects of combinations of food components are rare, but if they were to occur, they would be easily noticed. The main cause for concern is the not easily recognizable unspecific and chronic effects, such as growth retardation in children and poor state of health in adults. In addition, deficiencies may develop, resulting from interactions between non-nutritive components and nutrients.

Prevention of adverse combined actions is also not easy. Toxicological risk evaluation of food components, such as additives and pesticide residues, is based on the results of toxicity tests with single substances. Each food chemical, however, is a component of a complex mixture of many substances with the chance of interactions and toxic combined actions. It is impossible to test all combinations for potentiation, addition or antagonism. In specific cases, however, prediction of the possible interactions can be made on the basis of theoretical considerations of the underlying mechanisms. For that purpose, classifying combined actions according to the mechanisms involved is helpful.

13.2 Classification of combined actions

A useful classification of combined biological actions of chemicals distinguishes the type of site of action as well as the interactive potency. First, a distinction is made between combinations of substances with common sites of main action and combinations of substances with different sites of action. A second distinction concerns the occurrence of combinations of interacting substances and combinations of noninteracting substances. This leads to the following definitions:

- *simple similar action:* common site(s) of main action, and no interaction between the components. The action can be additive;
- *independent action:* different sites of action, and no interaction;
- *complex similar action:* common site(s) of action, and interaction;
- *dependent action:* different sites of action, and interaction.

Interactions can result in a higher intensity (potentiation) or a lower intensity (antagonism) of the effects of one of the components. If it is impossible to discriminate between potentiation and addition, the term synergism is used.

The next section deals with examples of the four types of combined action. In the subsequent sections, attention will be paid to interactions of food components with non-food factors and the consequences of combined action for food additive policy.

13.2.1 Examples of combined action

13.2.1.1 Simple similar action

Induction of effects by combining of substances with a common mode of action complies with the additivity rule as long as the receptors are not saturated.

Poisoning resulting from simple similar actions of food components has not yet been reported. However, it is beyond doubt that such combined actions occur. Biologically active secondary plant metabolites usually occur in food as mixtures of homologs and isomers. A number of these have similar actions. An example is the intoxication following the intake of green potatoes (see Part 2, Chapter 11). The toxic agents are solanine and chaconine. Their combined disturbing actions on biomembranes are assumed to be additive.

Another illustrative example is the combined toxic action of mixtures of polychlorinated dibenzodioxins, polychlorinated dibenzofurans, and polychlorinated biphenyls (particularly congeners with planar structures), occurring in, for example, mother's milk. The effects of these substances have been shown to be additive. Usually, the toxicity of such mixtures is expressed in terms of the concentration of 2,3,7,8-tetrachlorodibenzo-*p*-dioxin (TCDD), by adding the so-called TCDD toxicity equivalent concentrations of the individual components — concentration addition. Concentration addition is also used in the assessment of tolerances to "simple similar acting" pesticide residues in food products.

Difficulties in obtaining conclusive evidence of additivity (for the effects of food components) have not been encountered in studies on the toxicology of water pollutants. Therefore, it is important to look at the results of aquatic toxicological studies. The concentrations of the components to which the organisms under investigation are exposed can be maintained constant, and the effects of metabolite formation are minimal.

A study on the exposure of fish to a mixture of 50 different relatively stable lipophilic industrial chemicals may serve as an example. The toxicity of these substances is related to the general depressive action on the central nervous system. Their common sites of action are the nerve membranes. If each of the substances was added at 1/50 of its LC_{50}, the lethality of the mixture appeared to be approximately as high as that of one LC_{50}.

However, it should be noted that the induction of neurodepressive effects by combinations of lipophilic substances is only based on additivity for the general unspecific depressive action on the central nervous system. In addition, compounds may induce effects through interactions with specific receptors in the central nervous system. If food components have the same site of action, additivity is also possible for effects induced through interactions with specific receptors.

13.2.1.2 Independent action
Independency here means different sites of action and no interaction. However, different mechanisms can underlie the same effect and this may mean that the effects of some components of a mixture consisting of a large number of substances are similar, and are integrated into an overall effect (effect integration).

Acceptable daily intakes of food additives are estimated by dividing the NOAEL by safety or uncertainty factors. However, at the NOAEL a substance can still give rise to an effect, i.e., an unobserved effect. With regard to the validity of the NOAEL as basis for evaluation, it is important to know whether in the case of independency of action unobserved effects induced by the components of a mixture consisting of a large number of substances can be integrated into observable, or even adverse effects. The experience that lifelong daily intake of thousands of different food components with many independent actions is tolerated without any clear implication to health, provides insufficient evidence that this will always be the case. This problem was addressed in a recent study on the effects of mixtures of eight substances with different modes of action. The mixtures contained sodium metabisulfite, mirex (the insecticide decachlorooctahydro–methenocyclobutapentalene), loperamide (an antidiarrheal phenylpiperidine derivative), metaldehyde, di-*n*-octyltin dichloride, stannous chloride, lysinoalanine, and potassium nitrite. The mixtures were given to rats in the diet during 4 weeks. The dose levels were 0 (control), 0.1 and 0.33 of the NOAEL, the NOAEL and the marginal-observed-adverse-effect level (MOAEL) (see Section 18.3.5). After administration of 0.1 and 0.33 of the NOAEL, no effects were found that could result from the treatment. In the NOAEL group, on the contrary, effects were observed that might be attributed to effect integration: slightly decreased hemoglobin content and slightly increased kidney weight. The animals of the MOAEL group showed effects that were more serious as well as some that were less serious than the effects after administration of the individual components at their MOAELs.

The more serious effects included reduced food intake, impaired general state of health, growth inhibition, and liver damage. These findings provided no conclusive evidence for an increased risk from combined administration of chemicals at their NOAELs.

13.2.1.3 Complex similar action

Interaction can occur at the level of a common receptor, i.e., the site of main action. This can be competitive as well as noncompetitive by nature. Examples with respect to food are rare. Protection against goitrogens of the thiocarbamate type (e.g., goitrin) by iodine treatment through the diet can be regarded as a case of complex similar action. These goitrogens prevent the incorporation of iodine into tyrosine, the first step in thyroid hormone biosynthesis.

13.2.1.4 Dependent action

In the case of dependent action, interactions occur mainly at a pharmacokinetic/toxicokinetic level. This may lead to higher as well as lower intensities/toxicities.

In the medicinal field, there are many examples of enhancement of the effect of one drug by another, resulting from inhibition of the elimination of the latter. In general, however, this type of interaction is expected to be only of minor importance at the relatively low intake levels of nonnutritive food components. An interesting exception is the poisoning following consumption of the (edible) Inky cap mushroom (*Coprinus atramentarius*) in combination with alcoholic beverages. Characteristic symptoms are flushing, hypotension, headache, nausea, and vomiting. The combined action is similar to that of the combination of disulfiram and alcohol, which ends in inhibition of acetaldehyde dehydrogenase. This enzyme is involved in the elimination of acetaldehyde, the primary metabolite of alcohol. The toxic effects are attributed to acetaldehyde accumulation, resulting from inhibition of the dehydrogenase. The mushroom contains the precursor of an inhibitor of acetaldehyde dehydrogenase: the amino acid coprine. In the body, coprine is hydrolyzed under the formation of aminocyclopropanol or cyclopropanone (hydrate). The actual inhibitor of the dehydrogenase is probably the cyclopropaniminium ion. It is believed it reacts with thiol groups of the enzyme.

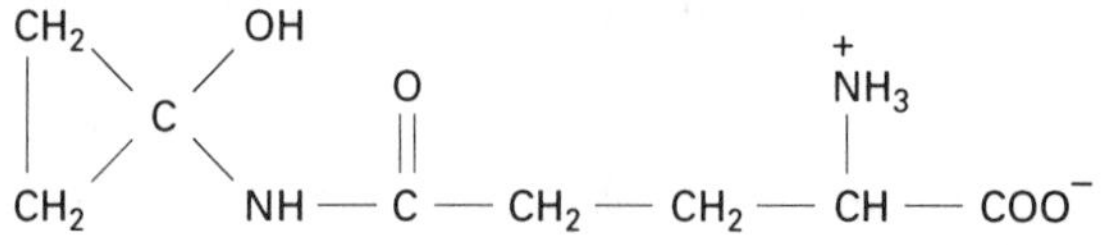

Coprine

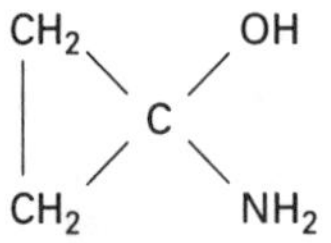

Aminocyclo-
propanol

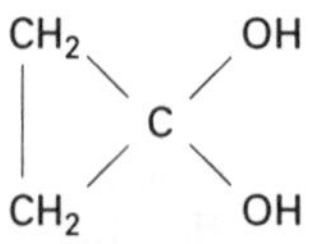

Cyclopropanone
(hydrate)

The fungicide thiram (tetramethylthiuram disulfide), the methyl analog of disulfiram, is also known to cause alcohol intolerance. Thiram may be present in vegetables as a residue, originating from its agricultural application.

Dependent action can also lead to a decrease in effect of one of the components of a mixture. This is, for example, the case if one component decreases the bioavailability of another.

An illustrative example is the decrease in absorption of metals as a result of the presence of phytic acid (inositol hexakisphosphoric ester) in the diet. Phytic acid occurs in cereal products and legumes. It forms insoluble salts with di- and tervalent metal ions. In that way, the absorption of zinc and calcium is inhibited by phytic acid. This is important if soybean proteins are used instead of animal proteins. The phytic acid content of soybeans is high and their zinc content is low. Consequently, dietary use of soybeans may lead to zinc deficiency. Intake of a balanced diet, i.e., with sufficient calcium and vitamin D, will prevent calcium deficiency.

Decrease in absorption of nutrients has also been reported to result from interaction without similarity of site of action after combined intake of thiamine (vitamin B_1) and the additive sulfite. Thiamine can undergo degradation by sulfite in the intestinal tract. In the case of thiamine deficiency, it is important to pay attention to the presence of antithiamines in food.

Another example of dependent action in relation to effects on the absorption of metals is the effects of interactions between heavy metals on their absorption. In industrial areas, the dietary intake of heavy metals can be high as a consequence of pollution. On the other hand, the absorption of heavy metals is usually low. In the case of lead, the average absorption is estimated at 10% of the intake.

Interaction between nutritional factors and heavy metals, i.e., effects of nutritional factors on the toxicity of heavy metals, have been the object of many studies. Synergism as well as antagonism have been reported. In the majority of cases, no conclusive evidence was obtained for the underlying mechanisms.

The complexity of the interactions is illustrated with the following example. It has been shown that calcium affords protection against the toxic effects of lead and cadmium. Further, calcium deficiency appeared to promote the absorption of both metals. The interaction between calcium on the one hand, and lead and cadmium on the other, is believed to be a competition for binding sites on a carrier protein which is involved in the uptake of the metals from the mucosal wall.

13.3 Toxic interactions after combined intake of food and nonfood chemicals

Toxic interactions need not only take place between food components, but can also involve food components and nonfood chemicals. In this context, interactions between drugs and food components are of particular importance. A well-known example is the inhibition of the metabolic inactivation of the pressor substance tyramine contained in food by antidepressant drugs like iproniazid. Tyramine is present in foods such as cheese, wines and coffee. It is detoxicated by monoamine oxidase (MAO). Iproniazid is an inhibitor of MAO.

Further, in mice a high intake of iron appeared to enhance the porphyrogenic activity of halogenated hydrocarbons such as the environmental pollutants hexachlorobenzene and polychlorinated biphenyls. The mechanism of this interaction has not yet been fully elucidated. In addition, an increased incidence of hepatic tumors has been found. This is believed to result from oxidative DNA damage by hydroxyl radicals formed from hydrogen peroxide by uroporphyrin in the presence of iron.

13.4 How does combined action affect food safety assessment?

The safety factors applied for the calculation of ADIs of food additives and food contaminants should not only account for inter- and intraspecies differences, but also for combined action. This means that the level of addivity should be known. However, to establish this level is very complex, if not impossible.

Potential consequences of combined action for food safety assessment are illustrated by the following arithmetic example. Assume a mixture consists of 10 components. The intake of each component is 20% of the ADI of the component. If the level of additivity is 100%, the total intake is twice the ADI. The safety factor for the mixture is cut by one half.

13.5 Summary

A classification of combined actions of chemicals in mixtures is given, together with examples of each type of combined action. While there have been many cases of acute or subacute poisoning due to (unexpected) combined actions of drugs, examples of combined actions of non-nutritive food components are scarce. Not easily recognizable unspecific effects, such as decreased growth in children and poor state of health in adults, are main causes for concern. The possibility of adverse effects from combined actions is one of the justifications of applying large safety factors in the calculation of ADIs.

chapter fourteen

Food allergy and food intolerance

T. Bruggink

14.1 Introduction

The cultural patterns of food consumption tend to change gradually with time. Although it is true that technology increasingly secures the safety of nutrients, this does not mean that each food product is safe. In fact, with the introduction of food additives such as coloring agents and preservatives, the number of substances that may generate adverse reactions has increased, and it seems that the incidence of allergic reactions has, too. This may be explained by developments in technology such as high-temperature processing and irradiation of food, leading to the creation of new antigenic sites. Other factors may be involved as well. It is known, for example, that allergy to inhalants is on the increase. There are many mechanisms underlying adverse reactions to foods. Because of their complexity, there has been confusion about the terminology that should be applied for the different kinds of adverse food reactions. A lack of consensus can easily lead to misunderstanding. Therefore, a discussion about this problem has been started and this has led to the much wanted result: the term adverse food reaction has been defined as any kind of abnormal response to a food (product). This can be an immunologically mediated response or *food allergy*, or a non-immunologically mediated response or *food intolerance*. The latter is a general term which can in turn be further divided into different subcategories (see Table 14.1 and Section 14.2.3). A third type of adverse food reaction is *food aversion*, meaning a pure psychological effect evoked by a food. Within the framework of this chapter, no further attention will be paid to this type of reaction. For a better understanding of the mechanisms underlying the first two types of reactions, the basic principles of allergic reactions and of the normal functioning of the digestive tract will be discussed (Sections 14.2.2.1 and 14.2.2.2 respectively). The main causes of food-allergic and food-intolerance reactions are mentioned in Sections 14.2.4.1 and 14.2.4.2, respectively. It is often difficult to discriminate between a food-allergic reaction and a food-intolerance reaction on the basis of clinical data, as the symptoms can be similar (Sections 14.3.1). Further, there is an extensive differential diagnosis, which renders the problem of diagnosis even more difficult (Section 14.3.2). However, it is important to come to the right diagnosis, because only then is it possible to institute an effective treatment (see Section 14.3.3). Proper treatment has to start at the root, and therefore it is necessary to know the factors which determine the development of a disease.

All together, this issue should be handled with care, due to the chance of overestimation as well as underestimation, with all the related problems.

14.2 General aspects of allergy and intolerance

14.2.1 Definitions

Allergy is defined as an abnormal reaction of the immune system to foreign (not infectious) material, leading to injury to the body that may be either reversible or irreversible. In

Table 14.1 Food-allergic and food-intolerance reactions

Term		Definition
I	Food allergy	Immunological reaction to food or a food product
II	Food intolerance	General term describing an abnormal physiological response to food or food additives that does not appear to be immunological in nature
	a Pharmacological reaction	Reaction resulting from the pharmacological effect of (a) food (component)
	b Metabolic reaction	Reaction resulting from the effect of (a) food (product) on a metabolic abnormality of the host
	c Toxic reaction	Reaction caused by toxic food components
	d Idiosyncratic reaction	Individual intolerance of a certain food or additive. The underlying mechanism is unknown.
III	Food aversion	A psychogenic reaction to (a) food (product).

general, four different types of immunological hypersensitivity reactions are recognized. In a *food-allergic* reaction, this abnormal immunological response is directed against a specific protein or part of a protein in food. *Food-intolerance* reactions are defined as reactions caused by an abnormal physiological reaction of the body to a specific food (component).

14.2.2 Allergy

14.2.2.1 Types of hypersensitivity

There are four types of immunological hypersensitivity reactions. Of the four types of hypersensitivity reactions, type I reactions are probably the most important, as will become evident in this chapter. This does not mean though, that other types of reactions or combinations do not occur. A food-allergic reaction takes place only if the immune system of the body reacts to food in a specific way. This is the case if the food has antigenic potency and has the opportunity to stimulate the immune system.

To protect the body against unwanted effects of food, like allergic reactions, several defense mechanisms are available. The sites where food first makes contact with the body are the mucosa of the oropharynx and that of the digestive tract. The most important defense mechnisms are also located at these sites. For a better understanding of the pathological reactions to food, the normal functioning of the digestive tract is described first (Section 14.2.2.2).

14.2.2.1.1 Type I hypersensitivity. After contact with an allergen, certain white blood cells (B lymphocytes) are triggered to produce antibodies of a special type, namely the immunoglobulin E (IgE) antibodies. These antibodies bind to cells (mainly mast cells and basophils). When there is a subsequent exposure to the same allergen, the allergen becomes bound to two adjacent IgE molecules, resulting in degranulation of the cell to which the IgE is bound. Several types of (preexisting or newly formed) mediators are released, which results in a complex reaction: muscle contraction, dilatation, and increase of permeability of blood vessels, chemotaxis (a mediator-triggered process by which other cells are attracted to the site of reaction), and release of other immune mediators. The reaction occurs mostly within 1 hour, and is sometimes followed by a so-called late reaction which starts hours later.

14.2.2.1.2 Type II hypersensitivity. Antibodies of the immunoglobulin G (IgG) or immunoglobulin M (IgM) class are generated against a cell-surface antigen or an antigen bound to a cell surface. This leads to an inflammatory reaction by which the cells are destroyed. Transfusion reactions due to blood incompatibility work according to this mechanism. There is no evidence that this type of allergic reaction plays a role in food allergy.

14.2.2.1.3 Type III hypersensitivity. Antibodies of the types IgG and IgM are formed against antigens that circulate in the blood. This results in the formation of antigen–antibody complexes which activate the complement system, followed by the release of different mediators from mast cells and basophils. When there is an optimal ratio of antibody to antigen, the complexes may precipitate at different sites in the body, e.g., the joints, the kidneys, and the skin. This type of reaction may play a role in some types of food-allergic reactions. However, it remains difficult to find conclusive evidence. Immune complexes may also be found in the bloodstream of normal individuals shortly after a meal. Type III hypersensitivity reactions also include a number of drug reactions and a few types of vasculitis.

14.2.2.1.4 Type IV hypersensitivity. In contrast to the above types of allergic reactions, no antibodies are involved in this type of reaction. After contact with an antigen, T-lymphocytes are sensitized. These T-lymphocytes then produce cytokines which activate other cells. An example is contact allergy of the skin due to cosmetics. In food allergy, this type of reaction is sometimes seen when food comes into contact with the skin in a person allergic to that specific food.

14.2.2.2 Defense mechanisms in the digestive tract

A major function of the digestive tract is to process food ready for absorption and to exclude harmful substances in the food. Proteins undergo enzymatic degradation to amino acids and dipeptides, fats to fatty acids and diglycerides, and carbohydrates to monosaccharides. Subsequently, absorption by gut enterocytes can take place. There are two types of defense mechanisms in the digestive tract. First, there is a *non-immunological defense* mechanism. The mucus membrane of the gut forms a protective barrier against penetration of pathogenic microorganisms and allergens. Also, the secretion of certain enzymes and gastric acid (which may lead to degradation of unwanted substances) and the enteric motility (which prevents excessive proliferation of bacteria in the small intestine as well as absorption of macromolecules through the digestive mucosa), contribute to the non-immunological defense.

Secondly, there is the *immunological defense*, formed by the gut-associated lymphoid tissue (GALT). The GALT consists of lymphoid organs (follicles, appendix, tonsils, and Peyer's patches) and solitary lymphocytes. Lymphoid organs contain B and T lymphocytes as well as antigen-presenting cells (APCs), mast cells, eosinophils, and basophils. Much research has been carried out on the anatomy and function of the Peyer's patches, which can be found in the small intestine. They are situated just beneath the mucous membrane and are covered by epithelial cells. Between the latter, the so-called M cells (microfold cells) are found, which transport the antigens from the gut lumen to the dome area by pinocytosis. The dome area consists of B cells, plasma cells, T cells and APCs. The APCs present the antigen to the B and T cells. The T cells produce cytokines which stimulate the B cells to switch from IgM production to immunoglobulin A (IgA) production, and activate the B cells to proliferate. The B cells migrate to other parts of the body, such as the respiratory mucosa. Meanwhile, they can receive other T-cell signals which stimulate differentiation to Ig-producing plasma cells. After this, a number of these plasma cells return to the GALT. Most of the plasma cells produce IgA (70 to 90%), some of them IgM (20%), and only a few IgG or IgE. The IgA binds not only to antigens, but also to microorganisms, to prevent infection. A small number of antigens, bound to IgA, are taken up and transported by the portal system to the Kupffer cells in the liver, and eliminated. Also, in healthy individuals, these immune complexes circulate in the blood shortly after a meal. Thus, both immunological and non-immunological mechanisms are involved in preventing food allergens penetration into the gut. In combination, they form the *mucosal barrier*.

Most food allergic reactions occur in infants. This can partly be explained by the fact that the permeability of the intestine in neonates is high, so that proteins can pass across the intestinal mucosa and interact with the immune system. Further, because of the immaturity of the immune system, defensive responses to antigens mediated by secretory IgA in the GALT are only poor. The complete development of both intestinal defense mechanisms takes months. In certain adults, allergic reactions to food occur, even though the defense mechanisms are present in the intestine. The defense can be insufficient at several levels. Mucosal factors as well as intraluminal factors may be responsible for an insufficient elimination of potentially harmful substances.

14.2.3 Intolerance

Various mechanisms may be responsible for food intolerance reactions. The different types of reactions, which are summarized in Table 14.1, will be briefly discussed.

Pharmacological reactions (Table 14.1, Food intolerance reaction IIa). The intensity of biological effects of substances may differ from individual to individual. A well-known example is *caffeine*, a methylxanthine derivative present in tea and coffee. Its biological action includes stimulation of the heart muscle, the central nervous system, and the production of gastrin. People who drink a large amount of coffee may experience restlessness, tremors, weight loss, palpitations, and alterations in mood. Another group of biologically active substances include the vasoactive amines such as histamine and tyramine, and histamine releasers. Excessive intake of *histamine* can cause headache, abdominal cramps, tachycardia, urticaria and, in severe cases, hypotension, bronchoconstriction, chills, and muscle pain. These symptoms appear within 1 hour after ingestion and may last for several hours. Histamine is normally present in food products such as cheese, wine, cream, fish (especially sardine, spatin), sauerkraut, and sausages. It can also be produced by bacteria in the gut. It is metabolized very quickly by enzymes in the gut mucosa and liver. *Tyramine* can be found in French cheese, cheddar, yeast, chianti, and canned fish, and can also be produced by microorganisms in the gut. Symptoms like migraine and urticaria can occur in sensitive persons. *Phenylethylamine* occurs in chocolate, old cheese, and red wine, and can provoke migraine attacks. *Histamine releasers* can function in a non-IgE-mediated way. Known histamine releasers are lectins, present in certain legumes, fruits, and oat. Also, chocolate, strawberry, tomato, fish, egg, pineapple, ethanol, and meat can cause histamine release. Symptoms following non-IgE-mediated histamine release resemble real allergic symptoms (Tabel 14.2). In children, the above foods may aggravate symptoms of atopic dermatitis.

Metabolic reactions (Table 14.1, Food intolerance reaction IIb). Several metabolic disorders in the recipient may result in adverse reactions to foods. The most important in this respect is enzyme deficiencies. The most frequently occurring, especially in Asian countries, is lactase deficiency, leading to intolerance of lactose. After lactose (a carbohydrate in milk or milk products) is ingested, it is not metabolized in the usual way and therefore not taken up by the gut mucosa. It is transformed by the intestinal microflora into a hyperosmolar product that causes diarrhea. Enzyme deficiency may also concern the enzymes disaccharidase and glucose-6-phosphate dehydrogenase, and the disorder phenylketonuria.

Toxic reactions (food poisoning) (Table 14.1, Food intolerance reaction IIc). The presence of toxic components in food has been discussed in Part 1, and their toxic effects in the preceding chapters of Part 2.

Idiosyncratic reactions (Table 14.1, Food intolerance reaction IId). The mechanism underlying this type of food reaction is unknown. It includes reactions to food and the majority of the reactions to food additives. Although the group of additives used in the food industry is very large, only a few have been found to be potentially unsafe for certain individuals. The most important additives in this respect are the azo dyes, sodium benzoate, sulfiting agents, monosodium glutamate, and annatto. They will be discussed separately. It is possible that in the future, some of these reactions will be considered metabolic or toxic, if more is known about their mechanisms.

14.2.4 Food components

14.2.4.1 Allergens
Allergens in food are either proteins, glycoproteins, or polypeptides. The allergenicity can be associated with the type of structure of the proteins and the peptides: primary, second-

ary, or tertiary. In the case of tertiary structures, allergenicity often disappears on denaturation, whereas in the case of primary structures allergenicity remains. Further, the protein has to be large enough to be recognized by the immune system as a foreign compound. In general, the allergenicity of molecules with a molecular mass lower than 5000 is low, unless they are bound to endogenous proteins. On the other hand, substances with a molecular mass higher than 70,000 are not absorbed, and do not come into contact with the immune system. There is a large number of foods wich may cause allergic reactions, but of only a few the allergens have been isolated and identified. The most common causes of allergic reactions are cow's milk, soy, fish, egg, nuts, peanuts, and wheat. These will be discussed briefly.

14.2.4.1.1 Cow's milk.   Cow's milk contains 30 to 35 g protein per liter, which include a large number of antigens. The main antigens are β-lactoglobulin, casein (about 30 g/l!), α-lactalbumin, serum lactalbumin and the immunoglobulines. β-lactoglobulin and α-lactalbumin are referred to as the whey proteins. Casein and β-lactoglobulin are the most heat-resistant. Cow's milk allergy (CMA) is most frequently seen in children. In 10% of the cases, the symptoms appear in the first week of life; in 33%, in the 2nd to 4th week and in 40%, during the following months. The main symptoms are eczema and gastro-intestinal complaints such as diarrhea, cramps, vomiting, and constipation. Also, rhinitis, asthma, and rash may develop. An often obvious feature is irritability and restlessness. There are some specific syndromes (protein-mediated gastroenteropathy and the Heiner syndrome) which are attributed to CMA, but these will not be discussed in this context. In the older child, rhinitis and asthma, and skin disorders such as urticaria and rash dominate. If the diagnosis is CMA, a few alternatives for cow's milk are available. One of them is soy milk as far as nutritional value and costs are concerned, although 20 to 35% of the children develop an allergy to soy. This may be partly explained by the fact that soy milk is often given directly after a period of cow's milk feeding. CMA causes an increase in gut permeability, possibly resulting in an increase in absorption of soy protein, and eventually in a more extensive interaction between soy protein and the immune system.

A second alternative to cow's milk is the protein hydrolysates, which may be considered hypo-allergenic as they contain no or few allergens. Goat's milk is sometimes also mentioned as an alternative. However, this should be discouraged, because goat's milk is strongly cross-reactive and deficient in folic acid. The prognosis of CMA is good; 50 to 90% of the children can tolerate cow's milk by the age of 2-3 years.

14.2.4.1.2 Vegetable allergy.   This kind of allergy may be provoked by beans (soy), peas, and peanuts. Especially, peanut allergy is well-known. Extensive reactions with urticaria, angioedema, nausea, vomiting, rhinitis, and dyspnea have been reported. Anaphylactic shock is not uncommon. The peanut allergen is very stable. It is resistant against all kinds of processing. In peanut butter and peanut flour (which is added to quite a few food products), the peanut allergen is still detectable. In peanut oil, the allergen is not or seldom present. Similarly, the soy allergen is rarely found in soy oil. Allergy to a particular legume does not invariably imply allergic sensitivity to all members of the legume family. Children with a peanut allergy seldom grow out of it.

14.2.4.1.3 Fish allergy.   Allergic reactions to fish are often serious. The cod-fish allergen is heat stable and resistant to proteolytic enzymes. In addition to symptoms such as rhinitis, dyspnea, eczema, urticaria, nausea, and vomiting following digestion of food, urticaria may occur after skin contact with fish. Also, shellfish can cause strong allergic reactions. Probably, fish families have a species-specific antigen as well as cross-reactive antigens. Allergies may be directed against a specific fish or multiple fish families.

14.2.4.1.4 Egg allergy. Egg white is the most frequent cause of egg allergy. Egg white contains about 20 allergens, the most important being ovalbumin, ovotransferrine and ovomucoid. The latter is heat-resistant. Other egg allergens that have been isolated are lysozyme and ovomucine. There is evidence that some cross-reactivity exists between the allergens of the egg white and the egg yolk. Egg allergy is more frequently encountered in children (appearing in the first 2 years of their life) than in older people. Children may eventually lose their allergy for egg.

14.2.4.1.5 Tree nut allergy. In several studies, a cross-reactivity has been reported between birch pollen, and nuts. This cross-reactivity shows itself in a syndrome that is known as the para-birch-syndrome. The complaints of people suffering from this syndrome result from a birch pollen allergy (sneezing, nasal obstruction, and conjunctivitis during the birch pollen season), and also from an allergy to nuts and/or certain fruits. The allergic reactions to these foods mainly cause symptoms such as itching in and around the mouth and pharynx, and swelling of the lips. In some cases, however, more severe reactions occur. Related fruits in this context are apple, peach, plum, cherry, and orange. Also, some vegetables such as celery and carrot have been shown to be cross-reactive with the birch allergen. The patient is probably allergic to bread. Other known cross-reactivity combinations are graspollen with carrot, potato, wheat, and celery. A graspollen-allergic person may become allergic to, for example, wheat as well, which may eventually lead to symptoms such as dyspnea. The exact mechanisms underlying these phenomena are not known. It might be that wheat allergens are inhaled during the ingestion of bread. Another explanation might be that an allergic reaction is caused in the gut, where mediators are released which, after absorption, may be transported to the lungs.

14.2.4.1.6 Wheat allergy. Wheat contains water, starch, lipids, and the proteins albumin, globulins, and gluten. Gluten consists of gliadin and glutenine. The various proteins in wheat can cause different symptoms. One example is the so-called baker's asthma in bakers allergic to wheat albumin. This reaction shows itself when wheat dust is inhaled. In food allergy, globulins and glutenine are the most important allergens. Allergic reactions can occur following the ingestion of wheat. In celiac disease, an allergy to gliadin plays an important role in the pathogenesis. After exposure to gluten infiltration of eosinophils and neutrophils, edema and an increase in vascular permeability of the mucosa of the small intestine can be observed. If the allergic reaction is chronic, the infiltration consists mainly of lymphocytes and plasma cells. Further, flattening of the mucosal surface is found. The disorder manifests itself typically 6 to 12 months after introduction of gluten into the diet. It is characterized initially by intermittent symptoms such as abdominal pain, irritability, and diarrhea. If not treated, anemia, various deficiencies, and growth failure may occur as a result of malabsorption. Improvement is seen about 2 weeks after elimination of gluten from the diet. In addition to the immunological reaction to gluten, a direct toxic effect may also play a role in causing the disease.

14.2.4.2 Additives

Food intolerance reactions can be caused by a variety of substances. The occurrence of metabolic or idiosyncratic reactions depends on the underlying disorder in the host. The different foods that may cause metabolic reactions have already been mentioned in Section 14.2.3. The additives that are most frequently involved in idiosyncratic food intolerance reactions will be briefly discussed.

14.2.4.2.1 Azo dyes. Tartrazine is a yellow dye which is, of all azo dyes, most frequently associated with certain symptoms. In Europe, it is admitted in lemonades, puddings, ice cream, mayonaise, sweets, and preservatives. Some authors claim that it can cause hyperreactivity in children. However, this remains controversial. Asthma has also been related to tartrazine intake, although recent studies have failed to identify sensitive patients in double-blind challenges. Other symptoms which are attributed to tartrazine are urticaria and angioedema, but these are extremely rare.

14.2.4.2.2 Sodium benzoate. This preservative is used in foods such as lemonades, margarine, jam, ice cream, fish, sausages, and dressings. Sometimes it is also added to flavorings. Benzoates can elicit asthmatic attacks in asthmatic patients. Further, benzoates may play a role in patients with urticaria.

14.2.4.2.3 Sulfites. Sodium and potassium bisulfite and metabisulfite are used in food products to prevent spoilage by microorganisms as well as oxidative discoloration. They are added to among others, salads, wine, dehydrated fruits, potatoes, seafood, baked goods, tea mixtures, and sugar products. Symptoms that may occur in sulfite-intolerant persons are airways constriction, flushing, itching, urticaria, angioedema, nausea, and in extreme cases hypotension. Different underlying mechanisms have been postulated. A conclusive explanation of the intolerance of sulfites has not been given.

14.2.4.2.4 Monosodium glutamate (ve-tsin). Salts of glutamic acid are used as flavorings, for instance in Chinese food, soup, meat products, and heavily spiced foods. The well-known "Chinese restaurant syndrome" was first described for a person who had consumed a Chinese meal. Symptoms such as tightness in the chest, headache, nausea, vomiting, abdominal cramps, and even shock, may show themselves. In asthmatic patients, ve-tsin may cause bronchoconstriction. The first symptoms may appear after 15 minutes, while an interval of 24 hours has also been described. The mechanism underlying this syndrome is not known.

If a person complains of dizziness, shortness of breath, nausea and vomiting shortly after consumption of a Chinese meal, this may have been brought about by a number of substances. The symptoms could be related to an allergic reaction, but also to an intolerance reaction. A Chinese meal often contains additives such as sodium glutamate, as well as many proteins and vegetables. For example, koriander (also a component of curry) and garlic are spices which may be responsible for this reaction, or also vegetables such as bean sprouts and cabbage. It should also be borne in mind that fish or fish products or peanut sauce could have been added to the meal. Only if a complete dietary recording and a medical examination have been carried out can the possible cause be identified.

14.2.4.2.5 Annatto. This is a coloring agent of natural origin that is added to cheese products, butter, dressings, syrups, and some types of oil. Some investigators have demonstrated that symptoms may worsen in patients with urticaria and/or angioedema, after ingestion of annatto.

14.3 Clinical aspects of food allergy and food intolerance

14.3.1 Symptoms

In the case of *food allergy*, late reactions seldom occur. The clinical symptoms of allergic food reactions are listed in Table 14.2. The oropharynx and gastrointestinal tract are the initial sites of exposure to food antigens. Symptoms such as edema and itching of the mouth often occur. However, these reactions may be transient and are not necessarily followed by other symptoms. In some people, certain fruits, nuts, and vegetables cause oral symptoms only, while in others a more extensive reaction is seen. The quantity of the offending food also plays a role in the gravity and extent of the reaction, although in

Table 14.2 Symptoms of food allergy

Skin symptoms	Itching, erythema
	Angioedema
	Urticaria
	Increase of eczema
Respiratory symptoms	Itching of (eyes,) nose, throat
	(Tearing, redness of the eyes)
	Sneezing, nasal obstruction
	Swelling of the throat
	Shortness of breath, cough
Gastro-intestinal symptoms	Nausea, vomiting
	Abdominal cramps
	Diarrhea
Systemic symptoms	Hypotension, shock
Controversial symptoms	Arthritis
	Migraine
	Glue ear
	Irritable bowel syndrome

principle a small amount of a certain food can readily cause a response. Sometimes the allergic reaction only develops if the food intake is followed by exercise. This is referred to as exercise-induced food allergy. Hypotension and shock are life-threatening consequences of a food-allergic reaction. Generally, the reaction is accompanied by other anaphylactic symptoms such as abdominal cramps, nausea, vomiting, diarrhea, dyspnea, urticaria, and angioedema.

Table 14.1 shows that *food intolerance* comprises many different clinical disease entities, with different symptoms. Often, the clinical picture is difficult to distinguish from an allergic reaction. The distinction intolerance/allergic cannot always be made on the basis of history alone.

14.3.2 Diagnosis

The manifestation of food allergy and food intolerance can vary from innocent symptoms, like rhinorhea, to life-threatening symptoms, such as shock. The diagnosis is made on the basis of clinical as well as laboratory data, according to the following procedure:

1. History of the patient (complaints, possible associations with food intake, family history, atopic manifestations);
2. Overview of food intake, recorded by a dietician. Often, people have already excluded food products of their own accord;
3. Physical examination (signs of eczema, asthma, rhinitis, abdominal disorders, and nutritional state);
4. Blood examination (eosinophils, total and specific IgE);
5. Skin tests (food allergens, inhalant allergens);
6. Exclusion of all potentially suspected foods (trial diet);
7. Challenge test, for one or a few food products or additives;
8. Gradual reintroduction of food products.

Some of the diagnostic tests are rather time consuming and costly. Also, they cause some risk or discomfort to the patient. The approach is modified depending on type of reaction involved, age, and other characteristics of the patient. Skin test results and specific IgE determinations may be unreliable. For many foods, the identity of the allergenic moieties is unknown and information about their stability is lacking. Examples include the allergens of some fruits and vegetables. It is known that most people allergic to apples can eat apple pie without any problems. The allergen in apple is not heat-stable, and is destroyed by baking. The question is how reliable the skin tests for apple are. Negative results do not rule out a possible allergy for the tested allergen. Positive results do not automatically imply that the particular food does indeed cause the symptoms. The golden standard for the diagnosis of food allergy remains reintroduction or challenge after a period of exclusion. If the diagnosis is correct and compliance is maintained, exclusion of the suspicious food(s) should result in improvement of the patient's condition, and challenge or reintroduction should lead to relapse of the symptoms. It should be realized that food intolerance is not IgE-mediated, and cannot be detected by skin tests and specific IgE determinations. Food intolerance can only be demonstrated by exclusion and challenge. Challenge tests should be carried out in a double-blind placebo-controlled way to prevent subjective interpretation of the results. However, it may be difficult to determine which food component is responsible for the patient's symptoms. An open food challenge with the natural product is then preferable. In addition, challenges with encapsulated foods also have disadvantages. The food is digested in a different way which may result in different symptoms. Further, the food must be pulverized before encapsuling, which may change its allergenicity. If the test results cannot give a conclusive answer about food allergy or food intolerance, other causes must be considered. There is an extensive differential diagnosis. For example, diseases of the stomach or gall bladder cannot be detected by these examinations. Further investigations, such as endoscopy of the stomach and X-ray have to be carried out. It is beyond the scope of this book to go more deeply into this matter.

14.3.3 Treatment

The preferred approach to the management of food allergy (as with any disease) is prevention. Prevention starts in the newborn. IgE synthesis begins before birth, during the 11th week of gestation. There is little evidence that the maternal diet during the last 3 months of pregnancy has any determining effect on the development of a food allergy in the child. If a child has atopic parents, the child has a greater chance of becoming atopic. If both parents have one of the above manifestations of allergy, the chance is about 70%; if only one parent is atopic, the chance is reduced to 30 to 40%, and if neither parent is atopic, the chance is 5 to 10%. For children carrying the risk of atopy, it might be advisable to recommend a special diet. As mentioned before, a dietary regimen in the first months of life can decrease the incidence of food allergy in early infancy. In view of the gut maturity, the introduction of solid foods should be postponed by 6 months. Breast feeding is preferred, as this has several advantages. First, the allergenic burden is less, although some allergens may be found in breastmilk. Secondly, breast milk contains IgA, to which bacteria, viruses, toxins, and also antigens are bound. The binding of antigens which are ingested by the mother herself and are excreted in the breast milk, to IgA is of particular importance. Binding of microorganisms to IgA probably diminishes the incidence of gastrointestinal infections in the infant, resulting in a reduction of the absorption of food allergens and a decrease of the chance of being sensitized. Substances have been found in breast milk which might promote the maturation of the gut. It should be noted that with regard to food allergy, breast milk is superior to current milk formulas, especially if eggs and milk are excluded from the maternal diet. Other alternatives to cow's milk have already been mentioned in Section 14.2.4. Preparations may be considered hypo-allergenic only if it has been proven that they rarely evoke allergic reactions in food-allergic subjects. Formulations based on amino acid solutions are the most hypo-allergenic. The disadvantages are high cost and bad taste. Recent formulations have almost solved these problems.

Once treatment has been started, strict avoidance of all offending foods is needed. It is the task of the dietician to provide a diet that guarantees optimal nutrition and at the same time excludes all hidden sources of offending substances. In the case of food often used as raw material for food products, e.g., milk, egg, and wheat, this can be very difficult. Even the most careful patient may ingest clinically significant amounts of the food to which he is sensitive. With every diet, the degree of sensitivity and the seriousness of symptoms should be taken into account. Some foods (e.g., certain vegetables and fruits) which are not tolerated raw, may lose their allergenicity upon cooking, and may be ingested without any problem if completely cooked. In case of food intolerance, a large number of food products often has to be excluded. For the patient, it is often difficult to know which foods contain additives. Since January 1990, however, this problem no longer exists since by law, all additives have to be listed on the food packaging. However, there are shortcomings to this requirement. One of these is that additives, which make up less than 25% of the end product, are not required to be listed on the packaging material. A second shortcoming is that additives which have no function in the end product also do not need to be listed. It can be concluded, therefore, that information on the composition of food products as given in Part 1 of this book is very important. In case of an intolerance of substances of natural origin, like histamine, avoidance of food products with high contents of such particular substances is often sufficient.

14.4 Summary

It is very important when dealing with adverse food reactions to use generally accepted terminology; this will avoid misunderstandings. As this chapter illustrates, many mechanisms underlying food allergy and food intolerance still have to be elucidated. It is often difficult to give an accurate diagnosis of food allergy. This is largely due to the limited reliability of the diagnostic means. Good therapy can only be started if the diagnosis is clear; herein lies an important problem.

If treatment is prescribed, the help of a dietician is essential, and good patient compliance is important. Some diets require considerable self-discipline on the part of patients. In the other extreme, patients may exclude many foods from their diet on their own accord, thus resulting in nutritional deficiencies. Other problems include the immense assortment of food products that are available, and the lack of knowledge of possible components, as well as the cost of specific diets, which may be quite considerable.

Reference and reading list

Allen, D.H., Delohery and G.J. Baker, Monosodium L-glutamate-induced asthma, in: *J. Allergy Clin. Immunol.* 80, 530–537, 1987.

Hattevig, G., Kjellman, N.I.M., and N. Sigurs B. Bjorksten, and N.I.M. Kjellman, Effect of maternal avoidance of egg's, cow's milk and fish during lactation upon allergic manifestations in infants, in: *Clin. Exp. Allergy* 19, 27–32, 1989.

James, J.M., A.W. Burks, Food hypersensitivity in children. *Curr. Opin. Pediatr.* 6, 661–667, 1994.

Kagnoff, M.F., Immunology of the digestive system, in: Johnson, L.R., (Ed.), *Physiology of the Gastro-Intestinal Tract*. Raven Press, 1987.

Metcalfe, D.D., Food allergens, in: *Clin. Rev. Allergy* 3, 331–349, 1985.

Sampson, H.A., Mechanisms in adverse reactions to food. The Skin. *Allergy* 50 (20 Suppl.), 46–51, 1995.

Simon, R.A. and D.D. Stevenson, Adverse reactions to sulfites, in: *Allergy, Principles and Practice*, Middleton *et al.*, (Eds.), St. Louis, C.V. Mosby Company, 1988.

Stevenson, D.D., R.A. Simon, W.R. Lumry and D.A. Mathison, Adverse reactions to tartrazine, in: *J. Allergy Clin. Immunol.* 78, 182–191, 1986.

chapter fifteen

Studies of adverse effects of food and nutrition in humans

W.M.M. Verschuren

15.1 Introduction

Studies in humans are indispensable for assessing and evaluating risks following the intake of food. The screening of substances for toxicity can be carried out in experimental animals. Extrapolation of the results to humans, however, is difficult. In laboratory experiments, the animals are locked up and the investigator regulates the exposure conditions (usually exposure to a single substance). In addition, the genetic background of experimental animals is often the same, as inbred strains are used. Except for the exposure, most conditions are maintained constant.

One of the problems with extrapolation of animal data to humans is that species can differ greatly in their sensitivity to a toxic substance. Further, in animal experiments, the exposure levels are often relatively high in order to detect possible effects following the exposure. On intake of food, humans are exposed to various combinations of substances and their biological effects can differ from what is expected on the basis of the effects of the individual components. Therefore, studies in humans are needed for the assessment of toxicological risks from the intake of foodstuffs. Studies in humans require a specific methodology; humans cannot be locked up for years, keeping all conditions but one constant; exposure of humans to carcinogenic substances is forbidden. Yet, relationships between exposure and adverse health effects have to be studied in humans.

Sometimes, humans expose themselves voluntarily to all kinds of harmful substances. In such cases, associations between exposure and adverse effects and diseases can be studied. An essential difference between human and animal studies is that humans, in observational studies, choose their own exposure. This may raise the problem that exposure is also related to other factors which may be important in relation to the disease. Thus, exposed and non-exposed subjects may differ in other factors, playing a role in causing disorders. An example is the observation that lung cancer occurs more frequently in people who drink alcohol than in those who do not. This can be attributed to the fact that among alcohol consumers the percentage of cigarette smokers is higher than in the group of alcohol abstainers.

This chapter is an introduction to the use of epidemiological methods in general, and to the use of epidemiology in studying associations between food intake and adverse health effects in particular. Section 15.2 introduces epidemiological methods. This is followed by a section on nutritional epidemiology in which pitfalls, possibilities, and limitations of nutritional methods are described. In order to circumvent the difficulties connected with studying nutrition, recently methods of identification of biological markers for the intake of particular food components are being developed. This will be dealt with in Section 15.4. Section 15.5 looks at the role of nutrition in the risk of cancer.

15.2 Epidemiology

15.2.1 Introduction

Epidemiology can be defined as the science that studies the occurrence and determinants of diseases in human populations. This section introduces the basic principles of epidemiology. It will enable the reader to evaluate critically the results of epidemiological research.

In epidemiology, the term "relationship" (between exposure and disease) is used if a disease is causally related to exposure. If causality has not (yet) been proven, the term "association" applies. Since a single study can never prove causality, the term association is generally appropriate. Associations between exposure and adverse health effects can be studied at different levels. Often, three steps can be distinguished in the etiology of a disease:

lifestyle habits	→	*physiological variables*	→	*disease*
nutrition		blood pressure, weight		cardiovascular
smoking		serum cholesterol concentration		diseases, cancer
physical activity		serum vitamin concentration		

Table 15.1 Cross-tabulation of subjects according to exposure
and disease state

		Disease		
		Yes	No	Total
Exposure	yes	a	b	a + b
	no	c	d	c + d
Total		a + c	b + d	a + b + c + d

In studying associations between nutrition and elevated blood pressure for example, food intake is the determinant or input variable (also called exposure variable) and elevated blood pressure is the (adverse) effect or outcome variable. In studies on possible relationships between blood pressure and coronary heart disease, blood pressure is the input or exposure variable and coronary heart disease is the outcome variable. Thus, a variable can be input variable in one study, and outcome variable in another one. Changes in physiological variables are sometimes referred to as adverse health effects, because they can be risk factors to diseases, e.g., elevated blood pressure to coronary heart disease.

The different ways to evaluate associations between exposure and disease are diagrammatically summarized in Table 15.1. Study designs and disease frequency parameters which can be related to the distinctions made in this diagram are discussed in the following sections. First, the diagram shows the simplest way in which a population can be divided with respect to exposure (yes/no) and disease (yes/no). One possibility is to select exposed (a + b) and unexposed individuals (c + d), followed by comparison of the number of diseased persons in the exposed (a) with the number of diseased persons in the unexposed (c). Another possibility is to select diseased (a + c) and non-diseased persons (b + d) and to compare the number of exposed persons among the diseased and the non-diseased (a vs. b).

15.2.2 Disease frequency parameters

For the description and quantification of the occurrence of a disease in a population, there are two important parameters: incidence and prevalence.

Incidence is defined as the number of new cases which arise during a specific period of time. An example is the yearly cancer incidence: the number of persons who during a year were diagnosed to have cancer for the first time. The significance of this parameter becomes clear if it is related to the number of inhabitants. Therefore, the *incidence rate* is often used. It is defined as the incidence divided by the number of persons at risk:

$$\text{incidence rate} = \frac{\text{number of new cases arising in a given period of time}}{\text{total number of persons at risk of the disease}} \text{ (per unit of time)}$$

The yearly cancer incidence rate can be expressed in terms of the number of new cancer cases during a year per 100,000 inhabitants. For example, if the incidence of cancer in country A is the same as that in country B, and country A has more inhabitants (persons at risk), the incidence rate is lower in country A. If a disease only affects a particular subpopulation, e.g., men, in the case of prostate cancer, the incidence is related to that subpopulation.

Prevalence is defined as the number of cases that are present in the population at a given point of time:

$$\text{prevalence} = \frac{\text{number of cases in a population at a given period of time}}{\text{total number of individuals in the population}}$$

As the equation shows, prevalence is dimensionless. Incidence and prevalence are related to each other. In a steady-state population (i.e., if the number of new cases equals the number of cases which disappear), the relationship between prevalence and incidence is given by:

$$P = I \times D$$

where P = prevalence, I = incidence rate and D = duration of the disease. This means that the prevalence is determined by the duration of the disease if the incidence rates of two diseases are equal.

15.2.3 Effect parameters

In epidemiological studies, biological effects are measured by comparing the occurrence of the disease of one subpopulation with that of another differing in exposure conditions. The differences in occurrence of a disease can be expressed in absolute or relative terms.

Differences in incidence rate between exposed and unexposed populations are *absolute* effects. They are calculated by subtracting the incidence rate in the unexposed group (I_0) from the incidence rate in the exposed group (I_1). The difference $I_1 - I_0$ is referred to as *rate difference*. The incidence rate in the unexposed group can be interpreted as the baseline incidence rate, and only the incidence rate exceeding this figure is due to the exposure. Therefore, the rate difference is also known as *attributable rate*. A difference in incidence rate of 0 means that the disease is not related to exposure ($I_1 = I_0$).

Relative effects are expressed in terms of the quotient I_1/I_0 which is called the rate ratio or relative risk (RR). Calculation of RR using the data given in Table 15.1 results in $(a/a + b)/(c/c + d)$. A relative risk of 1 indicates that the disease is not related to the exposure ($I_1 = I_0$).

The incidence can be estimated in a cohort study, but not in a case-control study (see Section 15.2.4.2 for an explanation). This is due to the fact that in a case-control study, cases and controls are selected at the same time. In such studies a measure can be calculated that is a good approximation of the relative risk: the so-called *odds ratio* (OR). This measure compares the ratio exposed/unexposed among the diseased with the ratio exposed/unexposed among the controls: $(a/c)/(b/d) = ad/bc$.

Intermezzo

In epidemiological studies, it is frequently observed that the relative risk (RR) in older age groups is lower than that in younger age groups. This is illustrated by the following example from the so-called Framingham Study (Figure 15.1). Diabetes is a risk factor for the development of cardiovascular diseases. The RR of coronary heart disease for diabetics is 2.7 in the age group of 45 to 54 years and 2.1 in the age group of 65 to 74 years. However, this does not mean that diabetes is a less important risk factor in the elderly. The absolute rate difference in the age group of 45 to 54 is 20, and 30 in the age group of 65 to 74. Since the rate in non-diabetics of the older age group is higher than that in the younger age group, the lower RR (2.1 vs. 2.7) leads to a larger rate difference.

The relative risk can be used to calculate another effect parameter. With respect to public health, it can also be important to know which proportion of diseased persons

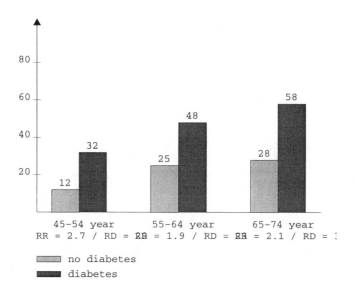

Figure 15.1 Annual cardiovascular disease incidence per 1000 individuals. Source: Kannel and McGee, 1979.

(cases) can be attributed to exposure, the so-called *attributable proportion* (*AP*). This proportion (*AP*$_e$) is obtained by dividing the rate difference by the rate among the exposed:

$$AP_e = \frac{I_1 - I_0}{I_1}$$

This equation can be converted to a function of the relative risk:

$$AP_t = 1 - \left(1/RR\right) = \left(RR - 1\right)/RR$$

There is also a parameter for the proportion of cases in the total population which can be attributed to exposure. The total population can be divided in a proportion unexposed individuals (*P*$_0$) and a proportion exposed individuals (*P*$_1$). The incidence rate in the total population (*I*$_t$) can be calculated from $I_t = P_0 I_0 + P_1 I_1$. The attributable proportion among the total population (*AP*$_t$) is defined as $(I_t - I_0)/I_t$. Substitution of $P_0 I_0 + P_1 I_1$ for I_t results in:

$$AP_t = \frac{P_1(RR - 1)}{P_1(RR - 1) + 1}$$

Intermezzo

Suppose the *RR* of liver cancer due to exposure to factor X is 2.0 and that of liver cancer due to exposure to factor Y is 10. To calculate which factor leads to the highest *AP*$_t$ it is important to note that the formula for *AP*$_t$ contains the *RR* for a particular exposure as well as the prevalence of the effect under investigation. Without information on the prevalences it is impossible to calculate *AP*$_t$. If 60% of the population is exposed to factor X and 0.5% to factor Y the calculation runs as follows:

AP_t for factor X is $0.6(2-1)/0.6(2-1)+1 = 0.38$
AP_t for factor Y is $0.005(10-1)/0.005(10-1)+1 = 0.04$

This example illustrates that RR only gives information on the strength of the associa-tion, and not on the contribution of the exposure to the public health risk for the total population.

15.2.4 Types of epidemiological studies

15.2.4.1 Experimental studies

In *experimental studies* the exposure conditions are chosen by the investigator, as in animal studies. If patients are the subjects, this type of study is often referred to as a *clinical trial*. For ethical reasons, exposure is bound to certain restrictions of which the most important one is that examining potentially toxic substances in humans is prohibited. This implies that potentially adverse effects of food components can only be investigated in *non-experimental studies*. For example, studying the beneficial effect of adding vitamin A to the diet of smokers in relation to the incidence of lung cancer would be permitted. In contrast, the effect of PCBs in mother's milk on the health of babies can only be evaluated in a non-experimental study design.

In experimental studies, two groups of subjects are compared with regard to the outcome variable: subjects exposed to the substance under investigation (intervention group), and subjects not exposed (control group). An essential condition of this type of study (referred to as an *intervention study*) is that the exposure is randomly distributed over the subjects. Maintaining all conditions constant except for the exposure has to be achieved by randomization of the study subjects, as lifestyle and genetic background differ greatly from one person to another. If, in an intervention study on the effect of vitamin C intake on lung cancer, the average number of cigarettes smoked by the intervention group is much lower than that of the control group, the incidence of lung cancer can be expected to be much lower in the intervention group, apart from the effect of vitamin C intake. This underlines the need for randomization of the exposure.

If possible, the study should be double-blind. This means that the investigator as well as the study subjects do not know whether they are in the intervention group or the control group. At the end of the study, the information on who received the substance under investigation and who did not is added to the information already available and the data obtained. In this way, the observations are not influenced by the investigator or the respondent.

15.2.4.2 Observational studies (non-experimental studies)

In *observational studies*, the exposure is "chosen" by the subjects themselves. The investiga-tor confines him/herself to observing the subjects and to collecting data on their exposure and disease, without interfering with their way of life. In the following, four types of observational studies will be discussed.

The various types of epidemiological studies are summarized in Table 15.2. The rank order from weak suggestions to strong evidence of a causal relation in the studies would be ecological studies, cross-sectional studies, case-control studies, cohort studies, and finally randomized controlled trials.

15.2.4.2.1 Cross-sectional studies. In *cross-sectional studies*, data on exposure as well as biological effects are collected at the same time. This kind of study is often used to describe the prevalence of certain exposures or diseases in a population. From an etiological point of view, an essential disadvantage of these studies is the problem of discerning effect from cause. For example, if the total cholesterol serum level is observed to be lower in persons with cancer, this does not allow the conclusion that a low cholesterol serum level causes cancer. It may just as well be that the opposite is true: cancer causes a low cholesterol serum level. In the case of an association between intake of saturated fatty acids and cholesterol serum levels, however, it is more likely that consumption influences the cholesterol serum levels, than the other way round. Knowledge of biological pathways is necessary for making valid inferences.

15.2.4.2.2 Follow-up studies (cohort studies). In a *follow-up study*, the subjects (also referred to as the cohort) are followed for some time (follow-up period). At the start of the study (also called the baseline), the cohort consists of people who are free of the disease under investigation and differ in exposure conditions. To begin with, all persons are examined and information on variables of interest is collected. During the course of the study the occurrence of diseases is recorded. From this, the incidence of the disease in the study population can be calculated. Based on these data, inferences on the association between exposure and occurrence of diseases can be drawn. An important advantage of a cohort study is that exposure is measured before the disease has set in. The appropriate follow-up period depends on the associations which are studied. In the case of salmonellosis following the consumption of raw eggs, a follow-up period of only a few days is sufficient. But, to study the associations between diet and specific types of cancer, a follow-up period of years or decades is necessary. Because the majority of follow-up studies concern chronic diseases, the follow-up period is usually long. Consequently, results are only available after many years. Further, for the assessment of associations between exposure and disease, it is necessary for the number of cases which manifest themselves during the follow-up period to be sufficiently large. This means that the cohort approach is not suitable for studying rare diseases. In order to assess the occurrence of diseases in the cohort in a reliable way, it is of great importance to keep track of all study subjects, and to prevent loss during follow-up as much as possible. Another advantage of follow-up studies is that a large number of both exposures and outcomes can be studied. At baseline, a large number of parameters are usually measured in all study subjects. For a cohort study on a chronic disease, for example, these parameters may include lifestyle factors such as diet, physical activity, smoking habits, and biological variables such as blood pressure, serum cholesterol concentration, height, and weight. Recent developments in the design of cohort studies include storage of biological material such as serum and white or red blood cells at $-20°C$ or $-70°C$. This can be very useful if during the follow-up period new hypotheses arise about the role of variables which have not been measured at baseline. In this way additional baseline information on the study subjects can still be obtained, for example, after 10 years of follow-up.

There are two special types of cohort studies. For a study on a particular effect of an industrial chemical, a cohort can be selected from groups of industrial workers who have been exposed to the chemical. Such cohorts are referred to as *special cohorts*. The prevalence of adverse health effects in such a cohort can then be compared with that among workers in the same industry who have not been exposed, or compared with adverse health effects in the general population.

Because a cohort has to be followed for many years after exposure has been measured, a *retrospective cohort study* is sometimes carried out. This means that a cohort is selected that has been exposed in the past. The investigator then has to establish the appearance of adverse health effects for all individuals of that cohort at the time of the actual study.

15.2.4.2.3 Case-control studies. While in a cohort study exposure is determined at baseline and the occurrence of diseases is followed after the exposure, a *case-control study* starts with identification of diseased subjects and then collects information on exposure in the past. In a case-control study, cases of a particular disease are selected and the patient's exposure in the past is compared with that of controls. This type of study is suitable for studying rare diseases. The numbers of subjects needed here are small compared to those needed in cohort studies. Since the cases are selected without knowing the size of the source population at risk from which they arose, no information on the incidence rate of the disease in the population is obtained in case-control studies. Consequently, the relative risk cannot be calculated. It is approximated by the so-called odds ratio (see Section 15.2.3). The advantage of this study design is that exposure and disease are both measured at the same time, and therefore one does not have to wait as long for results as in the cohort design. In this type of study, however, valid assessment of exposure may be a problem, since exposure in the past is measured after the disease has occurred. The disease may have affected recollection of the exposure by the subject. For instance, the occurrence of the disease may be a stimulus to search for an explanation, leading to a more accurate recollection of exposure. This may be the case for a woman who has given birth to a malformed baby, and who starts thinking about exposures during her pregnancy that may have caused the malformation. Mothers of healthy babies may not have such a stimulus. Therefore, on comparing exposure in complicated pregnancies with exposure in uncomplicated pregnancies, an artificial difference may be observed due to differences in recollection. Also, the disease may lead to denial of the exposure: people with lung cancer may underestimate the role of smoking in the past. For diseases with a long latency period and which influence the factor under investigation, information on exposure in the distant past is needed. This may well be impossible.

15.2.4.2.4 Ecological studies. In this type of study the unit of observation is not the individual but a group of people in a particular environment, such as workers in a factory or inhabitants of a city or a country. *Ecological studies* can be useful if information on individuals is not available; exposure is then an overall measure for the population under investigation. For example, nutritional data are sometimes only available per country as food balance sheets. The outcome variable under investigation in ecological studies is often mortality. For example, the mortality level due to cardiovascular diseases in different countries correlates well with the average saturated fat consumption per capita in those countries. This association has been supported by the results of intervention studies.

A well-known phenomenon occurring in this type of study is the so-called ecological fallacy. On comparing countries, it may be found that the higher the average level of a risk factor A for a country, the higher the mean level of mortality due to disease B, while within each country (based on individual measurements of A and B) risk factor A is negatively associated with disease B.

15.2.5 Precision and validity

Measurements are important in epidemiology. That refers to the input variables (determinants/exposure) as well as the outcome variables (adverse health effect/disease/mortality). For instance, in the case of a study on the relationship between magnesium intake and blood pressure, the investigator wants to know the blood pressure and magnesium intake of each subject. The measurements of these parameters are estimates of the "true" blood pressure and the "true" magnesium intake. These true values, however, are not known, and are therefore hypothetical. An estimate of a true value always contains some measurement error. This error may be random or systematic. Random errors can occur if the observer is not very accurate or the measuring device is not very easy to read. This results in values that are sometimes too low and sometimes too high. However, on average the over- and underestimation compensate each other, resulting in a group mean that is close to the true group mean.

Table 15.2 Summary of possibilities and limitations of epidemiological studies

	Experimental study	Cross-sectional study	Follow-up study	Case-control study	Ecological study
Possibilities	Strong indication of causal relation	Estimation of prevalence of exposure or disease	A large number of exposures and diseases can be studied Exposure is determined before onset of the disease Estimation of the incidence of a disease	A large number of exposures can be studied The number of study subjects may be relatively small Suitable for rare diseases	Can be used when information is only available on an aggregated level
Limitations	Only beneficial effects can be studied Only a small number of study subjects can be used for logistic reasons	Distinction between cause and effect is difficult	During the follow-up period the investigators must keep track of all study subjects Expensive (time and money) Only suitable for frequently occurring diseases	Exposure is determined after onset of the disease; reporting of exposure by the respondents might be affected by the disease	Ecological fallacy

A systematic error implies that all measurements are too low or too high. This can be the case if the measuring device is not properly calibrated. Apart from measurement errors, there is likely to be biological variability, as a result of which repeated measurements yield not exactly the same values. Biological variation can also be random (e.g., fluctuating around an average value) or systematic (e.g., height is largest at the beginning of the day and decreases slightly during the day).

Because of measurement errors and biological variability, the average value of repeated measurements usually gives a better estimate of the true value than the result of a single measurement.

Precision, also referred to as reproducibility, implies that there are no random errors in the measurements. The reproducibility of a measurement is high if there is good concordance between repeated measurements.

The *validity* of a measurement concerns the concordance between the value of a measurement and the true value, in other words: do you measure what you want to measure? A high reproducibility is a prerequisite for validity, but does not automatically imply validity.

Suppose a person weighs 65 kg. Balance A yields a value of 70.2 kg for each measurement, and balance B gives 65.3 , 65.6, and 65.7 kg on the first, second, and third measurement, respectively. In this case, balance A has a high reproducibility, but is not very valid. Balance B is reasonably precise and yields a valid estimate of the "true" weight.

With regard to the validity of the results of epidemiological studies, a distinction is made between internal and external validity. *Internal validity* is the validity of the inferences drawn for the population under investigation, while *external validity* refers to the ability to generalize of the results beyond the study population. It will be clear that if there is no internal validity, there can be no external validity either. In general, internal validity can be influenced by three types of bias: selection bias, information bias, and confounding. However, the distinction between these three is not always strict.

Selection bias can be defined as the fact that the effect measured is perverted due to the selection of the study subjects. This means that the association between exposure and disease in the study population differs from the association in the total population. Case-control studies are especially sensitive to selection bias. If subjects are systematically excluded from or included in the case or control group, the comparison of these groups can give biased results. Since cases are often recruited from hospitals, controls are sometimes also selected from the same hospitals. Since hospitalized persons are likely to differ from the general population, this may influence the study results. For example, if lung cancer cases are compared with controls which have been recruited from a hospital, the smoking habits between the two groups might not differ very much because smoking prevalence among hospitalized persons is higher than in the general population. This is due to the fact that smoking is a risk factor of a large number of diseases. Therefore, in the study design, special attention should be paid to the selection of controls. Often, several control groups are used to estimate the consequences of the choice of the source population of controls. Other source populations of controls that are used in addition to hospital controls are neighborhood controls (to control for socio-economic differences between cases and controls) or a random population sample, in order to compare the exposure in the cases with that in the general population.

Information bias is the term for errors in the necessary information, leading to errors in the classification of subjects. If the errors in the necessary information (e.g., in exposure measurement) are not related to the state of disease, the misclassification is called random or non-differential. This is the case if equal proportions of subjects in the groups which are compared, are classified incorrectly with respect to exposure or disease. *Random misclassification* dilutes the true difference and therefore always changes the observed effect towards the null hypothesis (i.e., no relationship between exposure and disease). If the measurement error in the exposure is related to the disease, the misclassification is called differential. *Differential misclassification* has more serious consequences. It can lead to either underestimation or overestimation of the effect.

A type of information bias that is of importance in case-control studies is *recall bias*, which means that cases differ from controls in the recollection of exposure. An example is that after giving birth to a malformed baby, mothers start thinking about potential causes for this malformation during their pregnancy (see Section 15.2.4.2). A way to solve this problem may be the selection of a control group of which the memory has also been activated. In a case-control study on congenital heart disease, for example, a control group can be selected with other congenital diseases. It should be noted that the congenital heart disease studied and the congenital disease in the control group should not have a common determinant.

The third type of bias is *confounding*, one of the most important concepts in epidemiology. Confounding can be defined as the combined effect of the factor under investigation and other (confounding) factors. An illustrative example of confounding is the finding that lung cancer is associated with alcohol consumption. However, this finding is caused by the fact that smoking is associated with alcohol consumption, and lung cancer is associated with smoking. In the association between alcohol consumption and lung cancer, smoking is called the confounding factor (the confounder). A factor can only be a confounder if the occurrence of the disease as well as the exposure under investigation is associated with it. There is an essential difference between confounding and information or selection bias. If information on the confounder is collected during the study, it can be adjusted for in the statistical analyses. Sometimes, however, an unknown or not measured confounder is present. Such a confounder cannot be adjusted for in the statistical analyses, and gives rise to biased study results.

External validity determines whether the results can be generalized beyond the study population. Internal validity is a prerequisite for external validity. If an association is not validly assessed for the population under investigation, it cannot be generalized to other populations. For external validity, a judgment must be made on the plausibility that the effect observed in the study population can be generalized. In this context, questions can be asked such as: Do associations found in men also apply to women?, Are associations found in young people also valid for elderly people?, and Are the results of an American study also applicable to the Dutch population?.

15.2.6 Causality

In epidemiological studies, associations of disease(s) with exposure may be found. This does not necessarily mean that the exposure caused the disease(s). The English statistician Hill introduced a number of criteria which should be met before inferences about causality can be made. Although only one of these criteria is imperative for a factor to be causal, all of them are briefly discussed below:

1. *Strength of an association.* Weak associations are more likely to be attributed to confounding than strong associations. On the other hand, weak associations certainly do not exclude causality. Particularly in nutritional research, the majority of the associations between food intake and adverse health effects can be classified as weak (meaning a relative risk of about 1.5 to 2.0);
2. *Consistency.* If an association is causal, it must be possible to observe this association in different populations under different circumstances. However, it is also possible that a factor causes a disease under one circumstance but not under another;

3. *Specificity.* This criterion means that the cause should lead to a specific effect. This can be easily proven to be wrong. For example, smoking causes not only lung cancer, but also several other lung diseases and ischemic heart diseases;
4. *Temporality.* This requires exposure always to precede the effect in order to be causal;
5. *Biological gradient.* In a number of cases, indeed, a dose–response relationship is found. Sometimes, however, all exposure levels measured are high enough to cause the disease. In that case, no dose–response relationship is observed;
6. *Plausibility.* The association should be biologically plausible. A problem with this criterion is that sometimes associations are found before the underlying biological mechanisms are elucidated;
7. *Coherence.* According to this criterion, associations are not incompatible with what is known about the etiology of the disease. It is closely related to plausibility;
8. *Experimental evidence.* An association should be confirmed in a controlled laboratory (animal) experiment. This cannot be done, however, if the toxicity of the substance under investigation in laboratory animals is extremely low;
9. *Analogy.* This means that if a substance causes a particular effect, a structurally related substance may cause the same effect.

In fact, only one of of these criteria is a "conditio sine qua non" to prove causality. The criterion of temporality should always be met: the cause must precede the effect! However, it is difficult to prove causality. In practice, this can only be achieved if information of a number of scientific disciplines is integrated. Sometimes, an association is indicated by epidemiological studies, and subsequently the mechanism is investigated in experimental animals or laboratory experiments. It can also be the other way around: an effect shown in experimental animals or laboratory experiments is confirmed in epidemiological studies.

15.3 Nutritional epidemiology

In the last few decades, the interest in the role of diet in the etiology of diseases has increased strongly. For the identification of the role of nutritional factors in the etiology of diseases, the methodology of food consumption measurement is of particular importance. Measuring individual food intake is difficult. In epidemiological studies, a number of methods are available to measure food intake. They will be dealt with briefly.

15.3.1 Methods for measuring food intake

15.3.1.1 Record method

The record method is used to obtain detailed information on food intake during a limited number of days, usually 1 to 7 days. During that period the subjects write down everything they eat, and measure the quantities. A problem with this method is that people tend to forget to write things down, or change their eating habits due to the fact that they have to write down everything they eat. A record method for 2 days cannot be used to obtain information on the usual diet of the study subjects. Due to the large day-to-day variability in the intake of foods, a 2-day period is too short to obtain a valid estimate of the usual food intake. If information on food consumption at the individual level is needed, the record method has to be repeated several times during a certain period of time. However, the 2-day record method can give a good estimate of food consumption at the group level, because then a large number of 2-day records is averaged to estimate the mean intake by the group.

15.3.1.2 *Interview method*

Two frequently used interview methods are the *24-hour recall method* and the dietary history method. In the 24-hour recall method, a complete description of the total food intake during the 24 hours preceding the interview is requested. As with the 2-day record method, a single 24-hour recall does not give a good estimate of food consumption by individuals, because of the large day-to-day variation in food intake. With the *dietary history method*, respondents are asked about their usual food intake during a specific period of time, usually the 2 to 4 weeks preceding the interview. This method gives a better indication of the usual dietary intake by individuals. Since a dietary history interview takes about 1 to 2 hours, this method cannot be applied in studies in which many thousands of people participate.

15.3.1.3 *Food frequency method*

If one wants to obtain dietary information from study subjects in a large-scale study, there is a need for a relatively quick and simple method. For this purpose, food frequency questionnaires have been developed. These questionnaires ask about the usual intake frequency (and sometimes also the quantities) of a limited number of food products. Only products which contribute substantially to the intake of the nutrients of interest are selected. A disadvantage of this method is that no information on total food consumption is obtained. Since food consumption patterns differ widely from one population to another, a new food frequency list has to be designed and validated for every study.

15.3.2 *Calculation of nutrient intake from food intake*

Once an estimate of the food intake has been made, associations between food intake and biological variables, diseases, or mortality can be studied. Information on the composition of the diet of an individual can be obtained from chemical analyses. Nutrients and other substances the investigator is interested in (e.g., contaminants) can be identified. However, this is usually expensive and laborious. Therefore, food tables are used which contain the average nutrient content of a number of frequently consumed foods. From these food tables, nutrient intake can be calculated. However, calculating nutrient intake from food intake introduces a source of error in the estimate of the true nutrient intake because the nutrient content of a particular food varies with the type of product, mode of cultivation, storage conditions, processing, and preparation.

Furthermore, no information on additives, contaminants, natural toxins, or products formed during preparation of foods can be obtained from the food tables. If one is interested in contaminants or natural toxins, for instance, special chemical analyses of foods have to be carried out. Particularly in the case of contaminants, the variability is high. One apple, for example, may have been sprayed with pesticides, whereas another may not. Therefore, it is not possible to give unequivocal averages for the amount of these substances in food tables.

15.3.3 *Analysis of dietary data*

Associations between food consumption on the one hand, and a biological variable or a disease on the other, can be studied on the basis of data on food as well as on nutrient intake.

Studies on food intake have the advantage that their results can be easily translated into preventive actions. In order to get insight into the etiology of a disease, it is important to know which food component(s) is (are) responsible for the effect. For example, a protective effect of the consumption of fruits and vegetables against lung, stomach, and colon cancer has been reported. For the prevention of those cancers, this can lead to the recommendation to eat more fruits and vegetables. However, the question remains which substances are responsible for the association. Possibly antioxidants, such as β-carotene and vitamin C play an essential role. Also, non-nutritive components with anticarcinogenic properties such as indoles, phenols, and flavones, may play a role.

When associations between dietary intake and diseases are studied, it should be borne in mind that the intake levels of many nutrients are strongly related to each other. For instance, a diet with a relatively high fat content will automatically have a relatively low carbohydrate content (see Chapter 12). This may lead to the problem that it is hard to distinguish the effect of a high fat intake from the effect of a low carbohydrate intake.

15.4 Application of biomarkers in epidemiology

15.4.1 Introduction

As described in the preceding section, measuring food intake is difficult, and in a number of cases almost impossible. An alternative would be to do it indirectly by measuring nutrient intake after consumption has taken place. For example, instead of estimating vitamin intake by measuring food consumption, the vitamin blood concentration can be used as an indicator of vitamin intake. The vitamin blood concentration is then called a *biomarker* for vitamin intake. The interest in biomarkers has increased greatly in the last few years, although they are not always the right solution. They have their limitations, as will become evident later in this section. Broadly, three categories of biomarkers are distinguished: markers of exposure, markers of effect, and markers of susceptibility. However, the distinction is not always strict. In this section, the use of biomarkers as a substitute for food intake (biomarkers of exposure) will be discussed.

In a number of cases, the biomarkers provide a more valid and precise estimate of food intake than food consumption methods. This is especially true for nutrients or contaminants of which the concentration in food may vary widely as a result of activities such as cultivation, storage, etc. (see Section 15.3.2). Errors as made by respondents in reporting their intake are prevented. Further, the use of biomarkers can provide information on micronutrients, contaminants, or substances formed during processing of foods. Another advantage is that biomarkers can be analyzed in retrospect, in frozen blood samples. However, it should be noted that if, for example, measured in serum, biomarkers do not only reflect interindividual differences in intake, but also in absorption, metabolism, and bioavailability. Since the human body keeps the concentration of many substances constant (homeostasis), levels measured in the body may not always reflect actual intake. Therefore, a requirement for a biomarker of intake is that there is a good relationship between the level of intake and the level of the biomarker. Biomarkers are most valuable if they reflect long-term intake. In that way, the biomarker is a good estimate of the usual intake that can be used for ranking individuals with respect to intake level. Not for all food components are suitable biomarkers available. A well-known example is the fact that the serum cholesterol concentration is a very poor marker of dietary cholesterol intake. On the other hand, the blood concentration of vitamin E is a fairly good indicator of dietary vitamin E intake.

15.4.2 Examples of biomarkers of dietary intake

As far as macronutrients are concerned, a well-known biomarker for protein intake is the 24-hour nitrogen (N) excretion. If subjects are in N balance, daily urine N excretion is strongly related to daily N intake. Also for a number of micronutrients, i.e., vitamins, biomarkers are available. In the case of vitamin E, the plasma concentration is well related

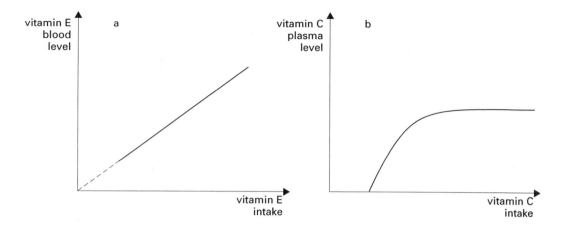

Figure 8.2 Blood/plasma level-intake curves for vitamin E (a) and vitamin C (b).

to intake. The relationship between dietary vitamin C intake and plasma vitamin C concentration is more complex. At high intake levels, plasma vitamin C levels reach a maximum (Figure 15.2).

Another example concerns selenium. The selenium concentration in toenails reflects long-term selenium intake. Biomarkers are also used for exposure to naturally occurring toxins. For example, on exposure to aflatoxins, the carcinogenic products of the fungus *Aspergillus flavus*, aflatoxin B1-albumin adducts can be measured in serum.

In comparison to the number of nutrients and other substances in foods, the number of biomarkers for intake is still small. Therefore, further development of the application of biomarkers in nutritional epidemiology is needed. When validated, the use of biomarkers can contribute substantially to nutritional epidemiology. In the future, a dietary questionnaire or interview in combination with the use of biomarkers may appear to be an adequate way to measure exposure. For some nutrients or substances, a questionnaire may provide reliable data, while for others the measurement of biochemical parameters may be a better or the only way to obtain reliable information.

15.5 Dietary factors and the risk of cancer

At present, much is known about the role of food intake in the etiology of cardiovascular diseases. The relationship between the intake of saturated fatty acids and the occurrence of ischemic heart diseases, for example, is now generally recognized. However, the role of food components in the induction of various types of cancer is less clear, although it is beyond doubt that dietary factors do play an important role. The types of cancer frequently occurring differ from one country to another. In Japan, cancer of the stomach occurs more often than in the US or Europe, while the incidence of breast and colon cancer is higher in the US and Europe than in Japan. The fact that in Japanese people who migrated to the US, the incidence of stomach cancer decreased whereas the incidence of breast and colon cancer increased, suggests that lifestyle and environmental factors are important.

As far as the role of dietary factors in the etiology of cancer is concerned, laypersons mostly think that contaminants and additives are the main risk factors. A well-known publication in which the contribution of dietary factors to the occurrence of cancer has been estimated is *The causes of cancer* written by Doll and Peto (1981). According to their estimates, the effects of contaminants and additives on the occurrence of cancer range from a decrease of 5% (due to a protective effect of antioxidants) to an increase of 1 to 2%.

Table 15.3 Summarizing of the conclusions about associations between food
components and cancer, based on literature data

Food component	Association[1]	Type of cancer
Fat	+	Colon, breast
	(+)	Prostate, pancreas
Alcohol	+	Mouth, throat, esophagus
Vitamin A and β-carotene	–	Lung, bladder
	(–)	Prostate
Nitrate, nitrite	+	Stomach
Vitamins C and E	–	Stomach
Products of pyrolysis		Recently, a number of these products have been found to be highly mutagenic and/or carcinogenic

[1]+: higher incidence of tumors is associated with higher intake of dietary factor.
(+): higher incidence of tumors is probably associated with higher intake of dietary factor.
–: lower incidence of tumors is associated with higher intake of dietary factor.
(–): lower incidence of tumors is probably associated with higher intake of dietary factor.

Epidemiological studies on the role of contaminants and additives in the induction of cancer are very cumbersome. Usually, exposure is very low and identification of exposed subjects is very difficult. Sometimes, there are large differences in effect between studies in experimental animals to which relatively high, single doses are given for a relatively short period of time, and human studies in which very low doses are ingested during long periods. An example is the long dispute about the safety of saccharin, a non-caloric sweetener. Saccharin has been used since its discovery in 1879. Studies carried out in the 1960s and 70s in rodents showed that high doses of saccharin caused bladder cancer. As a result of this finding a ban on the use of saccharin was proposed in some countries. To investigate potential effects on humans, different types of epidemiological studies were carried out. In descriptive studies, trends in the use of saccharin were compared with the occurrence of bladder cancer. In other studies, the incidence of bladder cancer in diabetics (from whom a rather large consumption of artificial sweeteners could be expected), was compared with that in non-diabetics. In case-control studies, bladder cancer patients and controls were compared for the use of saccharin. In a cohort study, the incidence of bladder cancer in saccharin users was compared with that in unexposed groups. The results of the various studies led to the conclusion that there is no increased risk of bladder cancer for humans from the use of saccharin. The composition of the diet with regard to macro- and micronutrients is of more importance for the occurrence of cancer than the intake of additives. Based on a large number of studies on micro- and macronutrients, Doll and Peto estimated that probably about 35% of all cancers are caused by an unbalanced nutrient content of the diet (with a confidence interval of 10 to 70%).

In 1986 the Dutch Nutrition Council reported that despite all the research that had been carried out, no definite conclusions could be drawn on the role of the different food components in the induction of cancer. Based on literature data, only general conclusions were presented about associations between dietary factors and several types of cancer. A few of these are listed in Table 15.3.

15.6 Summary

This chapter dealt with the basic principles of epidemiology, the ways in which epidemiological methods can be used for the assessment of food intake, the significance of the use of biomarkers for nutritional epidemiology, and the importance of epidemiology for the examination of the role of dietary factors in the risk of cancer.

Disease frequency can be expressed in terms of incidence rate or prevalence. Effects of exposure can be expressed in absolute terms, as incidence rate difference, or in relative terms, as relative risk or odds ratio. The proportion of the diseased which can be attributed to exposure, the attributable proportion, can be calculated for exposed individuals as well as for total populations. Epidemiological studies can be experimental (clinical trials/ intervention studies) or non-experimental (cross-sectional studies, follow-up or cohort studies, case-control studies, and ecological studies). Further, the concepts precision and validity, bias and confounding were introduced, and a number of criteria concerning causality were briefly discussed.

Food intake can be measured by using a record method, an interview method, or a food frequency method. Sometimes, the use of a so-called biological marker (biomarker) can give a more valid and precise estimate of the intake.

During the last decades, results of epidemiological studies have contributed substantially to the insight that dietary factors play an important role in the etiology of cancer. Nutritional imbalance of the diet with regard to macronutrients appeared to be the major cause. The risks due to the intake of food contaminants and food additives are minimal.

Reference and reading list

Doll, R., R. Peto, The causes of cancer, in: *J. Natl. Cancer Inst.* 66, 1195–1308, 1981.

Hill, A.B., The environment and disease: association or causation?, in: *Proc. R. Soc. Med.* 58, 295–300, 1965.

IPCS (Internationa Programme on Chemical Safety), *Biomarkers and Risk Assessment: concepts and principles.* Geneva: WHO, Environmental Health Criteria 155, 1993.

Kannel, B.W., and D.L. McGee, Diabetes and glucose intolerance as risk factors for cardiovascular disease: The Framingham Study. *Diabetes Care* 2, 120–126, 1979.

Kok, F.J., P. van 't Veer, *Biomarkers of Dietary Exposure.* London, Smith-Gordon and Company Limited, 1991.

Margetts, B.M. and M. Nelson, (Eds.), *Design Concepts in Nutritional Epidemiology.* Oxford, Oxford University Press, 1991.

Nutrition Council, *Nutritional factors in the causation of cancer.* Nutrition Council, Committee on Nutrition and Cancer, The Hague, March 1986.

Rothman, K.J., *Modern Epidemiology.* Boston, Little, Brown and Company, 1986.

Sturmans, F., *Epidemiologie. Theorie, Methoden en Toepassing.* Nijmegen, Dekker & van de Vegt, 1986.

Voeding en Kanker. Alphen a/d Rijn, Samsom Stafleu, 1984.

Taubes, G., Epidemiology faces its limits. *Science* 269, 164–169, 1995.

Willet, W., *Nutritional Epidemiology.* New York, Oxford University Press, 1990.

Part 3

Risk Management in Relation to Food and Its Components

chapter sixteen

Introduction to risk management

E.J.M. Feskens

16.1 Introduction

Part 1 of this textbook describes the ways in which consumers are exposed to dietary substances. Adverse effects and their underlying mechanisms of dietary substances are dealt with in Part 2. In Part 3, the information given in Parts 1 and 2 is integrated for managing toxicological risks due to food intake. Background information is provided to recognize and identify potential toxicological risks associated with dietary intake in present-day society. Some examples of health risks from food and its components will be given, and the importance of prevention, intervention, and control will be explained. In addition, a number of possibilities to reduce these risks will be described. Finally, an overview of the contents of the other chapters of Part 3 will be given.

Table 16.1 Ranking of food hazards, as perceived by the general public

1	Additives and food processing residues
2	Environmental contaminants (pollutants)
3	Nutritional imbalance
4	Naturally occurring toxins
5	Microbiological contamination

Table 16.2 Ranking of food hazards based on objective scientific criteria

1	Nutritional imbalance
2	Microbiological contamination
3	Naturally occurring toxins
4	Environmental contaminants (pollutants)
5	Additives and food processing residues

16.2 Public risk perception vs. expert risk opinion

Recently, the general public's interest in the quality of food has increased considerably. Also, more attention to this issue is paid by the press, in particular when hazards associated with contaminants and additives are concerned.

The five principal categories of food hazards are listed in Table 16.1 in order of importance according to the opinion of the general public.

The highest risk is believed to be associated with additives and contaminants, and the lowest with microbiological contamination. But is this ranking by the general public realistic? Section 16.3 discusses several examples which show that the ranking order of food hazards based on objective scientific criteria is completely different (see Table 16.2).

While, according to the general public, the toxicological risks from additives and contaminants are at the top of the list, the expert opinion scores them relatively low. The reverse applies to the toxicological risks from nutritional imbalance and microbiological contamination. The health risks acknowledged by experts are based on accepted scientific criteria. The next section describes the process of risk assessment in which the toxicological risks due to food intake are established on the basis of scientific data. Further, risk assessment will be discussed as the basic element of risk management.

16.3 Risk assessment, risk evaluation, and risk management

16.3.1 Risk management

Risk management is a complex process, based on information from various sources. A schematic overview of this process is presented in Figure 16.1.

First, the risk posed by a food component must be assessed, preferably in an objective and quantitative way. To do so, toxicological and epidemiological information is needed. Based on this information, guide values for components are determined. This is then followed by risk evaluation, in which the results of risk assessment are weighed against certain issues, such as those of socio-economical and political interest. Public perception also plays a role.

This process results in setting a standard. Such a standard is an important tool for risk management. Using the standard as a yardstick, the toxicological risks from food components are evaluated. If a standard is exceeded, the situation may become hazardous, and appropriate measures should be taken. These measures may concern risk intervention

Steps in the processes of risk assessment
and risk management

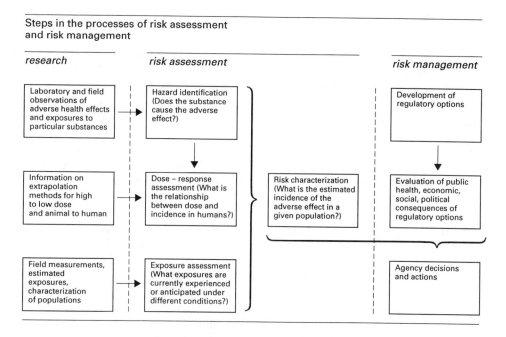

research *risk assessment* *risk management*

Figure 16.1 Framework for risk assessment and risk management. (Source: Grant and Jarabek, 1990.)

(relief of the risk situation) and risk prevention. In some cases, potential risks are regularly controlled or monitored by the authorities. It should be noted that risk management is also the concern of food producers, scientists, and consumers.

16.3.2 *Methods of risk assessment*

To assess the health risks from food components, information on the components, exposure to the components (see Part 1), the consumer, and the interactions between the components and the consumer (see Part 2) is needed. It should be noted that in many cases interactions with other dietary or environmental substances are also involved. Quantitative risk assessment therefore includes the following items:

- exposure assessment: daily intake of the components, duration and pattern of use;
- characterization of the relationship between exposure (dietary intake) and response (toxicity);
- elucidation of mechanisms;
- extrapolation of results from experimental animals to *humans*, and subsequently to *sensitive human populations*, the so-called high-risk groups;
- extrapolation from experiments, usually with high doses, to the real-life situation with exposure through the diet;
- extrapolation from short-term to *long-term exposure*;
- quantitative risk estimation, taking into account the estimated exposure (dietary intake) and the expected response (toxicity);
- estimation of the maximum allowable levels, guide values to be used in health policy. For foods, the Acceptable Daily Intake (ADI) (see Section 17.3.2 and 17.3.3) is applied.

Intermezzo

Calculation of guide values (ADI, TDI, PTWI), For non-carcinogenic components, the ADI is derived from the no-observed-adverse-effect level (NOAEL) (see Section 17.3.2 and Section 19.2.2) determined in experimental animals. NOAEL is divided by safety factors (e.g., 10 for taking into account the extrapolation from animals to humans, and 10 for taking into account a susceptible human subpopulation, such as infants), resulting in an integral safety factor of 100.

The guide value is not called ADI for all food components. For environmental pollutants, such as 2,3,7,8-tetrachlorodibenzo-*p*-dioxin, the guide value is known as the tolerable daily intake (TDI) (see Section 17.4 and 17.4.1 and Section 21.4.4.3): such pollutants are not added to food intentionally, and are therefore "tolerable" rather than "acceptable." Contamination with lead is another example. For this element, the guide value is called the provisional tolerable weekly intake (PTWI). The toxic effects of lead are measured in relation to the lead blood concentration and not to the intake. PTWI is based on the observation that at a certain level of weekly intake, intake is balanced by elimination, and therefore no accumulation in the body will take place. Provisional means that the available safety data do not warrant a final conclusion.

For *carcinogenic components,* ADI is not derived from a NOAEL. In the US, for example for substances indicated as carcinogens, and especially for components initiating cancer, a zero-risk approach is followed. To this purpose, a so-called "calculated mortality" procedure is used, involving linear extrapolation to a virtual low risk level (e.g., 10^6 over a lifetime). It is assumed that carcinogenesis starts with a cell mutation, and that the risk of cancer development is related to the daily dose of the component concerned to the power *m. M* corresponds to the number of hits of the carcinogenic component that is necessary for the initiation of cancer. Generally, for *m* the value 1 is used: this is a conservative model in which exposure to a component and cancer incidence are linearly related. Using information on the carcinogenic dose of a component in animal experiments, the daily dose can be estimated that would induce extra cancer incidence in humans. Currently, the maximal tolerable extra cancer risk is estimated at 10^{-6} per lifetime.

The information necessary for quantitative risk assessment should be provided by toxicological studies and epidemiological studies. For the toxicological screening of additives and contaminants, standardized protocols have been developed. It should be borne in mind that for many components not all necessary information on exposure, toxicity, and dose–response relationships is available. The next section gives a few examples of risk assessment, evaluation, and management.

16.3.3 *Examples of risk assessment, risk evaluation, and risk management*

The examples in this section underline the importance of risk assessment, evaluation, and management. Furthermore, they will show the lack of necessary information and its implications for risk assessment and ranking of risks. In fact, these examples explain the ranking order of food hazards as shown in Table 16.2. In each example, the following items will be addressed: toxicity of the substance concerned, its daily intake and the duration of use, the sensitivity of the consumer, and the existence of high-risk groups.

16.3.3.1 *Additives and processing residues*

According to the general public, the toxicological risks associated with the intake of food additives and food processing residues are high. In particular, sweeteners, antioxidants, dyes, and preservatives are substances that have recently received large negative publicity.

In contrast, more objective scientific criteria prove the toxicological risk from additives to be minimal. For most of these substances extensive information on toxicity is available; acute and subacute toxicity as well as chronic toxicity, including mutagenicity, carcinogenicity and teratogenicity, have been investigated. Epidemiological (human) data, however, are scarce.

16.3.3.1.1 *Saccharin. Risk assessment — Toxicity.* Saccharin is a sweetener which, at very high doses, has been shown to cause bladder tumors in experimental animals. Generally, for components suspected to be carcinogenic, the risk is estimated from so-called "calculated mortality" (see Section 16.3.2), the linear extrapolation to a virtually low risk level. However, the doses at which bladder tumors were shown to develop in experimental animals were very high. The carcinogenicity of saccharin appeared to be due to the formation of bladder stones, rather than to genotoxicty (interaction at DNA-level). Therefore, the use of saccharin has been approved in the U.S. and Europe, and the ADI calculation using the calculated mortality procedure was not applied. For safety reasons, the maximum daily intake was advised to be 2.5 mg/kg body weight. No epidemiological study has shown that cancer incidence and mortality are related to the use of saccharin.

Risk assessment — Intake. In general, the daily intake of saccharin is below the ADI. Soft drinks are allowed to contain a maximum of 200 mg saccharin/kg. The average use of soft drinks is 300 g/day, which means a maximum of 70 mg saccharin/day. However, the daily saccharin intake by diabetic patients may be several times higher than that of non-diabetics.

Risk assessment — Sensitivity. There is no evidence for an increased sensitivity of specific subpopulations. Diabetic patients may be a high-risk group owing to their extensive intake, rather than high sensitivity.

Risk evaluation. Saccharin most probably does not pose an important health risk to humans. It has been calculated that a cancer risk may only develop at a daily saccharin dose present in 800 cans of soft drink (equivalent to the sweetness of 25 kg of sugar). This agrees with the fact that no increased cancer risk for diabetic patients has been observed in epidemiological studies.

Risk management. Saccharin is already the subject of risk management. Its use in foods is regulated by several food acts. Saccharin-containing sugar substitutes should be labeled with the warning not to use more than 80 mg/day. Saccharin is prohibited in baby food. Authorities are also involved in risk management by checking the observation of food regulations and by giving proper dietary advice and information to diabetics. Scientists take part in saccharin risk management, for example, by investigating the need for diabetics to use sugar substitutes.

For most additives the situation is very similar: extensive toxicological information is available, and legislation on use is provided for. To this end, so-called "positive lists" are made up, i.e., an additive not on this list is not allowed to be used unless explicitly stated otherwise by law. In fact, additives are considered to be the safest food components. The toxicological risks from this category are believed to be minimal. However, there are a few examples of additives for which the evaluation of toxicological risks is more difficult. This is mainly due to interactions with other toxic substances, such as contaminants.

16.3.3.1.2 *Nitrite (and nitrate).* Nitrite is an important preservative. It is used in the production of cheese and meat products. Nitrite inhibits the growth and development of *Clostridia* bacteria. Exposure to the contaminant nitrate mainly occurs by drinking water and consumption of leafy vegetables.

Risk assessment — Toxicity. Toxic effects of nitrite include methemoglobinemia, leading to disturbances in oxygen supply, and hypertrophy of the adrenal zona glomerulosa.

Nitrite can also react with secondary amines to form N-nitrosamines, which have proved to be carcinogenic in several experimental animals. The toxic effects of nitrate originate from its bacterial reduction to nitrite in the oral cavity. Some epidemiological studies have suggested that in subjects with gastric lesions a higher risk of gastric tumors may be associated with a high nitrate intake. However, such effects have not been observed in non-patients, so that the evidence is limited.

The no-observed-adverse-effect level (NOAEL) of nitrite as calculated from the results of chronic toxicity studies in rats is 10 mg $NaNO_2$ or 6.7 mg NO_2^- per kg body weight. Since there is a difference in the reduction of nitrate to nitrite between rats and humans, the NOAEL of nitrate is calculated from the NOAEL of nitrite. Estimating the conversion of nitrate to nitrite at 5%, the NOAEL for adults is: $100/5 \times 10$ mg/kg body weight = 200 mg $NaNO_3$ per kg body weight or 146 mg NO_3^- per kg body weight. Currently, the ADI values are 5 mg $NaNO_3$ per kg body weight (3.65 mg NO_3^- per kg body weight) and 0.2 mg $NaNO_2$ per kg body weight (0.13 mg NO_2^- per kg body weight).

Endogenous nitrosamine (dimethyl- and diethylnitrosamine) formation has been demonstrated in human volunteers on a diet rich in fish and nitrate-containing products. Using the conservative one-hit model with linear extrapolation for carcinogenic substances (see Section 16.3.2) the acceptable daily dose for prevention of a lifetime tumor incidence of 10^{-6} can be calculated. For dimethylnitrosamine, this value amounts to $16–186 \times 10^{-6}$ mg/day, and for diethylnitrosamine to $11–14 \times 10^{-6}$ mg/day.

Risk assessment — Intake. According to the Dutch Food Act, addition of nitrate to foods other than cheese, melted cheese, and meat products is not allowed. For these products maximum acceptable limits are indicated, e.g., a maximum of 500 mg KNO_3 and 200 mg KNO_2 per kg meat. For baby foods, a maximum of 50 mg NO_3^- per kg dry matter is allowed. Leafy vegetables such as spinach contain high concentrations of nitrate by nature. Standards given in the Dutch Food Act are a maximum of 3500 mg NO_3^- per kg in summer, and 4500 mg NO_3^- per kg in winter.

A recent survey has shown that the average daily nitrate intake in the Netherlands ranges from 1.25 mg/kg body weight for men aged 65 to more than 3.6 mg/kg body weight for 1 to 3 year olds. These data were arrived at by combining information from a dietary survey in a large representative subpopulation with information on the nitrate content of various food products. Particularly among children, an excessive nitrate intake (higher than ADI) occurred quite frequently (20 to 40%). The intake of nitrite is probably lower than the amount of nitrite formed endogenously, and is estimated at 2.3 mg NO_2^- per day. Water accounted for 4% of the nitrate intake, while leafy vegetables such as spinach accounted for about 45% of the total estimated intake.

Risk assessment — Sensitivity. Infants are more sensitive to nitrite, resulting in nitrite-induced methemoglinemia, often leading to oxygen supply problems. Also in babies, nitrate is more extensively reduced to nitrite.

Risk evaluation and management. The toxicological risk of the preservative nitrite itself is probably low. Its use is regulated, as is the use of other additives like saccharin.

A major cause for concern is the fact that for many people in general and many children in particular the intake of nitrate is larger than the ADI. In the future, more attention should be paid to the reduction of nitrate emission, e.g., in the form of fertilizer, into the environment. This is the concern of the authorities; the agricultural sector in particular is responsible for this. The health effects of high nitrate intake by children as well as the validity of the current ADI levels need to be examined in more detail. The effects of food preparation on the nitrate content and the consumption of leafy vegetables, especially in winter, ask for attention. Public advice concerning this issue should be considered.

16.3.3.2 Environmental contaminants

In general, contaminants are believed by the consumer to pose high risks to health. According to the experts, however, environmental contaminants only rank fourth on the list of food hazards, as shown in Table 16.2.

Concerning polychlorinated dibenzo-*p*-dioxins and biphenyls — the subject of the next example — the public paid much attention to the high levels found in milk from cows grazing in the vicinity of waste incinerators. However, the guide values for contaminants are based on cumulative, life-long exposure. Therefore, the life-long duration of individual exposures should be taken into consideration when estimating the risks from such high levels of contaminants.

16.3.3.2.1 Polychlorinated dibenzo-p-dioxins and biphenyls.

Dioxins are emitted by waste incinerators. They are also known as by-products in pesticides. *Polychlorinated biphenyls* (PCBs) are well-known environmental contaminants, originating from their earlier use in transformers, and more recently in heat insulation.

Risk assessment — Toxicity. As far as the toxicity of dioxins is concerned, the congener 2,3,7,8-tetrachlorodibenzo-*p*-dioxin (TCDD) is known best. Wasting syndrome (weight loss) is a characteristic acute toxic effect of TCDD in animals. TCDD is not mutagenic. It induces to a large extent the biotransformation enzymes in the liver. Therefore, it is assumed to have a tumor-promoting effect. One epidemiological study reported an association between TCDD exposure and cancer occurrence in a group of workers in a chemical industry. In animals, also immunotoxic and teratogenic effects have been observed. Humans which were exposed to TCDD, e.g., as a result of an occupational accident, developed chloracne. Dioxins and PCBs can have similar biological effects. However, they differ in the intensity of their effects. Therefore, the so-called TCDD equivalent (TEQ) was introduced, relating the toxicity of all dioxins and PCBs to that of TCDD.

When rats were submitted to a lifetime exposure of 1000 pg TCDD per kg body weight, the effects in the liver were minimal. This dose was considered to be a "marginal-effect-level" from which the TDI was calculated using a safety factor of 250. Therefore, the Dutch TDI was 4 pg TCDD or TCDD equivalents per kg body weight. Recently, however, the WHO assessed the TDI at 10 pg TEQ per kg body weight. This value was obtained by using a toxicokinetic approach for humans, resulting in a NOAEL of 1000 pg TCDD per kg body weight. A safety factor of 100 was applied to calculate TDI. This TDI is identical to a lifetime maximum of 255.5 ng TEQ per kg body weight for 70 years.

Risk assessment — Intake. The daily exposure of the general population is estimated on the basis of data on the intake of foods by a representative subpopulation in combination with data on the dioxin and PCB contents of foods as determined by chemical analyses. Using this approach, the daily exposure is estimated to be about 130 pg TEQ, i.e., 2 pg TEQ per kg body weight for adults and 7 pg TEQ per kg body weight for infants. More than 95% of this exposure results from the intake of animal fat. Dairy products are estimated to account for 30 to 50% of the total exposure. Recently, life-long exposure was estimated at 70 ng/kg body weight for dioxins and structurally related substances.

Risk assessment — Sensitivity. About 1% of the children younger than 6 years are estimated to have an exposure of more than 10 TEQ per kg body weight per day. In this respect, the dioxin and PCB contents of breast milk are also of importance. Dairy milk has been shown to contain 2 to 4 pg TEQ per g fat. For breast milk, this is about 35 pg/g fat, implying that breast-fed infants are exposed to about 250 pg TEQ per kg body weight.

Risk evaluation and management. The exposure of breast-fed infants is only four times lower than the marginal-effect-level for rats. Therefore, the TEQ of breast milk certainly needs attention. On the other hand, it should be noted that the TDI levels are cumulative values, calculated on the basis of the results of a number of studies. This implies that conclusions are not allowed if, as is the case for breast-fed babies and children, these limits are exceeded during short periods of time. As far as risk evaluation is concerned, it should also be noted that from a nutritional point of view, for small babies breast milk has definite advantages over cow's milk. Therefore, breast-feeding should certainly not be discouraged. Other management measures, such as reduction of dioxin and PCB formation and emission, and checking of foods, are preferred.

Other groups with potentially higher exposures are, for example, industrial workers or individuals consuming milk and cheese from polluted areas near waste incinerators. Industrial safety, and food control should prevent toxic exposure of these groups.

16.3.3.3 Nutritional imbalance

As shown in Table 16.1, the consumer does not consider the risks associated with nutritional imbalance to be very important. However, since the beginning of this century it has become clear that the occurrence of several important chronic disorders, such as cardiovascular diseases and cancer, is affected by nutrients which form a substantial part of the diet. Epidemiological studies are particularly useful in bringing these risks to light. As a result, experts generally rate the risk of nutritional imbalance to be one of the highest of all food aspects. Several national nutritional councils have published extensive reports on the macro- and micronutrient contents of foods. For each nutrient, the so-called recommended dietary allowance (RDA) is given, also in relation to high-risk groups such as infants, children, pregnant women, and the elderly. These RDAs guarantee that the intake by 95% of the population is sufficient from a nutritional point of view. In addition, all major nutritional councils have prepared dietary recommendations for overall health maintenance. These guidelines are based on knowledge of the effects of nutrients and foods on the occurrence of chronic diseases, such as cardiovascular diseases and cancer.

Intermezzo

Guidelines for a healthy diet.

1. Pay attention to dietary variation.
2. Use dietary fats, in particular saturated fatty acids, in moderate amounts, and ensure a sufficient intake of polyunsaturated fatty acids.
3. Use dietary cholesterol moderately.
4. Ensure a liberate intake of complex carbohydrates and dietary fiber, and avoid frequent and high consumption of simple sugars.
5. Use alcohol in moderate amounts.
6. Use dietary salt in moderate amounts.

In the Netherlands, a large number of dietary surveys have shown that the macronutrient intake by the general population through the common diet does not match these guidelines. At present, the *average* daily energy intake by the population agrees with the recommended value, indicating that the energy intake of many people is too high (that is why the estimated prevalence of overweight individuals is about 20%!). The average fat intake is much higher than the recommended values, and this excess intake is estimated to be responsible for an extra 15% mortality due to coronary heart disease. In other Western countries, the picture as far as energy intake is concerned, is very much the same as that in the Netherlands. Also, sodium intake is high. The chronic "toxic" dose leading to hypertension in humans is about 60 g/day. Using a safety factor of 100, the ADI would be

10 mg NaCl per kg body weight. However, the usual intake by many Western populations is about 10 g/day, which is about 17 times the ADI. This means that the actual safety factor for salt (±6) is much lower than 100, the value commonly used for the determination of ADI levels for additives and contaminants (see Section 16.3.2). This is one of the reasons why the risks from nutritional imbalance are rated highest by the experts. The following example will show the toxicological risks from an important nutrient, dietary fat.

16.3.3.3.1 *Dietary fat.* Dietary fat is the main energy source in the human diet. The combustion of 1 g of fat results in the production of 37 kJ (or 9 Kcal). Dietary fatty acids are usually classified as follows:

- saturated fatty acids (SFAs), i.e., fatty acids with 4 to 18 C atoms. Well-known examples, occurring in large quantities in the diet, are palmitic acid (C16:0, i.e., number of carbon atoms:number of double bonds) and stearic acid (C18:0);
- monounsaturated fatty acids (MUFAs), fatty acids like oleic acid (C18:1), the main constituent of olive oil;
- polyunsaturated fatty acids (PUFAs), containing two or more double bonds, e.g., linoleic acid (C18:2). PUFAs are essential dietary components. The polyunsaturated fatty acids can be distinguished into n-6 and n-3 acids, referring to the location of the first double bond. Especially, fish oils are rich in n-3 PUFAs (see Chapter 6.2.1.1).

Risk assessment — Toxicity. Unsaturated fatty acids are susceptible to oxidation. The oxidation products may have several adverse effects, e.g., tumor induction. In addition, depletion of the anti-oxidant pool in the body may occur, and in some cases vitamin E deficiency may develop. Erucic acid (C22:1) is found in rapeseed oil and has been shown to induce cardiopathy (myocardial fibrosis) in experimental animals. For the combination of erucic and linolenic acid (C18:3, n-3), a NOAEL of 1% of vegetable oil intake has been suggested. However, it should be noted that the epidemiological evidence for these effects is limited.

A recent epidemiological study has suggested that PUFAs are associated with an increased risk of chronic non-specific lung disease (CNSLD). For CNSLD, a relative risk of 1.6 was observed comparing a linoleic acid daily intake equivalent of more than 5.6% of the total energy intake with an intake of less than 4% of energy. Since the prevalence of high linoleic acid intake was 25%, this results in a population attributable risk (see Part 2, Chapter 15) of 13%.

The recommended intakes of fatty acids are usually expressed in terms of energy percentages. For SFAs, a range of 0 to 10% of the energy intake is advised, while for PUFAs, a range of 3 to 7% of the energy intake is recommended. Since the total fat intake is recommended not to exceed 30% of the energy intake, the remainder can be provided by MUFAs.

Risk assessment — Intake. An extensive survey among a representative sample of the Dutch population has shown that in 1987/1988 the average daily intake of dietary fatty acids was 97 g/day. The intake of SFAs was 40 g/day (16.3% of the energy intake), and the intake of PUFAs 16 g/day (6.4% of the energy intake). The daily intake of erucic acid is estimated at less than 1% of the energy intake.

Risk assessment — Sensitivity. No information on the sensitivity of particular subpopulations to dietary fatty acids is available. On the other hand, it is known that infants may suffer from deficiency of the diet in essential fatty acids, possibly resulting in reduced neurological functions.

Risk evaluation and management. Toxic effects of several MUFAs and PUFAs have been observed, but at doses much higher than the usual intake. Therefore, these specific fatty acids probably do not cause a great health hazard, although their suggested role in the development of other chronic diseases such as CNSLD should be carefully considered in the future. In addition, it should be noted that this example clearly shows a dilemma with regard to risk management. One of the possibilities to reduce the intake of unsaturated fatty acids is to discourage the population's consumption of PUFA-rich food. This will not be useful if, instead, the consumption of SFA-rich foods increases, as this will enhance the risk of hypercholesterolemia and coronary heart disease. In fact, the adverse effects of SFAs on public health have been estimated to be more important than the potential detrimental role of unsaturated fatty acids in the induction of tumors and CNSLD. Also, positive effects of n-3 PUFAs have been reported.

Typically, there is an optimum for nutrient intake. Deficiencies as well as toxic effects need to be prevented. Public advice on nutrition in general, and fatty acids in particular, must be balanced and careful. This example illustrates a second difficulty in nutrition education: what is consumed is food rather than nutrients or food components. Fish, with the potentially beneficial n-3 PUFAs, may also contain small amounts of environmental contaminants such as dioxins and mercury. Therefore, pros and cons of fish consumption need to be weighed before univocal public advice can be given.

Besides authorities (guidelines, food labeling, public advice) and consumers (dietary habits), scientists and food producers also should be aware of these dilemmas. The recommended changes in dietary habits of the general public (see Section 16.3.3.3) will not result from public advice only. Food labeling may help, but also alternative food products should be developed and become available. A price policy would also contribute to behavioral change in people.

Intermezzo

Examination of cardiovascular toxicity, a necessity?   In general, the toxicological evaluation of substances includes acute, subacute, and chronic toxicity testing. Teratogenicity, mutagenicity, and carcinogenicity are studied. Only for a small number of substances (e.g., lead, cadmium) are cardiovascular effects considered. This is remarkable. For example, the total mortality in the Netherlands in 1989 was 9.1 per 1000 for men and to 7.9 per 1000 for women. For men, 30% of mortality was due to cancer, for women this was 25%. However, the contribution of cardiovascular disease to total mortality was larger: 41% for men, and 43% for women. For both men and women, acute myocardial infarction was the most important cause of death (17 and 13% of total mortality respectively). This indicates that for a complete evaluation of the health risk due to substances, a standardized cardiovascular or atherosclerotic screening is also necessary.

16.3.3.4 Naturally occurring toxins

For most naturally occurring toxins information is scarce. Also, information on their presence in food is usually lacking, which makes risk assessment difficult. Recently, the scientific interest in the potential toxic effects of naturally occurring toxins has increased. This concerns in particular mycotoxins, phytotoxins and phycotoxins. The lack of information may be one reason for the low ranking of hazards from these food components by the general public (Table 16.1). On the other hand, perhaps in combination with the fact that one of the most potent carcinogenic substances, aflatoxin, is a natural toxin, it may also explain the relatively high ranking of naturally occurring toxins by the experts (Table 16.2).

Aflatoxin is a mycotoxin occurring in peanuts and cereals originating from hot, humid countries. In developed countries, consumers may be exposed to aflatoxin as a result of international trade and the presence of the contaminated cereals in cattle feed.

Risk assessment — Toxicity. Many animals have been shown to develop hepatic tumors after exposure to aflatoxin B1. Several epidemiological studies on hepatic cancer have suggested that aflatoxin is involved in the etiology of this disease, in combination with hepatitis B virus. Aflatoxin M1 is also carcinogenic, but is less potent than aflatoxin B1. Such as for other carcinogenic substances, the cancer risk is derived from the "calculated mortality" procedure as described in Section 16.3.2.1. According to the Dutch Food Act, the content of aflatoxin B1 in foods is not allowed to exceed 5 µg/kg. The maximum content of aflatoxin M1 in milk is set at 0.05 µg/kg, because of its frequent use. Aflatoxin is not allowed to be present in groundnuts *(Arachis hypogaea)* at all or in any products prepared from them.

Risk assessment — Intake. A recent analytical survey reported that aflatoxin may be present in small quantities in peanuts, peanut products, buckwheat, and nutmeg. In baby foods, the average content appeared to be 0.06 µg/kg. Aflatoxin M1 was also found in cow's milk, but the standard level was not exceeded. Aflatoxin levels are higher in winter than in summer, due to the addition of cattle feed concentrate.

Risk evaluation and management. Apparently, in the Western countries the exposure to aflatoxins has increased due to the import of tropical products. Until now, however, no detrimental effects have been reported, and no elevated aflatoxin levels in food have been found. Surveillance of foods remains necessary, however, among others because the use of tropical products in cattle feed is expected to increase.

To prevent the occurrence of aflatoxin in these products, food processing in developing countries should be improved and controlled. In other words, Good Agricultural Practice (GAP) should be applied. The problem is monitored by organizations like the Food and Agricultural Organization (FAO). Unfortunately, however, local funds and equipment are often still lacking.

16.3.3.5 Bacterial contamination

The general public rates the risk due to microbiological contamination as minimal. According to objective scientific criteria, this risk ranks much higher on the list of toxicological risks from foods.

Many bacterial species produce toxins. These can be divided into two main groups:

- toxins formed after consumption of contaminated food, causing gastrointestinal disorders. These disorders have a long incubation period, as the toxins are only produced after multiplication of the microorganisms inside the host. Examples of this type of infection are *Salmonella* poisoning and cholera;
- toxins produced in the food before intake. The symptoms appear shortly after consumption, and the patients are not contagious. A well-known example is the induction of enterotoxic effects, caused by the consumption of food contaminated by *Staphylococcus aureus*.

Risk assessment — Toxicity. Staphylococcus aureus produces several toxins, classified as enterotoxins A to E. The toxins (mostly A) are responsible for acute food poisoning. The symptoms (diarrhea, vomiting) are mild, and occur shortly after the meal (1 to 6 hours). Therefore, most epidemics are not recognized.

Risk assessment — Intake. S. aureus can survive in foods with high salt concentrations and in briefly cooked protein foods. Epidemics are usually caused by contamination of ham, pastries with cream, or milk products.

Risk assessment — Sensitivity. Infants, sick people, and elderly people are groups whose reduced resistance may make them more sensitive to the toxins.

Risk evaluation and management. Based on the official data on the number of acute food poisonings, the risk would seem low. Only few cases are registered annually. However, the official records of microbiological contaminations do not represent the actual situation. A population survey has shown that in the case of diarrhea only 25% of the subjects consult a general practitioner. This means that identification of the pathogenic bacteria only takes place in a small subpopulation. The percentage of all cases of food poisoning that are officially recorded is estimated at only 1 to 5%. This suggests that microbiological contamination of food is a larger public health problem than generally assumed. Prevention of these intoxications is therefore important. Special attention should be paid to the production of foods. It is important to do this according to the guidelines known as Good Manufacturing Practice. Factors such as hygiene, temperature, pH, and water activity need to be controlled regularly by the industry as well as by governmental agencies. However, it should be noted that 75% of the contaminations occur where food is prepared, such as restaurants, hospitals, nursing homes, catering companies, and kitchens at home. Again, control by governmental agencies is necessary. In addition, it is important that the consumer is made aware of proper food handling. Especially, cooling and heating of food need attention, and public advice and education will be needed.

16.4 Important issues in risk management

The examples discussed in the preceding subsections serve two purposes. The first aim was to give an impression of the toxicological risks associated with food intake by the population. As shown in Tables 16.1 and 16.2, the perception of the toxicological risks from different food components differ between the public and the experts. The examples discussed show that in reality the highest risks do not originate from food additives and contaminants, as perceived by the public, but from nutritional imbalance and microbiological contamination. In fact, the risks due to additives are minimal. The difference in ranking between consumers and scientists is a cause for concern, especially as risk prevention and control is also partly the responsibility of the consumer.

As shown by the examples, the perception of toxicological risks by the public is different from the real situation. As will be discussed in Chapter 22, the public's perceptions of food risks are affected by information from the media. Also, psychological factors play a role. Self-inflicted risks, such as risks associated with food habits, are more easily accepted than risks coming from other sources (food producers). This may be due to the idea that risks posed from outside cannot be managed. These issues need to be taken into account, when public advice and behavioral health education are parts of the risk management process, as in the case of microbiological contamination.

The second purpose of the above examples is to introduce briefly the upcoming chapters. In Chapter 17, the basic requirements for risk assessment will be described in more detail. Attention will be paid to the standard toxicological protocols, and the nationally and internationally required toxicological data. Also, the importance of information on biotransformation and toxicokinetics will be stressed (see also the remarks on polychlorinated dibenzo-*p*-dioxins and biphenyls).

As described in the example on nitrite and nitrate, the calculation of the ADI values requires extrapolation from experimental animals to sensitive human populations. Chapter 18 deals with the factors affecting and hindering extrapolation, such as species differences and variation, and measurement errors, in more detail. New possibilities, involving toxicological modelling, will be discussed.

As shown in Figure 16.1, standard setting is a main step in the process of risk management. In Chapter 19, the principles, possibilities, and limitations of standard setting are described. Standard setting is not only based on risk assessment. As shown in the example of dietary fat, it also involves careful weighing against other issues, such as political and socio-economical interests. Special attention will be paid to harmonization of standard-setting procedures on national as well as international level. The latter has been shown to be important in the example of aflatoxin, as the standard setting in tropical countries affects the potential exposure to aflatoxin in other countries.

Up to now, epidemiological data are only rarely used as additional information for risk evaluation and standard setting. This is due to methodological limitations, such as low sensitivity and difficulties in characterizing the exposure of participants. As mentioned in the example of saccharin, an epidemiological study on cancer risk differences between diabetics, who are likely to use more saccharin, and healthy subjects revealed no statistically significant differences. As will be shown in Chapter 20, this study is one of a few examples of epidemiological studies that have contributed to risk assessment. It is expected that the input of epidemiology on risk management will increase in the coming years, as more possibilities become available for the use of so-called biomarkers to characterize exposure and disease. This is a fortunate development, as an advantage of epidemiological studies is their direct relevance to the human situation.

Chapter 21 provides a detailed overview of risk assessment, risk evaluation, and risk management.

16.5 Summary

The public perception of toxicological risks from foods differs from the experts' opinion. Several examples show that nutritional imbalance and microbiological contamination pose the highest food-related risks, followed by naturally occurring toxins and environmental contaminants. The risks from additives are only minimal.

These results are based on quantitative risk assessment, an important step in the risk management process. After identification of the risk, an evaluation against other interests (economical, social, political) takes place, before management measures are issued. The examples show that the responsibility for risk management, i.e., control and prevention of health risk, is shared by authorities, scientists, and food producers, as well as consumers.

Reference and reading list

Grant, L.D., and Jarabek, A.M., Research on risk assessment and risk management: future directions. *Tox. Indust. Health*, 6, 212, 1990.

Living with Risk. The British Medical Association Guide. New York, John Wiley and Sons, 1987.

Miller, S.A., Food additives and contaminants, in: Amdur M.O., J. Doull, C.D. Klaassen, (Eds.). New York, Pergamon Press, 1990.

Morgan, M.G., Risk analysis and management. *Sci. Am.* 269, 24–30, 1993.

Scala, R.A., Assessment, in: Amdur M.O., J. Doull, J., C.D. Klaassen, (Eds.). New York, Pergamon Press, 1990.

US Department of Health and Human Services. Public Health Services. *Nutrition monitoring in the US. An update report on nutritional monitoring*. DHHS Publication No. (PHS) 89-1255. Hyattsville, Maryland, 1989.

World Health Organization, *Diet, nutrition, and the prevention of chronic diseases*. Report of a WHO Study Group. Technical Report Series 797. Geneva, WHO, 1990.

World Health Organization, *Evaluation of certain food additives and contaminants*. 35th Report of the Joint FAO/WHO Expert Committee on Food Additives. Technical Report Series 789. Geneva, WHO, 1990.

chapter seventeen

Basic requirements of risk evaluation and standard setting

M. Smith

17.1 Introduction

Early man, in pre-Neolithic times, hunted for meat and gathered what food he could. Just like modern man, he had to balance his requirements of energy, protein, and other essential nutrients. He also tried to avoid consumption of toxic factors naturally present in certain foods, presumably achieved by careful observation and, of course, by trial and error. Clearly, it was possible for the fit to survive such a lifestyle. The population increased and eventually agricultural methods were adopted. As society developed, undergoing the industrial revolution, there was a further change in lifestyle to an urban existence, and the ensuing need for a variety of stable, nutritious and attractive foods.

The safety of the food supply is a topic of continual interest to the media and the public at large. There are many issues involved which include concerns about environmental contaminants, use of food additives, pesticide residues, microbial contamination, and nutritional quality. New developments in the food supply prompt discussions about the scientific evidence for safety and the use of suitable control measures. Governments fulfill their responsibilities for safeguarding the food supply through a variety of laws and regulations. These responsibilities include both the nutritional and the safety aspects of the food supply.

A distinction can be made between nutritional and toxicological mechanisms underlying adverse health effects from foods, although the endpoints, or outcomes may be similar, for example, illness, poor development, and possibly death. Nutritional changes can result from unbalanced intakes of the required nutrients, i.e., surplus or deficiency. They manifest themselves primarily as physiological changes.

Toxicological mechanisms depend on interaction of toxic substances with biochemical processes, primarily leading to definite disturbance of homeostases and ultimately to adverse effects.

Safety of the food supply can be defined in practical terms as the absence of toxicity following food consumption. However, from this chapter it will become clear that absolute safety is an unattainable goal for the food supply (and any other activity associated with human endeavor). Safety must therefore be defined in relative terms such that any dangers associated with food consumption are limited to an acceptable level. The dangers must also be weighed against the need for the consumption of a range of foods that supply nutrients sufficient for survival and good health. Toxicology is therefore more than just the study of poisonous chemicals, providing a method to assess the safety of the components which make up food. The application of modern toxicological methods improves the purely empirical observation of our ancestors and allows prediction of the possible toxicity of any new food or food component.

This chapter discusses the principles involved in the safety assessment of food components and how the information obtained is used by governments to ensure a safe and varied food supply. The nutritional evaluation of food and the mechanisms whereby governments can influence the quality and quantity of the consumed food is also covered.

17.2 Nutritional value of the food supply

Also in modern society, it is still necessary to balance the intake of nutrients to the requirements of growth and body maintenance. However, unlike in the times of our ancestors, at least in a large part of the Western world, there is the possibility of overnutrition from an abundant food supply. This, together with a sedentary lifestyle can lead to obesity and a number of associated diseases.

17.2.1 Nutritional considerations

As far as food intake is concerned, developed countries usually employ two types of recommendations: *dietary standards* and *dietary guidelines*. Dietary standards help to answer the question how much of a particular nutrient is adequate for the majority of the population. In 1943, the US Food and Nutrition Board of the National Research Council published a list of Recommended Dietary Allowances (RDAs). The list has been reviewed and reissued at regular intervals (the 10th edition was published in 1989) to incorporate new nutritional knowledge. The recommended allowances represented the quantities of certain nutrients believed to be adequate to meet the known physiological needs of practically all healthy persons in the US (see also Section 12.1). Their original use was as a guide for advising on nutritional problems in connection with the recruitment of healthy young people into the armed services.

Dietary standards are also used for:

- planning food supplies to subgroups in the population;
- interpreting food consumption records of individuals and populations;

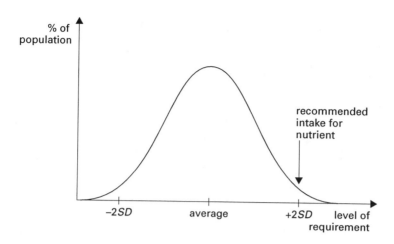

Figure 17.1 Distribution of the actual requirement of a given nutrient in a subpopulation. *SD* = standard deviation. Source: Beaton, 1985.

- evaluating the adequacy of food supplies to meet the national nutritional needs;
- designing nutrition education programs;
- developing new products in industry;
- establishing guidelines for the nutritional labeling of foods.

The recommendations are not meant to suggest that specific quantities of nutrients should be consumed every day. They are intended as a guide for intake levels averaged over a period of time (typically several days for most nutrients but over longer periods for others). This is necessary to take account of the day-to-day fluctuations in nutrient intake from the various foods consumed.

Intermezzo

RDA setting

The actual requirement of a given nutrient varies from one individual to another. If the requirements for all individuals in a given subpopulation are assumed to be normally distributed, the requirement for one particular person will be found on the characteristic bell-shaped curve (see Figure 17.1). If the RDA was set at the average requirement it would only satisfy 50% of the population. Therefore, the RDA is set slightly above the average requirement, typically by two standard deviations which, in a normal distribution, covers 98% of the chosen age or sex category. Special allowances are made for pregnant and lactating women whose requirements are unique in order to supply the fetus and suckling infant with the correct balance of nutrients.

An exception to the above is nutrient *energy* for which the RDA is set at the average requirement. The energy need varies from person to person. However, an additional allowance to cover this variation would be inappropriate because it could lead to obesity in the person with average requirements.

RDAs have been established by scientific committees in many countries but none of these can be applied globally as one single standard because of differences in diet and culture in the various countries. The task of setting RDAs is not an easy one, mainly due to a lack of basic information. As a consequence, different committees reach different conclusions, resulting in RDAs differing between countries.

It is important to understand the appropriate application of RDAs and the limitations of their use as stated in the 10th edition of the US RDAs:

- recommended allowances for nutrients are amounts intended to be consumed as part of a normal diet. Therefore, the RDAs are best met in diets composed of a variety of foods from a wide range of food groups rather than by *supplementation* or *fortification**. Such varied diets should also meet the requirements of other nutrients for which RDAs cannot currently be established;
- RDAs are safe and adequate levels of nutrient intake but are neither minimal requirements nor optimal levels of intake;
- RDAs are the amounts of nutrients which should be provided to particular groups of people. If the intake achieved by an individual is averaged over a sufficient length of time (to prevent an estimate being based on daily fluctuations of intake) and compared with the RDA, it will be possible to assess the risk of deficiency for that individual.

Dietary guidelines are recommendations for reaching an optimum nutrient balance in the diet. They aim to change the dietary pattern and thereby reduce the chance of chronic disease in a population. Such an approach is based on the results of studies on illnesses in populations or epidemiology, to enable the identification of dietary patterns associated with a low incidence of disease. The hypotheses developed from such studies may then be tested in animal models, assuming that a suitable model exists for man.

To date, dietary guidelines have not been very successful in so far that both the consumers' food choice and the industries' food supply are modified. The problem therefore appears to lie in the translation of qualitative recommendations into quantitative targets which may vary from country to country. In other words, the communication of these issues to the consumer and the food industry has not been very effective to date. The situation would be greatly improved if the European Union were to produce one set of quantitative recommendations for the whole of Europe. However, there is already a general consensus on dietary guidelines for the achievement of a healthy diet.

During the last half century, Western food supplies have become unbalanced. They now contain too much fat, too much sugar and salt, and not enough fiber. A healthy diet should be rich in vegetables and fruit, bread, cereals and other carbohydrate-rich foods, and may include fish and moderate amounts of lean meat and low-fat dairy produce. Such a diet is best to promote not only general good health but also to protect against the risk of heart and circulation problems, obesity, diabetes, some common cancers, and other Western physical disorders.

17.2.2 Nutritional evaluation of foods

A new or existing food can be characterized by its chemical composition. This can be achieved using chemical methods to analyze for:

macrocomponents: protein, fat, carbohydrate, and dietary fiber. These analyses can be
further subdivided to include the profile of amino acids, fatty acids, and fiber types;
microcomponents: vitamins and minerals, including trace elements.

Chemical analysis will establish the presence of a particular nutrient but will provide little information on its availability when consumed in food. Measurements of the

---

* Supplementation results from the consumption of nutritional supplements such as vitamin and mineral tablets. Fortification is the addition of nutrients to standard foods (e.g., breakfast cereals) particularly those which may have lost some nutrient content as a result of processing.

bioavailability of the nutrient demands testing in whole-animal or other biological assays. However, the results of the chemical analysis should indicate the types of biological testing required.

17.2.3 Strategy for nutritional testing

There are three aspects to be reckoned with when testing for nutritional values:

(a) Foods with a specific nutritional function require evidence that it actually fulfills its intended function both in experimental models and ultimately in man.
(b) Foods predicted to cause nutritional disturbance will need to be assessed to determine the qualitative and quantitative nature of the disturbance. For example, a fat replacer designed to provide the technical functions of fat without providing fat calories may lead to a reduction in the amount of essential fatty acids in the diet of certain consumer groups. Also, if a traditional food is produced by a new process it may be altered nutritionally. The nutritional equivalence of the food as produced by the old and the new process should be established.
(c) The nutritional properties of the food should be understood before toxicology testing is carried out so that nutritional disturbances can be distinguished from toxic effects.

17.2.4 Design of nutritional studies

The methods of nutritional research have not been standardized in the same way as toxicological studies, which are bound to internationally agreed protocols.

The following are some examples of study types to assess the nutritional properties of foods, carried out *in vitro*, in animals and, where appropriate, in man:

– on digestibility: *in vitro* and *in vivo* enzyme studies;
– on bioavailability: balance studies on intake and excretion of nutrients, growth, and carcass composition studies;
– on nutrient interaction: radioisotope and stable-isotope techniques;
– on physiological and biochemical effects: monitoring of blood and urinary composition, function tests; modification of the gastrointestinal microflora;
– on tolerance/adaptation: dose–response studies.

17.3 Toxicological factors affecting food safety

The presence of natural toxins and contaminants should be avoided whenever possible. Food production and processing should be carried out in such a way that their occurrence is minimized. Additives not only help to ensure maximum utilization and minimum deterioration of processed foods, but also facilitate the production of an attractive and wide range of food products.

To ensure a safe food supply it is first necessary to identify the *hazards* associated with the chemicals naturally present in food, i.e., the nutrients and any toxins of natural origin, and those chemicals added to food, either by accident (contaminants) or intentionally (food additives).

The next step is to assess the toxicological *risks* from the substances lacking nutrient properties, and thereby food safety. Although the terms hazard and risk will be defined elsewhere in this book (Sections 8.4 and 21.2), for a good understanding of food safety in the present context, it is crucial to recognize the difference between these terms and to ensure that some associated terms are used consistently.

Hazard can be defined as the intrinsic property of a substance that could lead to an adverse effect (e.g., cell toxicity or carcinogenicity). In other words, it is the toxic potential of the substance. *Risk* is a measure of the probability that a food component will cause an adverse effect as a result of human exposure. Therefore, risk is created by a hazard, but risk is not a necessary consequence of hazard. For example, a toxic chemical does not constitute a risk to man if, under the conditions of use, the target tissues are not exposed to the toxin. This may occur, for example if the toxicokinetic profile of the chemical in man is very different to that of the test species used to assess the toxicity of the chemical.

Hazard identification asks the questions: does a hazard exist?, and if so, what is it? A complex program of experimental techniques is often needed to answer these questions. Such a program could include analytical studies, *in vivo* animal studies, short-term *in vitro* cell culture tests and possibly epidemiological studies. *Risk assessment* is used to estimate the severity and likelihood of harm to human health (or the environment) from exposure to a toxic chemical. It must include an assessment of the source of that chemical and the characteristics of exposure (duration, dose, and dose response). The various factors are then integrated to give a measure of the risk. *Risk management* uses the information obtained from hazard identification and risk assessment. It also includes an assessment of the feasibility of taking action (together with a consideration of the political and economic impact) to determine the best course of action for reducing or eliminating the risk.

17.3.1 Safety assessment of new food components

Reviewing the process of safety assessment is useful from the point of view of a food company developing a new food or new food additive with a promising commercial application.

In the case of a new food, a detailed knowledge of its nutrient content is necessary for labeling on the packed food to inform the consumer. Such labelling enables the consumer to make a deliberate choice for the nutritional balance of his diet. If a traditional food is produced by a new process or a new variety is produced by selective breeding, analysis of the nutrient profile and the nutrients' bioavailability will indicate whether the novel food is equivalent to its traditional predecessor.

The new food may also contain natural toxins and it may be possible to detect and measure the levels of those that are known. In addition, the nutrient bioavailability studies and toxicological evaluation would indicate the presence of these natural toxins.

The possibility of contamination of the new food must also be considered by reviewing the processes used in its production, transport, and storage. If new contaminants are detected it will be necessary to assess the risk they pose and, as is also the case for known contaminants, to make sure they do not exceed the acceptable levels in the food.

The technial necessity of new food additives must be established to ensure that consumers are not exposed unnecessarily to the additional risk of a new chemical if it is of no particular benefit. Many people question the need for the many types of food additives presently available, but in practice, these additives are needed to ensure the availability of a range of attractive food products with a long shelf life. One only needs to look at the range of food products available in the supermarkets and consider how few would be possible without the use of some additives.

For the new preservative the support for its need should be based on its unique activity which will permit a new range of food products with an acceptably long storage life, which would not be possible with the existing preservatives. The next consideration is the safety of the new additive. It is the companies' responsibility to carry out the hazard identification. This information should be supplied to the regulatory authority which will, in conjunction with the company, carry out the risk assessment. If the risk associated with its use is deemed acceptable, the company will be granted permission to use the new additive. This approval will probably restrict the use to particular levels in certain food products or food categories. The regulatory authority can then incorporate the new chemical into its risk management programs to monitor its levels in food products and its intake by the population in general, and by certain high-risk groups in particular. This may take the form of post-marketing surveillance in which the occurrence of any unexpected effects may be monitored in certain groups. However, the approval for the use of any new food component is based on information currently available at the time of the safety review. If new information about the safety of a chemical emerges, its use must be reviewed. For this reason, the safety of food components is continually monitored by industry and government.

17.3.2 *Methods of hazard identification*

The type of toxic effect and the dose level at which it occurs are important issues in hazard identification (see Figure 17.2).

The test requirements are not necessarily the same for all food components. They will be influenced by properties of the substance such as:

- expected toxicity
- human exposure levels and pattern of use
- natural occurrence of the component in foods
- occurrence as a normal body constituent
- use in traditional foods
- knowledge of effects in man

It is impossible to give a detailed review of all the requirements for the safety assessment of a new food component here. Only a general outline of the approach will be presented.

The first step is to track the existing literature to find out whether the new chemical, or one which is structurally related, has been tested in the past. Once such information is collected, it is possible to design a program for safety testing to cover the pattern in which

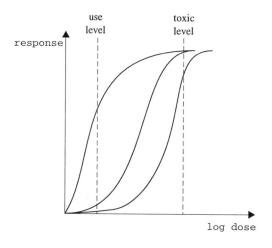

Figure 17.2 Dose–response curve. The steepness and shape of the dose–response curve indicate the size of the hazard as the dose or exposure is increased. Substances showing a steep curve with a low threshold before any toxic effect is detected, are of the greatest concern because their safety margin is very narrow.

the new food component is used. This would typically include studies of some or all of the following aspects:

- potential to cause mutagenesis in bacteria and mammalian cells
- absorption, distribution, metabolism, and excretion
- toxicity on repeated exposure for 4, 13, or 52 weeks in rodents
- effects on the reproductive systems and fetal development
- carcinogenic potential (e.g., 2-year feeding study in rodents)
- effect on the immune system
- effect on the nervous system
- effect on the endocrine system
- special studies on underlying mechanisms

The purpose of such tests is to build up a toxicological profile of the test material and to understand the dose–response relationship for any toxic response. A key determination is the assessment of the no-effect level from the feeding studies. Determination of the no-effect level depends primarily on the proper selection of doses for the study. Ideally, the highest dose should exert a toxic effect, whereas the lowest dose (a multiple of the human exposure level) should not show the effect. Additional dose groups are spaced between the top and bottom dose to define the dose–response curve further.

The risk assessment is carried out by determining the no-observed-adverse-affect-level (NOAEL) which is the highest dose in the most sensitive animal species which causes no toxic effects. The NOAEL is then divided by a safety factor to set an acceptable daily intake (ADI) level. The ADI is an estimate of the amount of a food additive, expressed on a body weight basis, that can be ingested daily over a lifetime without appreciable health risk. Substances that accumulate in the body are not suitable for use as additive. ADIs are only allocated to those additives that are substantially cleared from the body within 24 hours.

17.3.3 *Safety factors*

Safety factors are used to set an ADI that provides an adequate safety margin for the consumer by assuming that man is 10 times more sensitive than the test animal. A further factor of 10 is included which assumes that the variation in sensitivity within the human population is within a 10-fold range. The no-effect level, determined in an appropriate animal study, is traditionally divided by a safety factor of 100 (i.e., 10×10) to set the ADI. A food additive is considered safe for its intended use if the human intake figure is less than or equivalent to ADI. ADI is usually derived from the results of lifetime studies in animals and therefore relates to lifetime use in man. This provides a sufficient safety margin so that no particular concern is felt if man is exposed to levels higher than the ADI in the short term, provided that the average intake over longer periods does not exceed it. Higher safety factors may be used if the nature of the chemical's toxicity is of particular concern (e.g., if the substance is a carcinogen through a secondary mechanism, as is the case for bladder tumors following the formation of bladder stones caused by mineral imbalance), or if the chemical's toxicological profile is incomplete. Occasionally, lower safety factors may be used if there are human data to indicate that human sensitivity varies by less than 10-fold.

If a similar approach were applied to some essential nutrients (e.g., vitamin A, vitamin D, certain essential amino acids, and iron) it would become apparent that they may cause toxic effects at levels less than 10 times higher than those needed to satisfy the nutritional requirements for good health. This can be summarized as shown in the diagram below (Figure 17.3).

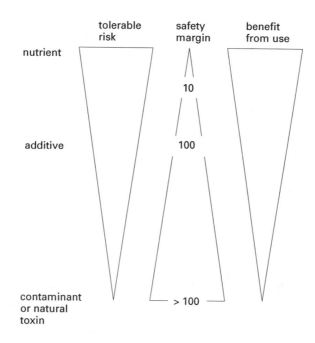

Figure 17.3 Use of safety factors. Small safety margins (2–10) are acceptable for essential nutrients e.g., selenium and vitamin A. Conversely, large safety margins (>100) should be set for contaminants. Additives will fall in-between (usually ~100). Source: ILSI Europe.

Using an ADI derived from a no-effect level found in an appropriate animal study and a suitable safety factor, implies an in-built conservatism reflecting the uncertainty of the extrapolation of experimental animal data to the diverse human population. In the case of contaminants, extrapolation is difficult from high-dose animal experiments to the human situation in which lower doses are consumed (see Section 17.4.2). The ADI also makes some allowance for the possible synergistic effects humans experience when additives are consumed together in foodstuffs. The effect of the interacting additives may be different from the responses to the individual additives.

17.3.4 Harmonization of safety testing procedures

Once a company has complied with its responsibilities for hazard identification and gained approval from the appropriate regulatory agencies, it may market its new food component worldwide. It is essential that any toxicological assessment is carried out to comply with internationally acceptable standards, to avoid the need for repetition of safety studies before gaining approval for use of the component in another country. The Organization for Economic Cooperation and Development (OECD) has developed guidelines for validated study protocols which provide acceptable basic standards for all member countries.

The World Health Organization (WHO) in combination with the Food and Agriculture Organization (FAO) through their Joint Expert Committee on Food Additives (JECFA) has also developed guidelines to improve the quality and general acceptability of food safety testing. Other organizations, including the Scientific Committee for Food (SCF) as the expert body within the European Union, EU, have also produced guideline protocols based on those of the OECD. Many individual countries have developed their own guidelines which may differ slightly from those of the OECD. Fortunately, most of these differences are disappearing as harmonization in standards increases, stimulated by the activities of the OECD, EU, and WHO.

17.4 Setting tolerable intake levels for natural toxins and food contaminants

Toxins of natural origin and contaminants are undesirable components which serve no nutritional or technical function (as do food additives) in the marketed food product. They constitute a large and diverse group of chemicals which man may consume in sizeable amounts. It is therefore essential that the toxicological profiles of the major natural toxins and contaminants are known so that their presence in food can be limited. The setting of acceptable intakes is based on an understanding of the toxicological profile of the component in question in a way similar to that described for food additives. In the case of contaminants, the term acceptable daily intake is changed to tolerable daily intake (TDI) to reflect the levels permissible in food to maintain a safe and varied supply (see also Section 16.3.2.1 and Section 21.4.4.3).

17.4.1 Assessment of toxicological risks from contaminants

Contaminants are often more toxic than additives. In the ideal situation, all toxic contaminants should be removed from food, but often the factors leading to their presence are difficult or impossible to control. Therefore, the unavoidable intake of contaminants should be limited to safe levels. As for food additives, limits for the presence of contaminants in food will need to be based on a no-effect level from an appropriate toxicity study. The safety factor applied to calculate the TDI of contaminants is frequently greater than the 100-fold factor used for additives.

For some contaminants that may accumulate in the body, the tolerable intakes are expressed on a weekly basis. The principle concern with respect to such contaminants is exposure for longer periods. This makes calculating intakes over a weekly interval more relevant as this eliminates daily fluctuations in intake.

In the case of genotoxic carcinogens (i.e., carcinogens which act directly by altering the genetic material), human exposure must be reduced to the lowest practically achievable level. The JECFA introduced the concept of "irreducible level," defined as "that concentration of a substance which cannot be eliminated from a food without involving the discarding of that food altogether and thereby compromising the ultimate availability of major food supplies." This level may in fact be the lowest detectable level and therefore the sensitivity of the analytical method is a key factor in defining the tolerable exposure.

Intermezzo

Sensitivity of measuring methods

The ability of analytical methods to separate and detect extremely low levels of contaminants in food has increased dramatically in recent years to the extent that detection of one part in a million (1 mg/kg) is routine. Methods for detecting 1 part per billion are commonplace and for some contaminants, additional orders of magnitude of sensitivity are achievable. However, it is very difficult to assess accurately the toxicological significance of exposure to such low levels.

Methods of hazard identification are usually based on feeding relatively high levels of the contaminant to animals to determine whether the substance has the potential to cause a toxic reaction, for example carcinogenesis. Therefore, the methods for safety assessment are frequently not as sensitive as certain analytical chemical techniques. As a consequence, it may be necessary to extrapolate the effects seen at relatively high dose levels (used in experimental animals) to the much lower exposure levels relevant to human risk. Such extrapolations are based on certain assumptions about the shape of dose–response curves. These assumptions are difficult to validate but the extrapolation may at least give an order of magnitude for the risk. Such techniques of quantitative risk assessment (QRA) are taken up with varying enthusiasm by the various regulatory bodies.

17.4.2 What are the toxicological challenges for effective risk assessment of foods in the future?

The available guidelines for the design of toxicity studies (for example those provided by the OECD) should be regarded as minimum standards for studies acceptable to the regulatory authorities. Test methods must be enhanced on a case-by-case basis to reflect the chemical nature and pattern of use or exposure to the new chemical. It is essential that the safety assessment of foods is based on the application of sound scientific principles rather than a checklist of toxicity tests to be completed prior to approval.

The guideline protocols and the safety factors employed for risk assessment are based on the assumption that new additives and ingredients will be used at levels well below 1% in the food. This assumption is not applicable to many of the current developments in food technology. The pace of development of new food additives has declined, but future developments will come from the area of biotechnology and the use of novel macro-components in the diet. The methods of genetic engineering are increasingly used to develop novel food sources with desirable characteristics. In addition, following the recognition of the need to modify the balance of macronutrient intakes to achieve a more desirable diet, materials such as fat replacers are actively developed.

These new developments pose an interesting challenge to the food toxicologist to combine his skills with the nutritionist's to develop test methods which will separate the toxic effects from those caused by altered nutrition in experimental models. Only a multidisciplinary approach will assure the continuous safety of our food supply.

17.5 Risk management

It is obvious then, that our food supply is composed of chemicals some of which are essential to nutrition, some of which serve no useful purpose, and others that are useful to maintain a varied food supply.

What can be done to ensure a safe food supply to the consumer? In other words, the food supply must contain sufficient nutrients to maintain health, while the levels of natural toxins, contaminants, and additives do not exceed those prescribed as safe. The two main weapons in the armory of a government are surveillance and enforcement.

Surveillance is concerned with estimating average and extreme intakes of foods by the general population or of high-risk groups within it, e.g., children, pregnant women, and the elderly. The intake of food components (both nutritious and potentially harmful) can be calculated and compared with recognized safety standards. The intakes of chemicals from food are measured from total diet surveys of standard food items purchased at regular intervals in different locations in the country. Amounts of foods consumed can be measured in diary studies where a record of the type and amount of a food or foods is kept. An extension of the diary study is a duplicate diet study in which a duplicated portion of each food item consumed is prepared and analyzed. Such studies are difficult and costly to carry out effectively, and therefore priorities have to be set and decisions to be made on the food components that are of the greatest concern and that need to be surveyed. Such techniques can provide an overview of the effectiveness of food control policies and a basis for their future development.

Enforcement is concerned with the compliance by agriculture and food producers with the legal limits for chemicals in food. Such procedures complete the chain from the safety assessment of foods and their components to the legal limits of chemicals in food established from the results of the assessment, in order to ensure consumer protection.

17.6 Summary

The challenges to early men who hunted and gathered their food were to obtain enough food to meet their nutritional requirements while avoiding those potential food sources that were acutely toxic. The challenges in present society are composed of some of the same elements, but the nature of the food supply is much more varied and complex. Nutritional standards and the safety of the diet are protected by regulatory processes laid down by governments. In addition, societies today are generally aware that certain food components may have effects on health in the longer term, for example on our cardiovascular health and the likelihood of developing some forms of cancer.

Safeguarding our food supply cannot be perfect, and risk evaluation and standard setting both have their problems, as illustrated in this chapter. Consumption of food, like any other activity, can never be entirely risk-free. Risks must be assessed and managed to protect the public from unsafe food components. There is a balance to be struck between the nutritional benefits of a varied diet and the low risk levels associated with food consumption.

Reference and Reading List

FAO/WHO *Evaluation of certain food additives and contaminants.* 22nd report of the Joint FAO/WHO Expert Committee on Food Additives. Technical Report Series No. 631. pp 14–15, 1978.

chapter eighteen

Extrapolation of toxicity data in risk assessment

H.J.G.M. Derks, C. Groen, M. Olling, M.J. Zeilmaker

18.1 Introduction

This chapter studies the role of extrapolation in the risk assessment of chemicals occurring in food. The concept of extrapolation is briefly introduced in Section 18.2. Both inter- and intraspecies (interindividual) extrapolation of data on chemicals showing a threshold in their dose–response relationship, and high-to-low dose extrapolation of chemicals which do not, are discussed. Further, some problems, which are in a certain sense specific to food chemicals, are presented.

Sections 18.3.1 to 18.3.4 investigate interspecies extrapolation and its limitations for four food chemicals from various classes. The relevance of extrapolation of toxicity data from one species to another in everyday life is explained in Section 18.3.5, using 2,3,7,8-tetrachlorodibenzo-*p*-dioxin as an example.

18.2 Extrapolation

One of the cornerstones of human toxicology is the assumption that toxic effects of chemicals in humans can be predicted from dose–response relationships established in experimental animals. In view of the countless biological and biochemical similarities between species, this assumption seems to be basically sound. Nevertheless, toxicological studies have proved beyond doubt that large interspecies differences in sensitivity to toxic chemicals do occur. In addition, it has been shown that between individuals of an outbred species, like man, similar differences in sensitivity may exist. Toxicology has responded to this problem with the introduction of safety or uncertainty factors which are applied whenever animal toxicity data have to be translated to safe human exposure levels. This process is referred to as *extrapolation* and is considered applicable to all chemicals exhibiting a threshold in their dose–response relationship. If chronic toxicity data have been collected in an experimental animal species, the human acceptable daily intake (ADI) is calculated by dividing the no-observed-adverse-effect level (NOAEL) in the animal by a standard uncertainty factor of 100 (see Chapter 17, Section 17.3.3). Larger extrapolation factors are used when only subchronic toxicity data are available or when the lowest dose tested in the animal still elicits slight toxic effects and repeating of the experiment with yet lower doses is not considered necessary in view of the type and severity of the observed effects. In both cases, the standard uncertainty factor is multiplied by an additional factor varying from 1 to 10. As mentioned above, the standard uncertainty factor is 100, which implies that in practice this additional factor often equals 1. However, extrapolation factors up to 2000 may be applied.

Genotoxic carcinogenic substances are assumed to exhibit no threshold in their dose–response relationship. Therefore, no absolute safe human exposure level can be defined. An important problem a toxicologist is confronted with in connection with this group of substances is that the dose levels needed to establish the dose–response relationship in experimental animals are many orders of magnitude higher than those likely to be encountered in human exposure situations. Simple linear extrapolation from these high doses to find the dose associated with negligible risk is considered to be safe, but rather conservative. Negligible risk is called the Virtually Safe Dose or Risk Specific Dose and is assumed to cause 1 extra tumor in 10^6 subjects after lifetime exposure. More often, mathematical models based on certain assumptions about the mechanism of carcinogenesis are used to fit the high dose data obtained in animals, and to predict effects at low dose levels. An often-used mathematical model is the multi-stage model which assumes carcinogenesis to be a multi-stage process, and tumor incidence to depend on the probabilities of transition of each stage into the next. The number of stages and the transition probabilities are estimated by curve fitting of the experimental data. Other mathematical models have been introduced which may differ in behavior in the low dose region, while they show no differences at high dose levels.

Usually, no adjustments are made to correct for interspecies differences in sensitivity to carcinogenic substances. However, in some cases dose levels have been normalized to body surface area rather than to body weight. Such a normalization may correct for interspecies differences in the pharmacokinetic behavior of xenobiotics.

For the assessment of the toxicological risks due to food chemicals, the same extrapolation methods are used as for other substances. However, chemicals present in food may behave differently from the same chemicals administered either in a pure form or as a solution. The food matrix may influence the extent and rate of gastro-intestinal absorption by several mechanisms. As a result, both the bioavailability and the maximum blood concentration may be affected. Since the route and method of administration of toxic substances to experimental animals are often chosen on the basis of convenience, the validity of animal data for human risk assessment may be compromised. Further, all food chemicals obviously enter the target animal via the gastrointestinal route and may therefore be affected by species differences in gastrointestinal physiology. Sometimes such problems can be prevented by careful selection of the experimental animal species. Other food-specific problems may arise from micronutrients with a small safety margin which do not allow the use of large extrapolation factors. Some of the above problems are dealt with in more detail in the next section.

18.3 Extrapolation and assessment of toxicological risks due to food chemicals

This section presents a detailed study on the role of extrapolation in estimating the toxicological risks from food chemicals. Examples are chosen from the four categories discussed in this textbook: vitamin A (nutrient), solanine (natural toxin), nitrate and nitrite (contaminants), and BHA and BHT (additives).

18.3.1 Micronutrients: vitamin A

18.3.1.1 Introduction
Vitamin A is required in small amounts in crucial biological processes such as controlling the differentiation and proliferation of epithelial cells, maintaining general growth and visual and reproductive functions. Therapeutically, vitamin A is used in dermatology for curing various skin diseases, and one of the metabolites of retinol, all-trans retinoic acid, is used topically to treat acne. Vitamin A, as retinyl esters, is also taken in various amounts as a food supplement.

The recommended average daily dietary intake of vitamin A was estimated by Sauberlich et al. (1974) at 600 retinol equivalents (RE) per day for adult men. Olsen (1987) estimated a total amount of 625 RE per day on the basis of metabolic turnover data. Adequate levels of vitamin A intake must be such that the concentration in the liver is maintained at 20 RE per g. From these data the regulatory authorities of the US and Canada recommended the following daily intake of vitamin A: male adults and pregnant women: 1000 RE per day, female adults: 800 RE per day, lactating women 1250 RE per day, and children of 1 to 3 yr 400, 4 to 6 yr 500, and 7 to 9 yr 700 RE per day.

18.3.1.2 Assessment of teratogenic risk
One of the most important toxic effects occurring after chronic and/or acute hypervitaminosis A is teratogenicity in early pregnancy. Rosa et al. (1986) described 18 cases of teratogenic effects in humans caused by hypervitaminosis A. Acute and chronic hypervitaminosis A may be caused by consuming vitamin A as a food supplement, or in liver. The vitamin preparations on the Dutch market, for example, contain from 300 RE to 15,000 RE vitamin A per dosing unit. Livers of calves may contain even more, 25,000 RE per 100 g!

From these data it is clear that women who are on a normal or rich diet with respect to the intake of vitamin A run a high risk by consuming liver or vitamin preparations in early pregnancy. Since only few human data are available, especially on the teratogenic effect, risk assessment is based on data obtained in animal studies. To estimate this risk a good interspecies extrapolation model is needed to extrapolate these data to the human situation.

Table 18.1 lists the lowest teratogenic doses of vitamin A in several species. The data clearly show a large interspecies variability in sensitivity to vitamin A.

The differences in route of administration (oral vs. intraperitoneal) and in *kinetics* of vitamin A may cause interspecies variability. Further, the fact that different effects were measured may play a role. There may also be interspecies differences in morphology of the uterus.

Table 18.1 Lowest teratogenic dose of vitamin A in various species after oral (p.o) or intraperitoneal (i.p.) administration

Species (body weight)	Dose (RE/kg/day)	Time after conception (days)	Effect on
Man (60 kg)	120 p.o	14–35	Cranium and face
Mouse (20 g)	3,300 i.p.	9	Cleft palate
Hamster (100 g)	30,000 p.o	8	Exencephalum
Rat (200 g)	50,000 p.o	9–11	Exencephalum

18.3.1.3 Toxicokinetics

As mentioned above, one of the causes of the interspecies' differences shown in Table 18.1 may be found in species' differences in toxicokinetic behavior of vitamin A. Therefore, the toxicokinetics of vitamin A and its precursors are briefly discussed here.

In the lumen of the gastro-intestinal tract, retinyl esters are hydrolyzed and the retinol formed is taken up by the enterocytes by means of passive diffusion. In contrast, carotenoids are taken up as such and converted to retinol in the enterocytes by cleavage.

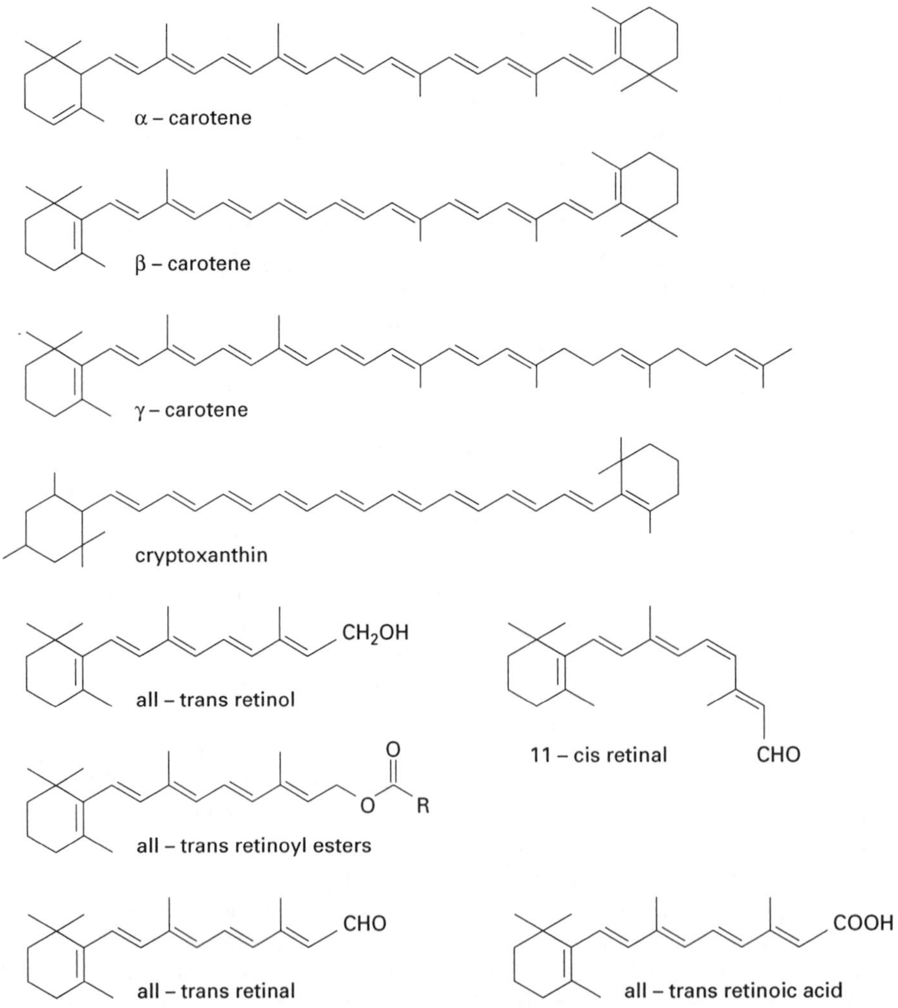

α – carotene

β – carotene

γ – carotene

cryptoxanthin

all – trans retinol   CH$_2$OH

11 – cis retinal   CHO

all – trans retinoyl esters

all – trans retinal   CHO

all – trans retinoic acid   COOH

In the enterocytes, retinol is re-esterified by two specific enzymes and the resulting retinyl esters are incorporated in the chylomicrons, followed by secretion in the lymph and transport to the liver via the thoracic lymph duct and systemic circulation. In the liver 10% of the total amount of the retinyl esters are stored in parenchymal cells and 90% in fat-storing cells. After hydrolysis and binding to specific proteins, retinol and retinoic acid are secreted into the blood and distributed to other organs. If the recommended amount of vitamin A is consumed, the amount in the liver remains constant and the blood concentrations of retinol and retinoic acid remain low. After intake of excessive amounts most processes, such as uptake, esterification, hydrolysis, and binding to proteins may become saturated, leading to an increase in the free retinol concentration and induction of toxic effects.

For the development of an extrapolation model to assess the teratological risk from vitamin A, the following toxicokinetic aspects must be examined in more detail in at least two species: linearity of absorption, bioavailability of retinol after administration of carotenoids or retinyl esters, and capacity of the liver to store retinol and to synthesize the relevant binding proteins. Also, concentration and form of vitamin A that the embryo is exposed to in the case of acute or chronic hypervitaminosis A are of high importance. If these toxicokinetic aspects of vitamin A are elucidated and can be related to physiological and biochemical characteristics, such as lymph flow, blood flow, and enzyme activities of the animals used, extrapolation to humans and an estimate of the risk can be achieved.

18.3.2 Natural toxins: solanine

18.3.2.1 Introduction

Solanine is just one of the countless substances of natural origin that may cause adverse effects in humans. The large number of natural substances known to possess potential toxicity probably only represents a small percentage of those that actually exist. This situation may be attributed to the fact that the available quantities of the substances are too small to use in toxicological experiments, and also because suitable analytical methods are not always available. This section shows the role of extrapolation in the assessment of human health risks due to solanine, and its toxicological evaluation.

18.3.2.2 Toxicological risk assessment

Symptoms of toxicity were recorded during an outbreak of potato poisoning among school children in South-East London in 1969. The peeled potatoes that were consumed contained 330 mg of glycoalkaloids per kg. In other cases, 410 mg and 430 mg of glycoalkaloids per kg potatoes have been reported to cause outbreaks of potato poisoning. From these casuistic data, a lowest-observed-adverse-effect level (LOAEL) (see Section 21.4.4.3) of about 2 mg/kg body weight was calculated. Generally, 200 mg of glycoalkaloids per kg potatoes is accepted as the upper safety limit. This value is based on an average daily intake of 300 g of potatoes by an adult, and includes a safety factor of 2. In a number of countries, this limit has been reduced to 100 mg/kg potatoes, as the safety factor 2 was considered to be inappropriate. Moreover, as compared to the assessment of risks from synthetic chemicals, there is a lack of data concerning long-term repeated intake of relatively small amounts of solanum alkaloids. There are indications that solanine and related substances can accumulate in tissues. This may lead to late toxic effects. Therefore, there is a need for at least semichronic toxicity studies. Summarizing, a more systematic approach is desired to come to a better estimation of the ADI of solanine and other natural toxins.

Table 18.2 Toxicokinetics of solanine in the rat and the hamster after intravenous as well as oral administration

Species	Dose (10^5 dpm[a])		$AUC_{0-\infty}$[b] ($\cdot 10^3$ dpm[a]⟨h/ml⟩)		Cl_m[c] (ml/h/kg)	F[d] (%)
	i.v.	p.o.	i.v.	p.o.	i.v.	p.o.
Rat	421 ± 6[e]	1240 ± 170[e]	1390 ± 230[e]	71 ± 33[e]	107 ± 17[e]	1.6
Hamster	364 ± 61	723 ± 8	3250 ± 310	239 ± 91	63 ± 14[f]	3.2

[a] dpm, disintegrations per minute; the solanine was radiolabeled.

[b] $AUC_{0-\infty}$, area under the plasma concentration vs. time curve from time zero to infinity.

[c] Cl_m, metabolic clearance of solanine.

[d] F, mean absolute bioavailability.

[e] Dose, $AUC_{0-\infty}$ and Cl_m are given as mean ±S.D.

[f] $p < 0.05$, compared to iv administration in rats.

18.3.2.3 Toxicokinetics

In the extrapolation of toxicity data from animal to man, interspecies differences in *bioavailability* are a factor to which special attention should be paid. This implies that in the case of solanine, blood levels rather than doses, should be used as a basis for extrapolation.

For most substances, there is a direct relationship between the blood concentration and the concentration at the site of action on the one hand, and between the concentration at the site of action and the intensity of the effect on the other. However, if the dose is used as a basis for extrapolation, the absorption from the site of administration into the general circulation, i.e., the bioavailability, is not accounted for. Lack of information on interspecies differences in bioavailability is an extra source of uncertainty in extrapolation.

Studies on the toxicity of glycoalkaloids have been carried out in different animal species. Severe gastric and intestinal mucosal necrosis was observed in hamsters receiving dried potato sprout material containing high concentrations of glycoalkaloids. Hamsters seem to be more sensitive to glycoalkaloids than rats and mice. However, little information is available on the underlying toxicokinetics.

Recent experiments suggest that the higher systemic toxicity in hamsters (and thus maybe also in man) is due to a higher bioavailability after oral administration. The difference in bioavailability of solanine between rats and hamsters is shown in Table 18.2.

It should always be kept in mind that not only the parent compound but also its metabolites can be toxic. For example, solanine is metabolized via different routes. Its metabolites are not toxic. However, for many other substances, it has been reported that the metabolites induce effects that are different from or stronger than those of the parent compound. In those cases, determination of the bioavailability of the parent compound solves only part of the problem.

Based on the difference in toxicokinetic behavior of solanine between rats and hamsters, and since after oral administration more disorders in the intestinal tract were observed in the hamster than in the rat, the hamster was chosen as a model for subchronic toxicity studies on this glycoalkaloid. The effects on toxicokinetics of factors such as dose level and food matrix have to be elucidated to enable a reliable estimation of the exposure to solanine. Matrix factors deserve special attention, since the public health authorities want to know whether additional requirements should be made for potato products, like starch, present in various types of diets.

18.3.3 Food contaminants: nitrite and nitrate

18.3.3.1 Introduction

As a naturally occurring substance, nitrate (NO_3^-) is a common constituent of the environmental compartments soil and water. From the soil compartment, NO_3^- may be taken up

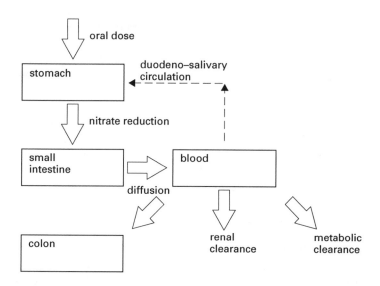

Figure 18.1 Disposal of nitrate following oral administration, including its duodeno-salivary circulation.

in drinking water. Since nitrate is the primary nitrogen source for plants, it enters the mammalian food chain by its ability to accumulate in plant materials. Consequently, the intake of food, especially leafy vegetables and drinking water, is the main route of exposure of humans to NO_3^-. The average intake of NO_3^- via food consumption is estimated at 100 to 150 mg/day and from water at 10 to 20 mg/day. This accounts for more than 99% of the total daily NO_3^- intake.

Nitrite (NO_2^-) is not a natural food component. It is used as a preservative in a number of meat products. In comparison with the daily intake of NO_3^-, the intake of NO_2^- by humans via food consumption is low, i.e., < 0.1 mg/day.

18.3.3.2 Toxicological risk assessment

In order to evaluate the toxicological significance of human exposure to NO_3^- and NO_2^-, the actual intakes of these substances need to be compared with estimated safe exposure levels. Based on a body weight of 70 kg, the actual intake of NO_2^- is <0.001 mg/kg/day, and of NO_3^- 1.4 to 2.5 mg/kg/day. Traditionally, safe exposure levels of humans to chemicals are obtained by extrapolating data from toxicity studies in experimental animals. In rats, the no-observed-adverse-effect level (NOAEL) of sodium nitrate was found to be 500 mg/kg/day and of sodium nitrite 20 mg/kg/day. Application of a safety factor of 100 to these values and correction for the differences in molecular mass between the sodium salts and the ions result in Acceptable Daily Intakes (ADIs) of 0 to 3.64 and 0 to 0.135 mg/kg/day, respectively. The actual intakes of NO_3^- and NO_2^- by the general population amount to 38 to 69% and <0.7% of the ADI, respectively. From this, it can be concluded that on average the actual exposure of humans to NO_3^- and NO_2^- via food and drinking water does not pose toxicological risks. However, mechanistic studies in experimental animals and man have shown that the traditional method of extrapolating data on the toxicities of NO_3^- and NO_2^- from animals to man is inadequate and deserves reconsideration. In order to show the inadequacy of the currently used extrapolation method, the fate of NO_3^- and NO_2^- in man should first be dealt with in more detail.

18.3.3.3 Toxicokinetics of NO_3^- and NO_2^- in man

Following oral administration, NO_3^- is almost completely absorbed (>98%). It is eliminated from the body in three ways (see Figure 18.1). First, NO_3^- is excreted via the kidneys. This accounts for approximately 60 to 70% of the total nitrate body clearance. Secondly, NO_3^- is metabolized to ammonium, urea and more reduced forms of nitrogen such as nitrous oxide (NO) and NO_2^-. This elimination pathway accounts for approximately 20 to 30% of the total body clearance of nitrate. Thirdly, NO_3^- may be excreted in the sweat which accounts for almost 11% of the total body clearance.

Once absorbed from the gastrointestinal tract, NO_3^- may circulate in the body by entering the so-called duodeno-salivary circulation. This circulation consists of the excretion of NO_3^- by an active transport mechanism from the blood into the saliva, followed by reabsorption of the excreted NO_3^-. Approximately 25% of the orally administered nitrate enters the duodeno-salivary circulation. Once NO_3^- has been excreted into the saliva, it may be reduced to NO_2^- by bacteria present in the oral cavity. In man an estimated 30% of NO_3^- is converted in this way. So, 8% of the orally administered NO_3^- may be converted to NO_2^- by bacterial reduction in the oral cavity. By combining these data with the estimated daily intake of NO_3^- (see Section 18.3.3.1) the daily intake of NO_2^- formed from ingested NO_3^- can be estimated at 7 to 10 mg/day, i.e., 0.09 to 0.14 mg/kg/day which is 0.69 to 1.07 times the current ADI for nitrite! This clearly shows that the setting of safe standards for nitrate and nitrite in food should not be based on the determination of standards for each individual substance but on an integration of knowledge of the disposition of both substances. Ideally, the standard setting should meet the following criteria:

1. standards for dietary nitrate should be based on *expected* derived nitrite toxicity;
2. the accepted intensity of nitrite-induced toxicity in man is equal to the NOAEL of nitrite in experimental animals divided by the product of inter- and intraspecies extrapolation factors. Currently, inter- and intraspecies factors of 10 are used in extrapolating data on nitrite toxicity from experimental animals to man.

If the above criteria are applied to the animal model most widely used in experimental toxicology, the rat, one is faced with the basic problem that the rat probably is not an adequate model for man with regard to nitrite-induced nitrate toxicity. The reason for this is that in literature it has been suggested that the duodeno-salivary circulation of nitrate does not exist in the rat. Hence, in this species nitrite-induced nitrate toxicity cannot be studied adequately. One way to solve this problem is to study nitrate toxicity in an animal species that does have a duodeno-salivary circulation, such as the pig. Alternatively, data obtained in the rat may be used for estimating safe human exposure levels for NO_3^-, i.e., the ADI. The procedure for obtaining the ADI for nitrate is then as follows.

First, a NOAEL for nitrite toxicity is determined in the rat. Dividing this parameter by an extrapolation factor of 100 then gives the ADI for nitrite. A total safe nitrite intake can be calculated by multiplying this value by body weight (BW), say 70 kg. If the direct dietary nitrite intake is negligible, the actual human intake of nitrite is solely determined by the conversion of NO_3^- to NO_2^- by bacteria in the oral cavity. The actual nitrite intake can then be calculated as follows :

$$\text{total safe nitrite intake} = \text{ADI}_{NO_2^-} \times \text{BW} \qquad (1)$$

$$\text{total safe nitrate intake} = \text{ADI}_{NO_3^-} \times \text{BW} \qquad (2)$$

$$\text{actual nitrite intake} = (\text{total nitrate intake} \times \alpha)/1.35 \qquad (3)$$

in which α equals the fraction of NO_3^- converted to NO_2^- by bacterial reduction in the oral cavity and 1.35 (= 62/46) is a multiplication factor for the difference in molecular weight between NO_3^- and NO_2^-. Combination of these equations gives the ADI for nitrate:

$$\text{ADI}_{\text{NO}_3^-} = \left(\text{ADI}_{\text{NO}_2^-} \times 1.35\right)\big/\alpha \qquad (4)$$

With the current $\text{ADI}_{\text{NO}_2^-}$ of 0.135 mg/kg/day and $\alpha = 0.08$, this would give an $\text{ADI}_{\text{NO}_3^-}$ of 2.28 mg/kg/day. In comparison with the currently used ADI for nitrate, i.e., 3.64 mg/kg/day, this means a decrease by more than 37%.

Once the ADI for a food contaminant has been calculated, its value may be used to set the dietary standard for that particular food component. This procedure can be easily illustrated by the calculation of safe drinking water levels for NO_3^-. For example, if drinking water accounts for 10% of the daily nitrate intake and the safe nitrate intake is set at a value found by multiplying $\text{ADI}_{\text{NO}_3^-}$ by BW, this nitrate intake can be estimated at approximately 250 mg/day (70 kg × 3.64 mg/kg/day). Assuming a daily water consumption of 1 l, drinking water is then allowed to contain 0.1 × 250 per l = 25 mg NO_3^- per l.

This procedure, valid for the *general* population, does not necessarily hold for high-risk groups, i.e., groups with expected high exposure levels and/or increased sensitivity. In the case of nitrate and nitrite, infants are such a group. In infants, the major toxic effect of nitrate and nitrite is nitrite-induced methemoglobinemia. Nitrite entering the blood circulation oxidizes hemoglobin (Fe^{2+}) to methemoglobin (Fe^{3+}), leading to reduced oxygen transport. Neonates are a high-risk group as they are methemoglobin reductase deficient. For the induction of methemoglobinemia, a NOAEL of 100 mg $NaNO_2$ per kg/day has been established in experimental animals. When combining the drinking water consumption of infants with the fractional conversion of NO_3^- to NO_2^- in the gastrointestinal tract and the infant's body weight, the safe NO_3^- concentration of drinking water used for the preparation of infant food can be calculated.

The safe concentration is 30 mg/l which is arrived at as follows. The ADI for sodium nitrite-induced methemoglobinemia can be calculated from the above NOAEL to be 1 mg/kg/day, i.e., for nitrite 0.67 mg/kg/day. If the conversion of NO_3^- to NO_2^- is taken into account and corrections are made for the difference in molecular mass between nitrate and nitrite, an $\text{ADI}_{\text{NO}_3^-}$ of 4.52 mg/kg/day can be calculated for infants [(0.67 × 1.35)/0.20] (see Equation 4). In combination with a body weight of 5 kg and a daily water consumption of 0.75 l, this results in a safe drinking water consumption of approximately 30 mg/l. If this concentration is compared to the calculated safe NO_3^- concentration in drinking water for the general population (25 mg/l, see above), it can be concluded that the generally supplied drinking water may be used safely for the preparation of infant food.

18.3.4 Food additives: the antioxidants butylated hydroxyanisole and butylated hydroxytoluene

18.3.4.1 Introduction

To preserve quality and to prevent loss of nutritional value, the addition of antioxidants to food containing fatty acids has a long tradition. Two well-known antioxidant food additives are butylated hydroxyanisole (BHA) and butylated hydroxytoluene (BHT) (see Figure 18.2).

Although highly lipophilic, BHA and BHT do not accumulate in mammals. The reason for this is the efficient elimination of these chemicals from the body. In the case of BHA and BHT, the discussion on setting the dietary standards has focused on the question whether or not these food additives have to be considered as non-genotoxic carcinogens, and consequently, on whether or not safe human exposure levels for these substances can be established.

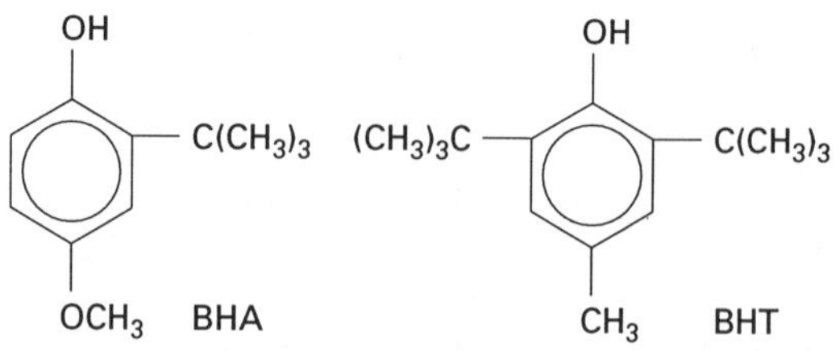

Figure 18.2   Structures of butylated hydroxyanisole (BHA) and butylated hydroxytoluene (BHT).

18.3.4.2 Toxicological risk assessment of BHA

In the rat, BHA induces epithelial hyperplasia and tumors in the forestomach. Since the forestomach is an organ specific to rodents (rat, mice, hamster) and not found in other animals, the question arises whether this effect can be used as the starting point for setting a dietary standard in man. To answer this question the Scientific Committee on Food of the European Commission Food-Science and Techniques asked in its 1983 evaluation of BHA for additional information on the following subjects:

1.   the induction of hyperplasia by BHA in the part of the gastrointestinal tract immediately preceding the stomach, i.e., the esophagus, and the glandular stomach in species without a forestomach, and
2.   the genotoxic properties of BHA.

On the basis of additional information from experimental studies, the Committee concluded in its reevaluation of BHA in 1989 that the effect of BHA on the forestomach epithelium is highly specific to rodents and does not occur in non-rodents. Furthermore, epithelial hyperplasia, qualified as a precancerous lesion, was found to be of a reversible nature and showed threshold characteristics, i.e., hyperplasia only occurred above a definite dietary BHA dose level. In species without a forestomach (guinea pig, dog, pig, monkey), BHA did not cause histopathological symptoms in the esophagus and the glandular stomach. All available mutagenicity data are negative, and BHA does not show any genotoxicity at all. Based on these data the Committee concluded that the induction of forestomach hyperplasia and tumors by BHA in rodents is of no significance in the assessment of human health risks from BHA exposure. Further, it was concluded that genotoxicity does not play a role in causing rodent forestomach tumors. Therefore, the Committee classified BHA as a rodent (and not human) carcinogen showing a threshold in the induction of effects. Consequently, the Committee accepted the calculation of an ADI for BHA to be relevant. In order to calculate this ADI, the NOAEL for the induction of hyperplasia in the rat forestomach was used as toxicity parameter for BHA. Experimentally, this NOAEL was found to be 50 mg/kg/day. Applying a standard safety factor of 100, this lead to an ADI of 0 to 0.5 mg BHA/kg/day for the safe chronic exposure level of the human population.

18.3.4.3 Toxicological risk assessment of BHT

As in the case of BHA, dietary standards for BHT were set at an expert meeting of the Scientific Committee on Food of the European Union. In its 1989 meeting, this Committee evaluated all available toxicity data on BHT. The toxicity profile of BHT was summarized as follows. In chronic toxicity studies, BHT induced liver carcinomas and adenomas in the rat at dose levels higher than 100 mg/kg/day. However, BHT was not found to be mutagenic or otherwise genotoxic. Therefore, the Committee considered BHT as a non-genotoxic carcinogen with a threshold in the induction of its carcinogenicity. In semi-chronic toxicity studies, BHT caused an increase in thyroid weight. In this type of study the lowest dose tested, 500 ppm BHT in the diet, still induced a significant increase in thyroid weight. However, the Committee concluded that "It is reasonable to assume that the likely NOAEL for thyroid weight change will be about 5 times lower than the lowest-observed-adverse effect level, i.e., 500 ppm." In subacute toxicity studies, BHT was found to interfere with blood clotting. The underlying mechanism is a reduction of the activity of vitamin K-dependent blood clotting factors. In the rat, the NOAEL for this effect was found to be 5 mg/kg/day. Taking all toxic effects into consideration, the Committee classified BHT as a non-genotoxic carcinogen in rodents. Likewise, the Committee recommended the determination of an ADI as a safe exposure measure for the human population. Since the NOAEL for the chronic toxicity (neoplasia in the liver) was about 50 times higher than the NOAEL for semi-chronic (increased thyroid weight) and subacute toxicity (hematological disorders), the latter parameter (5 mg/kg/day) was used for the calculation of the ADI of BHT. Applying a standard safety factor of 100, the Committee recommended an ADI of 0 to 0.05 mg/kg/day for BHT.

18.3.5 Extrapolation and standard setting for substances occurring in food

The choice of methods to extrapolate toxicological data from animals to man largely depends on the mechanism underlying the toxicity of the substance under investigation. Traditionally, the extrapolation of toxicity data of substances which give positive results in chronic carcinogenicity studies as well as in genotoxicity studies is carried out by using methods based on the assumption that there is no threshold dose (see Section 18.2). Toxicity data of non-carcinogenic substances are extrapolated by using methods assuming a threshold value mechanism (see also Section 18.2). Although the latter method offers a rather clear-cut possibility to extrapolate toxicity data from one species to another, its application in everyday safety evaluation procedures is often more ambiguous. This will be explained for the food contaminant 2,3,7,8-tetrachlorodibenzo-*p*-dioxin (TCDD).

TCDD is the dioxin with the highest toxicity. Dioxins are emitted into the environment by waste incineration and other combustion processes. They may enter the food chain. Chronic toxicity studies in experimental animals showed that TCDD is a liver carcinogen in the female but not in the male rat. Further studies revealed that TCDD was not capable of inducing genotoxic effects in *in vitro*-genotoxicity assays and that its carcinogenicity is probably associated with an altered function of female steroid hormones. These findings were used as a starting point for the evaluation of the toxicological risk from TCDD. In practice, however, different authorities took diverging scientific standpoints for the extrapolation of TCDD toxicity data to man. As a result, quite different estimates of the toxic potential of TCDD were reached. In the Netherlands, for example, an expert panel was of the opinion that the experimental data on the toxicity of TCDD provided sufficient evidence to classify this substance as a non-genotoxic carcinogen in experimental animals. The panel concluded that in the case of TCDD safe exposure levels, i.e., an ADI could be calculated in a valid way. For the calculation of the ADI, liver carcinogenicity (in rats) was chosen as the critical toxic effect. For this effect, a marginal-observed-adverse-effect level (MOAEL) of 1 ng/kg/day was established in a chronic experiment in rats. The MOAEL is the lowest found concentration of a substance which causes a marginal adverse effect. The MOAEL is between the NOAEL and the LOAEL. From this effect level a NOAEL was calculated by applying an extrapolation factor of 2.5. The panel considered this value for the MOAEL–NOAEL extrapolation factor adequate in view of the type of effect observed at the MOAEL. Application of inter- and intraspecies extrapolation factors of 10 then gave an ADI of 4 pg/kg/day. In contrast to the Dutch Health Authorities, the US Environmental Protection Agency (US EPA) concluded that the available information on the toxicity of

TCDD did not give conclusive evidence with regard to its carcinogenicity mechanism. The US EPA decided to consider TCDD as a genotoxic carcinogen and to base its safety evaluation on acceptable rather than safe exposure levels. To calculate the acceptable exposure level 1 extra liver tumor incidence per 10^{-6} after lifelong exposure to TCDD was taken as an acceptable risk level. The calculation of the exposure level was based on a quantitative dose–response relationship between the daily TCDD intake and liver tumor incidence; the relationship was assessed by using a multi-stage carcinogenesis model. This relationship was then used for the calculation of the risk specific dose (RSD, see Section 18.2). This calculation resulted in an acceptable exposure level of 6.4 fg/kg/day.

Whether or not TCDD is considered a genotoxic or a non-genotoxic carcinogen, the extrapolations mentioned above were based on the so-called *external dose concept*. This means that the toxic potential of a substance is proportional to the amount of the substance to which an organism is exposed. The exposure levels were expressed in terms of units of weight of the substance per kg body weight. The external dose concept has been used for the interspecies extrapolation of TCDD toxicity data up to 1991. In 1991, however, this concept was abolished. In that year a World Health Organization Expert Committee decided to use the actual concentration of TCDD in its target organ, i.e., the liver, rather than the ingested amount for the calculation of the safe human exposure levels to TCDD *(internal dose concept)*. The reason to replace the external dose by the internal dose lies in the widely accepted view that the toxicity of a substance is best characterized by the following two factors: the disposition of the substance in the organism (toxicokinetics) and the mechanism underlying its toxicity. In order to assess the disposition of TCDD in mammals as a function of the dose, the Committee used a one-compartment model. Toxicokinetic analyses showed that, at equal exposure levels, TCDD concentrations are expected to be 10-fold higher in the human liver than in rat liver. On the basis of a NOAEL of 1 ng/kg/day for TCDD carcinogenicity in the rat, this analysis predicted a NOAEL of 100 pg/kg/day in man. By dividing this (estimated) NOAEL by a safety factor of 10 (for intraspecies variation) the ADI of TCDD was obtained, i.e., 10 pg/kg/day.

The kinetic extrapolation method used by the WHO Expert Committee is an example of classic toxicokinetic modeling. A limitation of this type of modeling is its inability to give a physiological interpretation of the various compartments forming part of the model. The classic toxicokinetic modeling does not allow organ-specific toxicokinetic and toxicodynamic processes to be taken into account in safety evaluation procedures. To obviate this limitation, alternative kinetic approaches have been developed in the last decade. These so-called physiologically based pharmacokinetic (PBPK) models describe the disposition (absorption, distribution, metabolism, and excretion) of substances in the organism on the basis of blood flows through the organs instead of distribution over compartments. Figure 18.3 gives a diagrammatic representation of a PBPK-model of TCDD disposition in the rat.

A system of five blood flows is shown: blood circulation, and four flows through the liver, fat tissue, slowly perfused organ system (SPO, mainly skin and muscle) and richly perfused organ system (RPO, mainly kidneys, lungs and spleen). After a physiological flow diagram as shown in Figure 18.3 has been defined, absorption and elimination of the substance concerned are included in the model. For TCDD, this refers to absorption, elimination by hepatic metabolism, and biliary excretion of the metabolites formed. The model also includes a toxicodynamic parameter, viz. the induction of hepatic P-450 mixed-function oxidase (MFO), a well-known effect of TCDD and structurally related chlorinated aromatic hydrocarbons. The mechanism underlying this induction has been found to consist of a sequence of events: uptake of TCDD by the liver, binding of TCDD to a cytosolic receptor protein (the aryl hydrocarbon or Ah receptor), and stimulation of the *de novo* P-450 MFO synthesis. The determination of the exposure level of the liver to TCDD

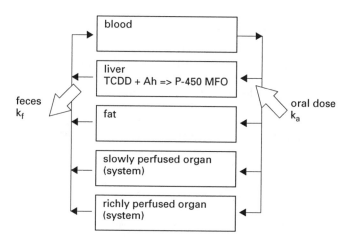

Figure 18.3 Flow diagram for a PBPK model of TCDD disposition in the rat.

is based on this mechanism. The PBPK model has been used to analyze the disposition of TCDD in experimental animals (rat, mouse) and man. The results showed that the disposition of TCDD in these species could be described by *one* PBPK model, irrespective of the dose level (high to low dose extrapolation), the route of administration (route to route extrapolation) and the dosage schedule (acute, semi chronic or chronic exposure conditions). Further, these analyses showed that TCDD-induced *de novo* synthesis of P-450 MFO was the primary factor determining the disposition of TCDD (and thus its toxicological risk) in rodent liver but not in human liver. This underlines the importance of taking interspecies differences in toxicity into account in toxicological safety evaluation. In contrast to classical toxicokinetic modeling, PBPK models can predict the disposition and toxicity of substances in mammals on the basis of a common physiological approach of the organism. PBPK models enable the incorporation of detailed knowledge of toxicity mechanisms as well as variations in the physiological state (growth, pregnancy, sex, disease, age) into toxicological safety evaluation. These possibilities make PBPK models suitable for quantitative and physiologically valid interspecies extrapolation of toxicity data. In this connection, PBPK models are continuously the subject of extensive scientific research.

18.3.6 Concluding remarks

The preceding sections have shown that in principle, extrapolation of data on the toxicity of food chemicals is carried out in the same way as that of chemicals in general. Specific problems may be related to particular subcategories such as micronutrients and using inappropriate methods for the administration of toxic substances to experimental animals. In general, the application of uncertainty factors to establish safe human exposure levels may provide a sufficiently large safety margin to compensate for inter- and intraspecies differences, if in the choice of animal models for toxicity studies, mechanistic aspects are taken into account. In exposure situations above the established safe level, however, the quantitative basis of this methodology is insufficient to allow reliable risk evaluations. Knowledge about the toxicokinetics of a substance in both experimental animal and man reduces the uncertainty in the extrapolation step by enabling the calculation of the quantitative relationships between external and internal dose levels. A more fundamental approach would be the incorporation of toxicokinetic as well as toxicodynamic differences in the extrapolation step. This approach, which is the objective of advanced modeling techniques, may ultimately lead to more adequate quantitative extrapolation methods.

Reference and reading list

Blomhoff, R., Transport and Metabolism of Vitamin A. *Nutr. Rev.* 52, 513–524, 1994.

Chappel, W.R. and J. Mordenti, Extrapolation of toxicological and pharmacological data from animals to humans, in: B. Testa, (Ed.), *Adv. Drug Res.* 20. London, Academic Press, 1991.

European Union, Scientific Committee on Food, Opinion on Nitrates and Nitrites, 98th Meeting of the Scientific Committee on Food/EU, 1995.

Friedman, M., Composition and safety evaluation of potato berries, potato and tomato seeds, potatoes and potato alkaloids, in: Finley, J.W., S.F. Robinson and D.J. Armstrong, (Eds.), *Food Safety Assessment*, ACS Symposium Series No. 484, 429–462, 1992.

Jadhav, S.J., R.P. Sharma, D.K. Salunkhe, Naturally occurring toxic alkaloids in foods, in: *CRC Crit. Rev. Toxicol.* 9, 21–104, 1981.

Joint Expert Committee on Food Additives (JECFA), FA/WHO Technical Report Series, 44th Meeting, Rome, 859, 1995.

Olsen, J.A., The storage and metabolism of vitamin A, in: *Chemica Scripta* 27, 179, 1987.

Rosa, F.W., A.L. Wilk, F Kelsey, Teratogen update: vitamin A congeners, in: *Teratology* 33, 355, 1986.

Sauberlich, H.E., R.E. Hodges, D.L. Wallace, H. Kolder, J.E. Canham, J. Hood, N. Raica, L.K. Lowry, Vitamin A metabolism and requirements in human studies with the use of labeled retinol, in: *Vit. Horm. (Leipzich)* 32, 251, 1974.

Sennes, F.J. de and A. Hollaender, Nitrates and nitrites: Ingestion, Pharmacodynamics and Toxicology, in: *Chemical Mutagens, Principles and Methods for their Detectrion* 7, 1982.

Shapiro, K.B., J.H. Hotchkiss, D.A. Roe, Quantitative relationship between oral nitrate-reducing activity and the endogenous formation of N-nitrosoamino acids in humans, in: *Food Chem. Toxicol.* 29, 751–755, 1991.

Slanina, P., Solanine (glycoalkaloids) in potatoes: Toxicological evaluation, in: *Food Chem. Toxicol.* 28, 759–761, 1990.

Das Nitrosamin Problem. Weinheim, Verlag Chemie, 1983.

chapter nineteen

Setting toxicological standards for food safety

F.X.R. van Leeuwen

19.1 Introduction

Toxicological standard setting is a process carried out by legally qualified national authorities to protect the public health or the quality of the environment. A toxicological standard for a substance can be defined as a limit value for its content in food, (drinking) water, soil, or air. These toxicological standards are not only based on toxicological knowledge, but are also the result of a thorough risk–benefit analysis. In the process of standard setting, toxicological guide values or health-based recommendations are weighed against technical feasibility and check possibilities, and socio-economical and political interests (Figure 19.1). Thus, standards are based on scientific as well as practical considerations. It should be noted that standards are only of value if they can be implemented and enforced.

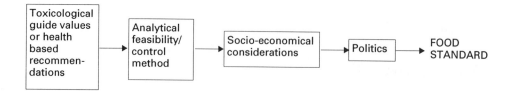

Figure 19.1 Process of standard setting.

A guide value can be defined as a limit value with the aim to maintain or protect the quality of human life and ecosystems, and to minimize risks. A guide value is an estimate of the highest acceptable or tolerable exposure level. It is based on an objective evaluation of all available toxicological information, reflecting the state of the art, including application of appropriate safety (or uncertainty) factors. In practice, guide values are maximum daily (or weekly) doses or maximum concentrations in food, drinking water, or environmental compartments.

This chapter will answer such questions as: what is a standard with regard to food safety? How are standards set? Who is responsible for the setting of standards? The general principles of recommendations for the protection of human health are discussed and the role of international bodies, such as the World Health Organization and the European Union in setting harmonized standards and the effects of these standards on national regulatory measures will be elucidated.

19.2 General principles

Usually, health-based recommendations or guide values are based on data obtained from toxicological studies in experimental animals, and only sometimes on observations in man.

It is the aim of safety evaluation to identify the type of adverse effect and to establish and quantify the dose–response relationships over certain periods of time. Therefore, adequate toxicological data are essential to determine the level at which human exposure to a substance can be considered as safe. For food additives, it was decided a long time ago by the Joint FAO/WHO Expert Committee on Food Additives (JECFA) that an acceptable daily intake (ADI) should be established that would provide "an adequate margin of safety to reduce to a minimum any hazard to health in all groups of consumers." Thus, the ADI was defined as an estimate of the amount of a substance in food, expressed on the basis of body weight, that can be ingested daily over a lifetime without an appreciable health risk. Guide values or standards based on this ADI should minimize the probability of the occurrence of adverse effects in man, if exposed to a particular substance. Crucial in this approach is the establishment of a threshold dose above which any functional or structural disturbance shows itself as a pathological effect of which the intensity increases with increasing exposure (due to both dose and duration). In evaluating the toxicological potential of substances (present in food), it is essential to distinguish between genotoxic substances, for which it is assumed that no thresholds exist, and non-genotoxic substances, which can be evaluated according to the threshold approach.

19.2.1 Determination of a threshold

For non-genotoxic substances, a deviation from the statistical mean of a normally distributed value must be reached before a particular effect in an organism can be observed. The threshold dose for the most critical effect in a test is the highest exposure level without adverse, i.e., toxicologically relevant, effects. It is called the no-observed-adverse-effect level (NOAEL). In practice often more than one method will be available for the toxicological evaluation. In general, the critical test is the most sensitive one, carried out in the most sensitive animal species, assuming that man is at least as sensitive as this particular animal species. The results of the various methods are compared. In a toxicological evaluation, the following points are examined: the relevance of the effect as well as the animal model, both in view of the extrapolation to man; the validity of the tests, and the quality of the report.

19.2.2 Determination of the NOAEL

For the determination of the NOAEL, a series of doses is used. In order to establish the dose–effect relationship, the dose levels are chosen in such a way that the highest dose causes an adverse effect that is not observed after the lowest dose.

Ideally, in a long-term toxicity study, the highest dose should evoke symptoms of toxicity without causing excessive mortality, and the lowest dose should not interfere with development, normal growth, and longevity. In between, doses should be selected sufficiently high to induce minimal toxic effects. The determination of an adverse effect in a particular study depends not only on the doses tested, but also on the types of parameters measured and the ability to distinguish between a real adverse effect and a false positive finding. In long-term toxicity tests, the average value of a specific parameter at a particular dose level is compared with the average value of the parameter in control animals. An effect can then be defined in purely statistical terms as a significant deviation of a control value. However, in determining an adverse effect, the biological relevance of this deviation should be taken into consideration. If, for example, a slight but significant alteration is only observed at the highest dose level, it is difficult to define it as a real adverse effect. More weight should be given to a particular change in a parameter, if a dose–response relationship can be established, or if the observed change is related to changes in other functional or morphological parameters. If an effect is irreversible, the relevance of the effect is unquestionable. In some cases, however, the biological relevance of an effect must be interpreted in relation to historical control values. This is often the case when the value of the particular parameter is highly variable among the control animals used in a number of different toxicological studies. The historical control data should originate from the same species, strain, age, sex, supplier, and laboratory to enable proper comparison.

There are many sources of uncertainty in toxicity testing. For example, effects may not show themselves if the number of animals is too small (to discriminate between various test groups), the time of observation is too short for the manifestation of a particular effect, or the experimental design is too limited to obtain conclusive evidence. In addition, the differences in sensitivity to a particular substance between man and experimental animals may not be known. Therefore, safety factors are applied in the setting of guide values for man based on animal data to compensate for these uncertainties.

19.2.3 Application of safety factors

In the extrapolation of animal data to the human situation, safety factors are applied to provide an adequate safety margin for the consumer. Usually, most national as well as international regulatory bodies traditionally apply a safety factor of 10 for interspecies variation and 10 for intraspecies variation, resulting in an overall safety factor of 100. If toxicity data in human beings are available, such data take precedence over animal data, and, generally, in such cases a safety factor of 10 is appropriate. A lower safety factor may suffice if the substance under investigation is identical to traditional food components, e.g., nutrients such as vitamins and amino acids, if the substance is metabolized into

Table 19.1 Organizational differences between WHO standards and EU standards

	WHO standard	EU standard
Impact	Worldwide	European Union
Status	Advisory	Imperative

endogenous compounds, or if it lacks overt signs of toxicity. For substances serving as essential sources of energy in the human diet, the safety factor 100 is not applied either.

Although safety factors are employed to protect the health of the consumer, they reflect all the uncertainties in the process of extrapolating animal data to health-based recommendations for man. Therefore, the term "uncertainty factor" may be more appropriate.

19.2.4 High-risk groups

As mentioned in the previous section, when establishing guide values an uncertainty factor of 10 is applied to account for interindividual variations in the sensitivity to a particular substance. In some cases, however, specific human subpopulations can be identified as being particularly at risk. These groups may consist of young children, pregnant women, elderly persons, or specific groups of patients, for example those suffering from chronic non-specific lung disease, cardiovascular diseases, or renal deficiencies. If such a group can be clearly identified, the guide value for the general population may be based on this group.

19.3 Who is responsible for standard setting?

Within the framework of public health legislation, national regulatory authorities are responsible for standard setting with regard to food safety. The authorities can carry out the process of standard setting as a separate national affair, or adopt standards set by international bodies such as the World Health Organization and the European Union. To achieve harmonization in food standards, many countries adopt standards set by the WHO. However, since 1992 the member countries of the EU are required to accept the decisions taken by the European Commission and enforce Union standards into their own national legislation. The difference between WHO standards and EC standards are summarized in Table 19.1.

19.3.1 Role of the World Health Organization

The World Health Organization is an international advisory body with the overall aim of protecting human health. As far as toxicological risk assessment is concerned, it is not a legislative body. It backs national authorities in setting standards for the protection of human health. The International Program on Chemical Safety (IPCS) plays a guiding role in the international procedure of evaluating risks from chemicals and setting tolerances for residues of chemicals in food. Through the IPCS, the WHO participates in two joint committees of the WHO and the Food and Agricultural Organization (FAO). The Joint FAO/WHO Expert Committee on Food Additives (JECFA) and the Joint Meeting on Pesticide Residues (JMPR) serve as scientific advisory bodies of the Codex Alimentarius Commission, a joint FAO/WHO commission that sets standards for chemicals in food. The Codex Alimentarius Commission is responsible for the implementation of the Joint FAO/WHO Food Standards Program, that is intended to:

(a) protect the health of the consumer and ensure fair practice in food trade;
(b) promote coordination of all food regulatory activities carried out by international governmental and non-governmental organizations;
(c) establish priorities, and initiate and give guidance to the preparation of provisional standards by and with the aid of appropriate organizations;
(d) finalize provisional standards and, after acceptance by governments, publish them in a Codex Alimentarius;
(e) amend published standards, after appropriate survey, in view of certain developments.

Although the Codex Alimentarius and FAO/WHO do not have any legal authority and the standards they propose are not standards as defined above, the Codex standards have been shown to be of great value in the harmonization of food standards.

It is the aim of Codex to offer proposals for Maximum Residue Limits (MRLs) to national governments for acceptance into the prevailing national registration or standardization system. There are Codex Committees on food additives and contaminants, on pesticide residues and on veterinary drug residues. The membership of the Codex Committees is open to all nations, and their meetings are attended by formal national delegations. While the considerations of JECFA and JMPR are purely scientific (as these bodies consist of experts or advisory members speaking as private persons), the proposals of the Codex Committees are partly based on national politics.

Regional differences in the use of additives, pesticides, or veterinary drugs are a problem in the harmonization of (worldwide) MRLs. Officially recommended use rates for pesticides are usually higher in those countries where extreme climatic conditions favor the development of pests or diseases than in more temperate climates. Further, countries which are important exporters of foods such as grains and meat, tend to favor relatively high MRLs, while countries that are importers tend to favor low MRL values.

In tackling these differences, the Codex Commission follows a thorough stepwise procedure, leading to the acceptance of a formal Codex Standard (see Figure 19.2).

The above procedure gives members an opportunity to participate in the decision process and to use the final result for their own national standard setting. However, national or regional policy sometimes disturbs this ideal in standard setting, for example, when the European Union uses other MRLs, based on the recommendations of one of its own Scientific Committees.

19.3.1.1 Role of the Joint FAO/WHO Expert Committee on Food Additives

The Joint FAO/WHO Expert Committee on Food Additives evaluates food additives, food contaminants and residues of veterinary drugs. JECFA first convened in 1956 with the mandate to:

- formulate general principles governing the use of food additives, with special reference to their legal authorization, on the basis of considerations such as innocuousness, purity, limits of tolerance, and the social, economic, physiological, and technical reasons for their use, taking into account work already done on the subject by national and other international bodies;
- recommend, as far as practicable, suitable uniform methods for the physical, chemical, biochemical, pharmacological, toxicological, and biological examination of food additives and of any degradation products formed during the processing, for the pathological examination of experimental animals and for the assessment and interpretation of the results.

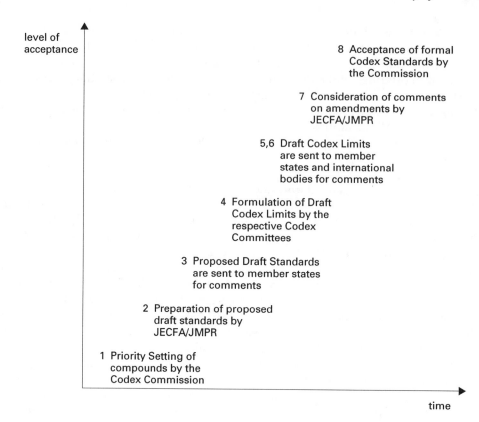

Figure 19.2 Procedure, leading to the acceptance of a formal Codex Standard.

This mandate was later (1987) expanded to include food contaminants and veterinary drugs.

For food additives, ADIs or provisional ADIs (when the available information does not warrant a final conclusion) are calculated. This parameter indicates the safe daily dietary intake of a substance. The actual daily dietary intake should not exceed the ADI. Therefore, information on dietary intake is necessary. This can be obtained from market-basket or total diet studies. In the case of major food components and some novel foods (modified starches, polyols, modified celluloses), it is often not necessary to calculate an ADI, since the effects observed in toxicity experiments concern the nutritional value. In such cases, no numerical value for the ADI is given (ADI not specified). These products are believed to be acceptable.

For residues of veterinary drugs, the WHO panel of the Joint Expert Committee evaluates the toxicological information and establishes, if possible, ADIs (or provisional ADIs). The FAO panel proposes limits (MRLs or provisional MRLs) for residues of veterinary drugs in products of animal origin, based on the WHO ADIs and on information about the distribution of the residues in tissues of the target animal. In setting the MRLs, the maximum theoretical intake should not exceed the ADI. This maximum theoretical intake is estimated using the (exaggerated) consumption package for products of animal origin as compiled in Table 19.2.

Veterinary drug residues include parent drugs as well as their metabolites. The metabolites are taken into account if they are toxicologically relevant, i.e., present in a considerable quantity or having a toxicological or pharmacological potential. The MRL is

Table 19.2 Average daily consumption of animal products

Cattle/swine		Poultry		Fish	
Muscle	300 g	Muscle	300 g	Muscle	300 g
Liver	100 g	Liver	100 g	Liver	100 g
Kidney	50 g	Kidney	40 g		
Fat	50 g	Skin	60 g		
Milk	1.5 l	Eggs	100 g		

expressed in terms of parent drug levels or in terms of levels of a marker metabolite, if the percentage of the marker metabolite formed from the parent drug is known.

Intermezzo

Example. In 1990 JECFA evaluated the antibiotic oxytetracyclin and calculated an ADI of 0 to 0.003 mg/kg body weight (0.2 mg per person) based on the results of a study on the antimicrobial activity of tetracyclin in human volunteers. JECFA established the following MRLs: 0.6 mg/kg for kidney, 0.3 mg/kg for liver, 0.2 mg/kg for eggs, 0.1 mg/kg for milk and muscle, and 0.01 mg/kg for fat.

However, using the data presented in Table 19.2, the estimated maximum theoretical daily intake for oxytetracyclin residues in beef, eggs, and milk is: 150 µg in milk, 30 µg in muscle, liver and kidney, 20 µg in eggs, and only 0.5 µg in fat. In total, this is approximately 260 µg. This exceeds the ADI of 200 µg per person.

However, JECFA concluded that application of these recommended MRLs does not pose a risk to the consumer, since the NOAEL used for the calculation of the ADI was very conservative, and the consumption data used in Table 19.2 are at the upper limit of the range for the individual intake of animal products. Thus, in practice, the safety rules are interpreted with a certain flexibility though strict rules are applied for the derivation of health-based recommendations.

19.3.1.2 Role of the Joint Meeting on Pesticide Residues

In 1963 The Joint Meeting on Pesticide Residues convened for the first time. The WHO panel of the JMPR evaluates pesticide residues on the basis of toxicological and biochemical data. If the data are inadequate, the JMPR allocates an ADI for each individual pesticide under investigation. The FAO panel of the JMPR evaluates disposition of residues and resulting residue levels under conditions of Good Agricultural Practice, on the basis of data on patterns of use.

In order to evaluate the acceptability of a proposed MRL, it is necessary to compare the dietary intake of pesticide residues calculated on the basis of the MRL with the ADI. The dietary intake is calculated by multiplying each MRL with the quantity of the corresponding diet component, followed by summation of the residue quantities obtained. It should be noted that the use of the MRL in the calculation of total intake may lead to a higher value than the actual intake, since the actual residue levels will often be lower than the recommended MRLs.

Food consumption patterns vary considerably from one country to another, and from one culture to another. At the international level, the total intake is calculated on the basis of a hypothetical average global food consumption package, composed according to the recommendations in the FAO Food Balance Sheets, i.e., consisting of components of each cultural diet. At the national level, the total intake is calculated on the basis of actual consumption data. In practice, these are cultural diet data.

These three ways of calculating the daily intake of pesticide residues are summarized below.

1. Theoretical maximum daily intake (TMDI):
    TMDI $= \Sigma Fi \times MRLi$
    Fi $=$ the hypothetical average intake of a diet compoent
    MRLi $= ...$

2. Estimated maximum daily intake (EMDI):
    EMDI $= \Sigma Fi \times Ri \times Pi \times Ci$
    Ri $=$ the actual residue level in the diet component
    Pi $=$ adjustment factor taking into account reduction (or increase) in residue quantity due to industrial processing
    Ci $=$ adjustment factor taking into account reduction (or increase) in residue quantity due to preparation of the food (boiling, frying etc.).

3. Estimated daily intake (EDI), which is a refinement of the EMDI at national level, based on adequate actual data.

The procedure in which the dietary intake of pesticide residues is compared with the ADI starts with the intake parameter that can be the highest, TMDI. If TMDI does not exceed ADI, it is highly unlikely that the ADI will be exceeded in practice, and therefore the MRL proposals can be considered to be acceptable. If TMDI is higher than ADI, a parameter concerning the *actual* situation (EMDI) should be used in order to eliminate the pesticide from further consideration.

For veterinary drugs, another procedure is applied. MRLs for veterinary drugs are based on theoretical maximum consumption data. Furthermore, veterinary-drug-residue limits are set for the fresh animal product, and effects of industrial or in-house processing on the residue content are not taken into account.

19.3.1.3 International Program on Chemical Safety

Within the framework of the International Program on Chemical Safety (IPCS), WHO has drawn guidelines for the protection of drinking water quality. Recently, a revision of these guidelines was carried out for a large number of organic and inorganic substances, including disinfectants and pesticides. It is the WHO's intention that these guidelines should be applied in setting national standards, not only for community piped-water supplies but for all sorts of drinking water except for bottled mineral waters. Adoption of these worldwide guidelines is dependent on national priorities and socio-economic factors. Since water is one of the primary needs for life maintenace, it must be available even if the quality is not entirely satisfactory. This implies that setting standards that are too stringent could limit the availability of water. This is considered unacceptable, in particular in regions with water shortage. On the contrary, it is WHO's opinion that this consideration is never allowed to lead to guide values posing health risks.

The WHO states that the established guide values protect health for lifelong consumption. The quality of drinking water should always be maintained at the highest level. On the other hand, short-time exposure above the guide value does not necessarily imply a health risk, but it should be a signal to competent authorities to consider certain measures. The information used for drawing guidelines for drinking water does not only include toxicological data but also data on the occurrence of contaminants in drinking water, physical properties like solubility, and aesthetic and organoleptic aspects. In cases where threshold doses were exceeded, ADIs were calculated, or adopted if they were available from other international bodies. For genotoxic carcinogens, which may be present as contaminants in drinking water, the risks were assessed on the basis of an acceptable risk of one additional case of tumorigenesis per population of one million lifelong exposed persons.

Since exposure to the substances of which the guidelines are under revision not only occurs via drinking water but also via other routes (food, air), the ADI may be partly ingested. In general, intake via drinking water amounts to 10% of the ADI. Since for most pesticides exposure via other routes is extensive, an intake value of 1% of the ADI is employed. For disinfectants used for the purification of drinking water, exposure via other routes is negligible. Therefore, higher intake values (up to 50%) are applied. The toxicological guide values calculated according to the above procedure were compared with taste and odor thresholds. If the latter values were lower, the standards were based on organoleptic quality.

Intermezzo

Example. Drinking water may be contaminated by monochlorobenzene as an indirect result of its use as an organic solvent in pesticide formulations, or as a degreasing agent in industry. Based on chronic toxicity data, WHO established a tolerable daily intake (TDI) of 0.09 mg/kg body weight (see also Sections 16.3.2.1, 17.4, 17.4.1, and 21.4.4.3). For calculation of the guide value, a body weight of 60 kg and a consumption of 2 l of drinking water per person per day are used. If the intake via drinking water amounts to 10% of the TDI, the total acceptable intake is 0.54 mg and the guide value 0.27 mg/l.

It should be emphasized that this toxicological guide value far exceeds the lowest reported taste and odor threshold for monochlorobenzene, being about 10 µg/l. Therefore, the latter value will probably be used by national authorities as standard for monochlorobenzene in drinking water.

19.3.2 Role of the European Union

The European Community was founded as a free-trade association for its member countries. One of the objectives was to achieve harmonization in setting food standards. Since January 1992, however, all member countries have to accept the products produced in other member countries without any restriction, and have to apply identical criteria for quality and safety. In practice, this means that member countries cannot approve a marketing authorization for substances used in the production of foods without the agreement of the European Community. The safety evaluation of food additives or substances present in a food as a result of their use in its production process, is formally carried out by the Commission of the European Communities.

Within the Commission, several scientific working groups are involved in food safety evaluation (see Figure 19.3). Proposals made by these working groups for the safe use of food additives and for maximum residue limits are, if adopted by Regulatory Committees, enforced by the Council of Ministers. Enforced proposals are published in the Official Journal of the European Union and are, from that time on, imperative for the regulatory authorities in the member countries.

19.3.2.1 Activities of the European Scientific Committee for Food

The Scientific Committee for Food (SCF) advises the Commission with regard to directives for food additives, flavoring substances, solvents, materials in contact with food, contami-

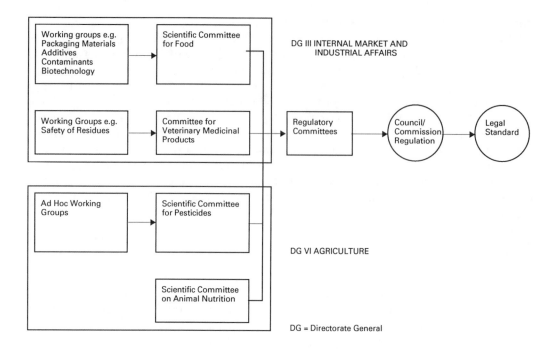

Figure 19.3   Scientific working groups involved in food safety evaluation.

nants, novel foods, and foods for particular nutritional use. Consultation of the SCF is obligatory in all cases concerning public health.

The SCF evaluates the available toxicological and analytical information in order to estimate the maximum limits for the safe ingestion of the substances under investigation, and designates these guide values using the following classification:

- Acceptable Daily Intake (or provisional ADI if more data are required) for lifetime exposure, to be used to set standards for the use of particular food components;
- ADI not specified, if the technological limits are believed to provide a sufficiently large safety margin;
- Acceptable, limited and well-defined use;
- Not Acceptable: the intentional use is considered unsafe. Particularly in the case of carcinogenic substances, no acceptable values can be given;
- Tolerable Daily Intake, for lifetime unintentional exposure (e.g., environmental pollutants and contaminants originating from packaging materials).

According to the present EU regulation, any new request for the admission of a new substance that is covered by the Food Directive should no longer be addressed to the member state concerned, but directly to the Commission.

19.3.2.2 Activities of the European Committee for Veterinary Medicinal Products
On behalf of the Committee for Veterinary Medicinal Products (CVMP), the Scientific Working Group Safety of Residues evaluates the food safety aspects of veterinary drugs used in animal production. Since January 1992, the decisions made by this Working Group and authorized by the CVMP, overrule the national safety evaluation of veterinary drug residues. At this moment, no admission of a new veterinary drug in a member country is possible if a Union Standard has not been set. In contrast to the members of the other scientific committees, the members of the CVMP and of the Working Group are national

representatives. This means that not only scientific judgments contribute to decisions, but also national policy arguments. In order to establish ADIs and MRLs, the Working Group follows a procedure as used by JECFA. If possible, the Working Group adopts ADIs and MRLs already established by the Codex Commission on Veterinary Drugs, but sometimes the scientific judgment of the Working Group differs from those of JECFA and Codex, resulting in a different conclusion. In JECFA, the uncertainty with respect to the toxicological evaluation and the lack of sufficient data often lead to a number of questions to be answered by industry, and no ADI or MRL is established in such a case. The EU, however, is entitled to set residue levels for all veterinary drugs. Before 1997, about 400 biologically active substances present in veterinary drugs have to be evaluated and MRLs have to be established. This means that if there are not sufficient data available for an appropriate safety evaluation, a pragmatic approach has to be chosen which enables the establishment of provisional ADIs by applying larger uncertainty factors, resulting in the establishment of provisional MRLs. If the use of a particular component is a serious reason for concern, the MRL is also based on the detection limit.

Recently, the Codex Commission on Veterinary Drugs published the first regulation on MRLs for residues of veterinary drugs in foodstuffs of animal origin. In this regulation, for each biologically active substance the animal species for which the MRL is applicable, the marker residue on which the MRL is based, and the target tissue for which the MRL should be used, are listed.

19.3.2.3 Activities of the European Scientific Committee for Pesticides

For safety evaluation, the Scientific Committee for Pesticides (SCP) follows a procedure similar to that of JMPR. In general, this means that carcinogens are not acceptable as pesticides, and for other substances ADIs have to be established. The ADIs are compared with the estimated intakes of the residues through the consumption of various agricultural products. Based on this comparison, residue standards are set.

19.3.2.4 Activities of the European Scientific Committee on Animal Nutrition

Additives used in cattle, swine, and poultry feed to prevent the outbreak of diseases have already been evaluated in the past by the Scientific Committee on Animal Nutrition (SCAN) as an accepted Union procedure. Following the evaluation of all available toxicological data, conditions of use were described, which were safe for the consumer, and these conditions were included as an annex to the veterinary drug acts in several countries. However, SCAN is now in the process of developing procedures for standard setting of feed additives, a process that, in the light of the ongoing harmonization, needs to be comparable to the procedures used by the CVMP and by JECFA.

19.3.3 National regulations

Nowadays national standards appear to be of minor importance in relation to EU regulation. In the past, the responsible national regulatory authorities were obliged to evaluate substances with regard to consumer safety, and to set residue standards in foodstuffs within the framework of the local Food Acts, the Pesticide Acts, or the Veterinary Drug Acts. As was mentioned before, this responsibility is now taken over by the respective scientific and regulatory committees of the European Union. The decisions reached in the EU with respect to food standards should now be implemented in the national legislations, and standards should be adopted in the national Residue Regulations. This implies, as mentioned before, that no new marketing authorization can be granted in a member country without a Union Standard.

However, this process does not necessarily mean that all EC member states have exactly identical standards. If a member state, for whatever reason, sets a different standard, it has to accept that if this standard is higher than the Union Standard it can not sell the particular product in other member countries, and consumer organizations certainly will question this decision. If a country sets a standard that is lower than the Union Standard, it has to accept food products from other member countries coming up to the Union Standard. If such a lower standard will lead to additional restrictions in the use of the particular substance, one can expect the industry to complain, and to seek its rights via the European Court.

19.3.4 Role of industry

Although industry in general has no formal responsibility in the process of standard setting, it still plays an important role. First, industry provides the necessary information about the identity and purity of the substance, conditions of use, analytical methods for detection of residues, efficacy, and toxicological data that are essential for the safety evaluation. During evaluation in JECFA, JMPR, or EU Committees, hearings take place at which the industry is offered the opportunity to clarify existing problems or to comment on decisions taken by these bodies.

The Codex system, as described before, is unique in its possibility for industries to participate in pre-Codex meetings and to be members of the national delegations. In these delegations the industry representatives, however, have no voting status. Further, the International Group of National Associations of Manufacturers of Agrochemical Products and the International Animal Health Industry participate as observers in the Codex meetings without voting rights but with a limited opportunity to join in the debate. During the process of drafting a new EU regulation, the Commission or the respective Working Groups inform the industries about new proposals and offer them the opportunity to respond.

Reference and reading list

Codex Alimentarius Commission, Joint FAO/WHO, Food Standards Programme, 1989.

Eden, C., Setting the standard. *Food Rev.* 19, 17–18, 21, 1993.

Gardner, S., Food safety: an overview of international regulatory programs. *Eur. Food Law Rev.* 6, 123–149, 1995.

Hathaway, S.C., Risk assessment procedures used by the Codex Alimentarius Commission and its subsidiary and advisory bodies. *Food Control* 4, 189-201, 1993.

Truhaut, R., The concept of the acceptable daily intake: a historical overview, in: *Food Additives and Contaminants* 8, 2, 151–162, 1991.

World Health Organization, *Principles for the Safety Assessment of Food Additives and Contaminants in Food*, Environmental Health Criteria 70. Geneva, 1987.

World Health Organization, *Principles for the Toxicological Assessment of Pesticide Residues in Food*, Environmental Health Criteria 104. Geneva, 1990.

chapter twenty

Epidemiology in health risk assessment

A.E.M. de Hollander

20.1 Introduction: Why epidemiological data in health risk assessment?

The identification and quantification of human health risk associated with exposure to chemicals is a complex process in which a variety of disciplines are involved, such as toxicology, epidemiology, clinical medicine, chemical subdisciplines (analytical chemistry, organic chemistry, biochemistry), and biostatistics. All contribute, but none provide a complete picture. Among these disciplines epidemiology is becoming increasingly important. As will be pointed out in the next sections of this chapter, this is largely due to the growing scientific awareness that the relevance of results obtained in experimental animals to human health is limited. In Section 20.2, some important methodological limitations of epidemiology in studying environmental (including nutritional) risk factors are discussed. Section 20.3 indicates the prospective role of epidemiology in risk assessment and the way in which methodological limitations may be overcome.

20.1.1 A safer world for rats?

The information on toxicological risks from food contaminants and additives (both natural and man-made) is mainly derived from toxicological studies in animals. Compared with epidemiological studies, these studies have the advantage of an experimental design. All conditions are maintained constant except for the factor of interest: exposure to a certain substance. In toxicological studies, exposure and exposure conditions (such as housing, diet, and climate) can be carefully monitored and controlled. Histopathological and bio-chemical methods offer possibilities to study adverse responses with high sensitivity. However, toxicological research is not meant to make this world safer for rats and mice. It is also improper to deal with humans as if they were rats weighing 70 kg.

Translation of results obtained in experimental animals to human populations as a step in quantitative risk assessment requires three important assumptions:

– animals under laboratory conditions and human populations respond alike;
– the response to (high) experimental exposure is relevant to human health and may be properly translated to environmental exposure (including food intake) levels which are often orders of magnitude lower;
– (standard) experiments in animals reveal all responses to a substance which are potentially adverse to humans.

All three assumptions may be challenged. Consequently, they may give cause to substantial uncertainty in quantitative risk assessment.

The sensitivity of species (or strains or even individuals) to a toxic substance can differ dramatically. For instance the LD_{50} (see Section 8.9.1) of 2,3,7,8-tetrachlorodibenzo-*p*-dioxin (TCDD) differs several orders of magnitude from one species to another. In hamsters, it is almost 10,000 times higher than in guinea pigs. There are not only differences in intensity. TCDD-induced tumorigenicity may also involve different organs in different species.

Although TCDD is believed to be a non-genotoxic carcinogen in rodents, results of human studies do not give rise to great concern. Epidemiological research on Vietnam veterans exposed to the TCDD-tainted defoliant Agent Orange did not reveal any effect. Even among a population of highly exposed people living near the chemical plant that exploded in Seveso in Italy in 1976, the only irrefutable effect was chloracne (rash). Only recently (and not undisputed in the scientific press) an increase of 50% in the cancer risk was found in a large cohort of workers occupationally exposed to high levels of TCDD for long periods. It mainly involved soft tissue sarcomas, while in experimental animals liver tumors were predominant. The exposure levels were about 500 times higher than the exposure levels human populations are likely to experience. In a low-exposure group, no increased cancer risk was shown, although the exposure was still 90 times the average environmental level. These results suggest that humans are by far less sensitive to TCDD than laboratory rodents. However, there is no guarantee humans will be the least sensitive species if other toxic substances are involved.

Quantitative risk assessment based on animal experiments means the translation of results obtained in genetically homogeneous experimental animals under well-controlled laboratory conditions to a free-living, heterogeneous human population exposed to a wide variety of risk factors affecting the state of health. In addition to the fact that experimental animals are roughly equally sensitive to toxic effects of the substance they are exposed to, they do not smoke, and do not drink, do not have dangerous occupations or hobbies, and do not have unhealthy dietary or sexual habits which may obscure the effects of the substance under investigation.

In most experiments, the test animals are only exposed to one toxic substance at the same time, while humans are generally exposed to a variety of chemicals. Exposure to mixtures or combinations of different substances may have unpredicted (and unpredictable) health effects as a result of all sorts of interactions between the components. This has been an issue of concern among toxicologists for more than a decade now. However, there is still no satisfactory answer to the question how to deal with combined actions in health risk assessment.

Since the (statistical) sensitivity of animal experiments is limited by the number of animals in the exposure groups, toxicologists are often forced to use relatively high exposure levels to ensure the detection of potentially adverse effects. As Theophrastus Bombastus von Hohenheim, better known as Paracelsus, already stated five centuries ago: "Alle Dinge sind Gift und nichts ist ohne Gift. Dasz ein Ding kein Gift ist, macht allein die Dosis" (only the dose determines toxicity). Substances humans cannot live without, such as oxygen and water are toxic at doses lower than ten times the normal intake. Thus, one may query the significance of effects which are observed at "unphysiologically" high exposure levels. Illustrative for this dilemma is the recent discussion in the American scientific press in which the relevance of animal experiments for carcinogenicity is called into question by several prominent toxicologists. In order to avoid false negative results chemicals are tested in chronic animal studies at maximum levels which are tolerated without clear signs of toxicity. Some toxicologists argue that these chronic exposures almost by definition lead to severe disturbances of homeostases. Responses associated with tumor promotion, such as excessive cell proliferation to compensate for cytotoxicity and disturbance of hormonal balances, have been observed at these levels. The authors pointed out that it would only be surprising if those lifelong disturbances would not be expressed in altered tumor incidence rates. In view of the fact that almost half of the tested chemicals appeared to be carcinogenic, some authors suggest that carcinogenicity revealed by these studies may often be an artifact of the experimental design, which is of no relevance to most environmental exposures.

Another subject of scientific dispute is the way in which results of animal experiments should be translated to real-life exposure levels. In Chapter 18, several models for extrapolation to low dose levels have been described. All lack information on what happens at the low doses to which humans are actually exposed. For instance, estimates of bladder cancer risk from saccharin, made on the basis of data obtained in rats, varied by as much as six orders of magnitude, depending on the assumptions used to translate from high to low dose levels. Estimates ranged from 0.001 cancers per million exposed, using the multi-hit method, to 5200 cancers per million exposed, using the single-hit method. In contrast, a review of 13 case-control studies in humans led to the conclusion that there was no consistent association between saccharin intake and the incidence of bladder cancer.

Only in human studies it may be verified whether environmental (including dietary) exposure to so-called rodent carcinogens indeed increases the risk of cancer. When assessing human health risk based on animal studies, the question ought to be asked whether these studies will reveal all relevant responses. One has to consider the possibility that the toxicological methods are "blind" to more subtle responses, which may have great impact on public health in the long term.

The case of oral contraceptives is very illustrative. Before their introduction, toxicity experiments in rodents revealed that female sex hormones could induce breast tumors. Considering the underlying mechanisms and low-dose extrapolation, it was concluded that no such effects were to be expected in women using oral contraceptives. However, the most important side effect of "the pill," the disturbance of blood coagulation, has never been found in experimental animals. Other responses which may not be easily revealed in animal experiments are minor neuropsychological disorders, such as chronic headaches, concentration disturbances, forgetfulness, and depressiveness. These symptoms have been reported in painters exposed to organic solvents during long periods of their lives.

20.1.2 Risk communication

Epidemiology may contribute to a rational public and political awareness of the risks of daily life. Descriptive epidemiological studies may help health authorities to see the state of public health and its relationship with environmental problems in true perspective. With relatively simple statistical parameters of the health impact of serious diseases, one can inform the public and policy makers on the importance of certain risk factors. This may be useful in setting priorities for the funding of research, prevention, and control programs. For instance, one may rank diseases in terms of potential years of life lost and then conclude that cancer is by far the most serious threat to public health, followed by coronary heart disease, and traffic accidents (see Figure 20.1, based on Canadian health statistics). The diagram in Figure 20.2 shows the lost life expectancy for an individual, caused by several risk factors.

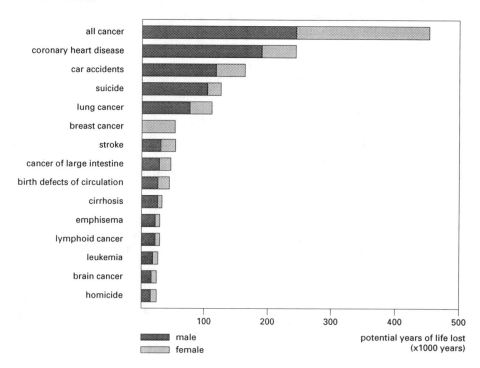

Figure 20.1   Health problems ranked by potential number of years of life lost.

A recent review of worldwide trends in age-related cancer mortality showed rapid increases in the prevalance of various types of cancer (e.g., of the brain, the central nervous system, breast, kidney). These could not be explained by increased accessibility of health care records, changes in disease registration, improvements of diagnostic technology or by life-style trends. The investigators suggested that these trends reflect an increase in environmental or occupational exposures to carcinogenic factors. Such descriptive studies may stress the need for more research and preventive measures to reduce exposure to carcinogenic agents. However, public health is a complicated subject that may be looked upon from many different angles. To find the right method to measure health impact is not easy. During the last decades public health science and management have focused on the

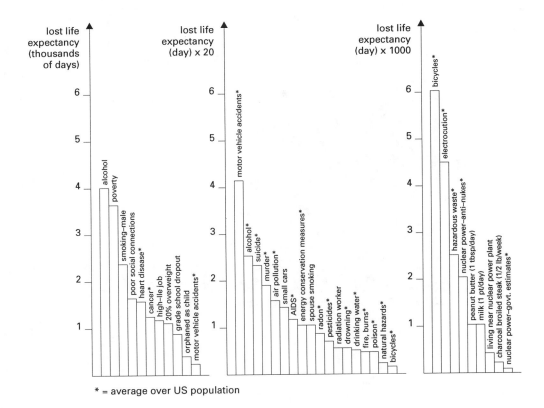

* = average over US population

Figure 20.2 Comparison of risks. Asterisk designates average risk spread over the total US population: others refer to risks of those exposed or participating. The ordinate scale is shown at the left. The heights of the bars are multiplied by 20 in the center section and by 1000 in the right section. The first bar in each of these reproduces the last bar in the previous section, showing the effect of scale change. (Source: Cohen, 1991.)

prevention of (early) death. Nowadays many have recognized the fact that there is more to life than dying, as the saying goes. The area of special attention for public health policy is shifting from prevention of fatal disease to improvement of quality of life by reducing the period of dependence and disability in elderly life. Prevention should aim at chronic morbidity as in the cases of rheumatoid arthritis, chronic obstructive pulmonary disease, diseases of the eyes and ears, diabetes, dementia, and other multi-factorial diseases of old age which cause severe disability. The diagram in Figure 20.3 shows the principles of reduction of morbidity.

The great challenge for public health care is to make the disease-curve move faster to the right than the death-curve in order to reduce the black area in the diagram that represents severe disability. It is the concept of the ideal car that disintegrates completely after exactly ten years of loyal service without any prior mechanical problems. Some authors argue that for the purpose of health risk assessment, attention should be focused on risk factors associated with chronic morbidity rather than on pursuing every carcinogen that shows itself in animal experiments or occupational epidemiology.

It should be noted that there is no scientific consensus yet on how one should measure public health. The above examples were given to show that public health is more than just counting bodies, or the sum of years of life lost.

The public's perception of the threats of daily life may sometimes be seriously biased. Often the public opinion is fixed on the effects of popular risk factors such as

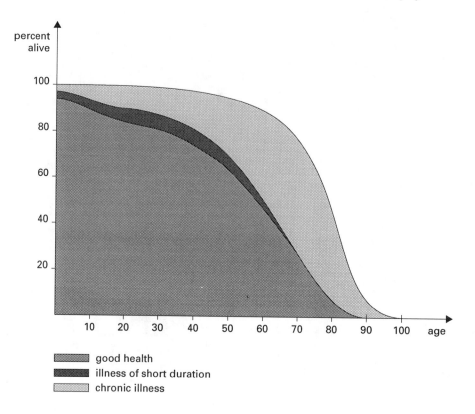

Figure 20.3   The burden of disease.

asbestos, TCDD, benzene, and radiation. However, the risks from these factors are, if verifiable, in fact often trivial compared to the dramatic health effects of factors such as traffic, smoking, dietary habits, occupational exposures, and poverty (wealthier is healthier).

Epidemiological data may give a good notion of the real risks of daily life by transfering information on health risk from the expert sphere to the public sphere in a way that appeals to lay-people. This can be achieved, for instance, by comparing the effects of different risk factors on public health. Although an oversimplification, the risk yardstick in Figure 20.4 might be a first step.

20.2   Limitations of epidemiology in risk assessment

In contrast to toxicology, epidemiology is an observational science. This is both its strength and its weakness. No extrapolation from one species to another is required. However, the study of free-living populations has important methodological limitations which have to be taken into account if the results are to be applied in health risk assessment.

20.2.1   Time interval between exposure and response

Epidemiological studies in which food components are tested for their adverse health effects are considered unethical in Western society. Furthermore, some health effects of major concern need a considerable amount of time to manifest themselves. In the case of chronic diseases like cancer, the induction period may span almost one generation. As in the case of exposure to industrial asbestos, it takes several decades until carcinogenicity

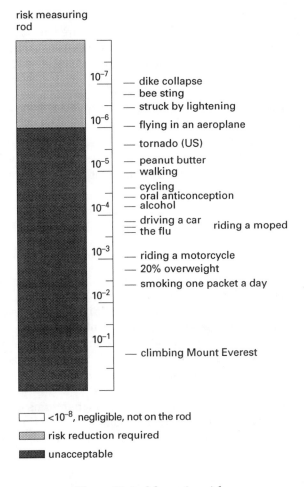

risk measuring rod

- 10^{-7}
  - — dike collapse
  - — bee sting
  - — struck by lightening
- 10^{-6}
  - — flying in an aeroplane
  - — tornado (US)
- 10^{-5}
  - — peanut butter
  - — walking
  - — cycling
  - — oral anticonception
- 10^{-4}
  - — alcohol
  - — driving a car   riding a moped
  - — the flu
- 10^{-3}
  - — riding a motorcycle
  - — 20% overweight
  - — smoking one packet a day
- 10^{-2}
- 10^{-1}
  - — climbing Mount Everest

⬜ $<10^{-8}$, negligible, not on the rod

🔲 risk reduction required

⬛ unacceptable

Figure 20.4 Measuring risks.

can be observed. Meanwhile, many people may already have suffered from hazardous exposures. Thus, for the screening of new chemicals or new applications, epidemiology is a useless tool and one has to rely on toxicological research.

On the other hand, epidemiology may provide possibilities to detect health risks after a certain food additive or contaminated food has found its way to the market. This so-called post-marketing surveillance is sometimes applied when new drugs are introduced.

20.2.2 Low sensitivity

Since World War II numerous associations between potential causal factors and major diseases have been found and confirmed. These were strong associations with large relative risks (ratio risk in persons exposed to the factor of interest/risk in the unexposed). Examples are smoking and lung cancer, excercise or dietary habits and coronary heart disease, urban pollution and lung disease, industrial asbestos exposure and mesothelioma. The results hardly needed statistical analysis to be convincing. Most of the evident risk factors have been revealed and confirmed. The associations epidemiologists are looking for nowadays are likely to be more subtle. Owing to computer programs, the design and statistical analysis of studies have become more sophisticated. Nevertheless, the results often fail to be convincing.

One of the problems epidemiologists have to deal with is the low sensitivity of their methods. Slight increases in the risk of exposed compared to unexposed will remain unnoticed, because of chance, lack of heterogeneity of exposure, inadequate parameters of exposure and/or response, insufficient time elapsed since onset of exposure, or bias and confounding.

If the prevalence of a disease is high, a slight increase in relative risk has great public-health impact. If approximately 40% of the male population dies of cardiovascular disease, even a slight increase in the relative risk caused by a certain risk factor leads to a significant increase in the number of deaths per year.

In the next subsections, the most important aspects of the sensitivity of epidemiological methods are discussed. For a more detailed description of epidemiological studies, the reader is referred to Chapter 15 of Part 2.

Intermezzo

Example

In the summer of 1972, a chemical factory accidentally distributed ten 25 kg bags of polybrominated biphenyls (PBBs, a flame retardant with carcinogenic potency) to dairy farms throughout Michigan for use as cattle feed additive. The chemicals caused widespread deaths among cattle, calf loss, and decreased milk production, but the mistake was not discovered until 9 month later. Since PBBs persist in tissue, it was possible to measure levels of the contaminants in cattle and dairy food products. This monitoring led to widespread quarantining and destruction of animal food products in which PBB levels exceeded certain limits.

Reports on various diseases in families on contaminated farms led to epidemiological studies designed to detect possible PBB-related health effects. Cases of PBB-syndrome consisted largely of subjectively reported symptoms, such as headaches, joint pain, loss of appetite, and skin rash. Epidemiological studies failed to show associations between PBB serum levels and reported symptoms in PBB exposed family members. Low levels of PBB were demonstrated throughout the whole population of Michigan, but not in other neighboring states or in Canada. The chemicals were found to concentrate in mother's milk. Continuous surveillance of health effects was carried out in exposed families. No evidence of increased disease incidence, including cancer, has yet been revealed. However, it may take decades before the carcinogenic potency of a substance can be observed in epidemiological research.

20.2.3 Chance

In epidemiological studies, the effects of differences in exposure to a risk factor are compared. The sensitivity of this type of study is limited, because groups may differ in the outcome of response variables by chance alone. The role of chance can be illustrated by calculating the confidence intervals for an observed relative risk equal to 1 as estimated with the standardized mortality ratio of a cohort study or the odds ratio from a case-control study (Tables 20.1 and 20.2). Table 20.1 shows how the relative risk estimate improves with an increasing number of cases. Table 20.2 shows that the reliability of the relative risk estimate depends on the number of cases as well as on the proportion of controls exposed. Given an estimated relative risk of 1, with a certain confidence interval (95% is normally used), these intervals give the range of results which may be observed by chance alone.

Even if there is no effect of exposure, an industrial medium-sized cohort study may show an increased risk of 20 to 30% by chance alone. If one compares cases and controls

Table 20.1 95%-Confidence intervals for an estimated standardized mortality ratio of 1, based on the results of a cohort study

Number of responders in the exposed group	Comparison with an unexposed group of the same size
100	0.750–1.334
200	0.816–1.225
1000	0.915–1.094
5000	0.959–1.042

Source: Krewski et al., 1989.

Table 20.2 95%-Confidence intervals for an estimated odds ratio of 1, based on the results of a case-control study (unmatched analysis, equal number of cases and controls)

Percentage of controls exposed	Number of cases			
	100	200	1000	5000
5%	0.222–4.498	0.354–2.822	0.656–1.524	0.832–1.202
25%	0.502–1.992	0.620–1.612	0.812–1.231	0.913–1.096
45%	0.551–1.815	0.661–1.512	0.835–1.197	0.923–1.083

Source: Krewski et al., 1989.

Table 20.3 Total sample size required to detect an increased risk in epidemiological cohort studies with same-size exposure groups

Relative risk (%)	Risk in unexposed cohort (%)		
	1	10	20
1.05	1,032,609	114,096	64,497
1.10	266,433	29,569	16,796
1.25	46,546	5,233	3,014
1.50	13,231	1,519	894
2.0	4,083	487	298
4.0	789	107	73
5.0	538	77	55

Source: Krewski et al., 1989.

for exposure to a risk factor (in so-called case-control studies), the role of chance is even more prominent. In a case-control study with as many as 200 cases, one might find an increased risk ranging from 50 to 200% even if there is no real association between risk factor and response. This variation due to chance decreases with the size of the population. An important implication of this phenomenon is that negative results of epidemiological studies offer no guarantee that effects are actually absent.

The effect of chance implies that very large cohorts are necessary for the detection of low risks. Table 20.3 shows that cohort studies involving several thousands of subjects are required to detect relative risks lower than 2 if the disease concerned is rare in the unexposed cohort.

20.2.4 *Exposure*

To identify and quantify risk associated with environmental exposure, one needs to design a natural experiment in which identifiable populations differ distinctly in exposure level. This is often difficult. The case of smoking and lung cancer was clear-cut, owing to the fact that there was an extreme heterogeneity in exposure levels between populations of smokers and non-smokers. However, if everyone in the Western world would have smoked a pack a day no epidemiological study would have been able to identify smoking as the predominant cause of lung cancer. Lung cancer would probably have been believed to be genotypical. If everyone is exposed to the predominant risk factor, the distribution of lung cancer is largely determined by genetic predisposition for the disease. In general, one might argue that the wider a particular environmental risk factor is spread, the less it explains the distribution of a disease. If a cause is ubiquitous, no epidemiological method will be able to detect it, because all methods rely on differential exposure to a risk factor.

This implies that epidemiological studies on health risk from food components will become increasingly difficult since there is a tendency towards buying mass-produced foods from supermarkets. In addition, increasing mobility eliminates the effects of differences in local environment. Health effects of certain food contaminants can only be studied if populations distinctly differ in exposure. Examples of such natural experiments concern populations living in areas where the water or soil is contaminated with nitrate, arsenic, or cadmium, resulting in a relatively high intake of these contaminants with food or drinking water. In epidemiological research one can compare the incidence of a particular disease in a certain area with national statistics or with the incidence in uncontaminated regions. Comparing people who use artificial sweeteners or drink coffee with people who do not, is another example of studying distinct differences in exposure. If the exposure conditions differ little from each other, the sensitivity of epidemiological methods is often too low.

It is generally recognized that exposure of defined populations is often hard to quantify. Exposure assessments based on questionnaires, as in the case of dietary patterns, are inherently inaccurate. Exposure levels in ambient and indoor air, food, or drinking water are often poorly documented, if monitored at all. If the disease under investigation follows chronic cumulative exposure, as in the case of cancer or cardiovascular disease, exposure patterns may be determined by factors, such as time-activity patterns, housing, occupation, and dietary habits. These may change drastically in the period between the start of exposure and the onset of the disease. In general, continuous monitoring of exposure is not feasible.

Often, the exposure variable is a poor surrogate, because it does not properly reflect the actual exposure. In studies on the association between passive smoking and lung cancer, non-smokers were classified as belonging to the group of passive smokers, depending on the smoking habits of their spouses. This classification may reflect the exposure to environmental tobacco smoke inaccurately, because exposure takes place in a variety of environments, at work, during public transport, or in all sorts of public places such as pubs and theaters. The relevant latency period for lung cancer ranges over several decades. During such a period, changes may occur in smoking habits, working environment or even in the spouses involved.

Intermezzo

Example

In the city of Antofagasta, Chile, the water supply contained high concentrations of arsenic during the period between 1958 and 1970. In this desert area, the water has to be brought from the Andes, where arsenic is found in naturally high concentrations, over a distance of 300 km. Copper mining was another probable source of arsenic contamination. In the early 1960s, a large number of citizens (children in particular) were found to have an abnormal skin pigmentation. Clinical investigations revealed that this pigmentation was accompanied by other symptoms such as chronic cough, chronic diarrhea, abdominal pain, and blood vessel disorders.

Using arsenic concentrations in hair and nails as exposure markers, a clear association was shown between arsenic exposure and intoxication symptoms. Further, a significant higher prevalence of abnormal pigmentation was found in comparison with a control population.

In retrospective studies, epidemiologists have to rely on the memory of participants to make an assessment of exposures. Unwittingly, this information can be seriously flawed. In case-control studies, a well-known type of information bias is caused by the fact that cases (people with the disease) or their relatives tend to overreport their exposure to a risk factor compared to controls, especially if an association between the risk factor and the disease concerned has been suggested in the media. This may then become a self-fulfilling prophecy.

In epidemiology, measures or "surrogates" of exposure are often of a qualitative nature. At best, they discern the less exposed from those who are more exposed. Sometimes, several groups differing in exposure can be defined depending on the quality of the exposure data. Misclassification of exposed as unexposed or the other way round tends to dilute the association between exposure and disease, as it diminishes the differences in exposure between the groups to be compared.

This can be illustrated by the results of three large cohort studies (including about 90,000 workers in rubber or petrochemical industries) on the relationship between benzene exposure and leukemia. With relatively high statistical power due to the large sizes of the cohorts, all three studies failed to show any effect of long-term benzene exposure. However, further analysis of the job characteristics revealed that many workers classified as exposed, had been exposed to negligible concentrations in the past. If subpopulations of workers with the highest exposures were defined and compared with unexposed, an effect became apparent.

Quantitative health risk assessments are based on quantitative relationships between exposure and response. This requires quantitative exposure data. Often such data are unreliable or lacking, and consequently have to be estimated by "educated guessing." This is very important, because apart from the extrapolation model, a quantitative risk assessment is as reliable as the exposure data. The latest episode of the continuing debate in the US on the occupational exposure standard for benzene is illustrative. This standard is based on a quantitative evaluation of an increased leukemia risk that was found for cohorts of workers who where followed since the 1950s and '60s and for which historical exposure data were provided by chemical industries. Recently, it was revealed that the actual exposure levels in the past had been much higher. As it appeared, the information on exposure levels provided by the industry had been carefully brought in accordance with occupational exposure standards which applied in the '60s. Because the exposure levels used in the quantitative risk extrapolation were too low, the true risk may have been overestimated. Consequently, the discussion on adjustment of the current occupational exposure standard has started again.

Intermezzo

Example

In April of 1956 a 6-year old girl was taken to a hospital in Minamata, Japan, with gait and speech defects as well as delirium. Many cases followed in all age groups. After several years, it was found that methylmercury was the cause of the epidemic of neurological disorders. Mercury derivatives from various small industries had accumulated in sediment and fish in Minamata Bay. Several hundreds of families of fishermen incurred injury. Among the exposed families there was at least a tenfold increase in birth defects, in particular severe cerebral palsy.

20.2.5 Response

Most diseases can be looked upon as multi-causal. A variety of factors contribute to the incidence. When comparing the outcomes of responses in different exposure groups, epidemiologists try to make adjustments for risk factors such as age, sex, socioeconomic status, smoking, and dietary habits. In practice, this implies that exposure groups are selected in such a way that these factors are equally distributed. If known risk factors are equally distributed over the exposure groups, comparison may reveal an association between exposure to the risk factor and the response concerned. The more a risk factor contributes to the development of a disease, the more epidemiological studies are likely to identify it and be able to quantify its influence.

Intermezzo

Example

In Iraq an outbreak of intoxications due to the consumption of fungicide-treated grain occurred in 1971–72. Of a total of 6530 hospital admissions, 459 patients died. The symptoms were typical for organic mercury intoxication. Analytical data showed a definite association between the consumption of bread baked from fungicide-treated grain and the mercury concentrations in hair and blood.

Examples of predominant risk factors are smoking (historical obduction data suggest that lung cancers were relatively rare before the "smoking era" started), asbestos (almost 80% of mesothelioma incidence can be traced back to occupational asbestos exposure in the past), and vinyl chloride (only two cases of angiosarcoma, a blood vessel tumor of the liver, in a cohort of workers alerted a medical company doctor to suspect the PVC-monomer). Toxic effects following the intake of food are often associated with accidental contamination. In the history of epidemiology there are many examples of this kind. In 1956 the accumulation of methylmercury in fish from Minamata Bay in Japan caused severe neurological disorders in many people as well as neurological birth defects (see also Section 20.2.4). In 1981 contaminated cooking oil caused an epidemic of acute lung disease in Madrid, Spain, with more than 20,000 cases over a period of three months. The contaminant caused severe immunotoxic effects. Toxic effects are often caused by components of vegetable origin. In India many people suffer from paralysis of the lower limbs, caused by the consumption of chick peas. In Africa a syndrome of severe neuropathological symptoms is caused by chronic hydrocyanic acid intoxication as a result of the poor man's diet consisting only of cassava. However, this kind of one-to-one associations between risk factors and diseases (disease specificity) is rare. In most cases, the risk factor under investigation is one of several, which reduces the sensitivity of epidemiological studies substantially.

If an association between the risk factor under investigation and another risk factor is found, one has to take into account the possibility of confounding. For example, comparison of an urbanized population with a rural population will show that the lung cancer incidence rate is higher in the first group. Thus, one might conclude that exposure to the urban ambient air induces lung cancer. However, the difference in lung cancer incidence may well be explained by differences in smoking habits and occupation, two major risk factors for lung cancer associated with urban life.

Inaccurate response parameters may also dilute the association between a risk factor and a disease and thus decrease the sensitivity of epidemiological methods. This can be best illustrated by studies on the association between radiation exposure and leukemia. A substantial risk increase could have escaped observation in epidemiological studies, if all types of leukemia had been combined. Only by studying the various types of leukemia individually, the association between exposure and acute nonlymphatic leukemia was revealed.

Intermezzo

Example

In 1955 the first cases of an "epidemic" disease causing severe bone pain and multiple bone fractures were reported by a local general practitioner in a rural area in Japan. The disease, referred to by the Japanese as Itai-Itai disease ("ouch-ouch"), was clinically diagnosed as osteomalacia in most cases. Epidemiological studies showed that the disease occurred throughout the region. Only when it was found that the disease was associated with tubular kidney damage, a link with cadmium poisoning could be demonstrated. Local food and drinking water appeared to contain high levels of cadmium.

20.3 Prospectives

The earlier mentioned limitations imply that the role of epidemiology in quantitative assessment of toxicological risks from food intake is often relatively modest:

- epidemiological studies are only feasible after human exposure has taken place and sufficient time has elapsed for disorders to develop;
- it may not always be possible to define exposure groups that differ substantially and consistently in their exposure to the risk factor of interest;
- the sensitivity of the available methods is often low; small health risks may not be noticed due to chance, lack of heterogeneity of exposure, inaccurate parameters of exposure and/or response, bias or confounding;
- negative results, even from impeccable studies, cannot prove the absence of an effect.

In spite of these limitations many investigators claim that the advantages of epidemiology are not yet fully exploited in toxicological risk assessment. In this section, several aspects of optimizing the application of epidemiological data in risk assessment will be discussed.

20.3.1 Toxicology and epidemiology as complementary disciplines

In identifying, quantifying, and verifying human health risk, toxicology and epidemiology are complementary. Both disciplines have their limitations. To put it bluntly: animal studies may only predict what will happen to other individuals of the same species under the same circumstances, while epidemiology often tries to compare incomparable free-living human populations. On the other hand, using the biblical metaphor: 'the lame leading the blind,' together they may bring us where we want to be. If toxicological and epidemiological data are evaluated in coherence, most of the individual shortcomings may be overcome. Animal studies can provide a first screening of hazardous chemicals used in food production, while the existence of toxicological risks at realistic exposure levels is most adequately studied in humans. Further, explorative epidemiology may reveal unknown associations between environmental factors and diseases, but only toxicological research can elucidate the underlying mechanisms.

20.3.2 Analysis of all available data, meta-analysis, and publication bias

Considering the limitations of the disciplines involved, human health risk assessment should comprise all available data, human and animal, positive and negative results. As far as epidemiological data are concerned, especially the validity of the study design should be carefully scrutinized. A number of important questions have to be addressed. What is the validity of measurements of exposure and response variables? How can the pitfalls of epidemiology, i.e., confounding, selection, and information bias be circumvented? Did the study generate a hypothesis by computerized analysis (some investigators use the metaphor "torturing") of existing data sets or did it test an a priori hypothesis by an appropriate study design?

Lack of statistical power can sometimes be overcome by combining (pooling) data from different studies through so-called meta-analysis. While in primary epidemiological research, populations of exposed and unexposed are the subjects, in meta-analysis the results of these primary studies function as such. In meta-analysis, weight can be assigned to the results of the individual studies, according to the reliability or the sample size of the primary studies. Meta-analysis offers the opportunity to study how variables affect study results. Such variables may be date of publication, exposure levels, study design, definition of exposure, definition of disease, methods of statistical analysis, and geographic location of studies.

In analyzing the results of a number of studies, one methodological pitfall of great concern remains: publication bias or bias to positive results. Editors of scientific journals are more likely to publish results indicating positive associations, even if they are only weak and from a methodological point of view inadequate, than results of impeccable studies indicating no association at all. (The book on train accidents which never happened on railroads which were never built, was never written.) Authors may (subconsciously) be inclined to refrain from completing work that does not offer the promise of positive findings, publication, and peer recognition. The consequence may be that a literature study comprising all available animal and human data still gives a distorted picture of human health risk.

20.3.3 Causality

An old truth is that by itself, epidemiology will never be able to prove the causality of an association between a risk factor and a disease. The likelihood of causality is often evaluated using the following criteria:

- an appropriate *time sequence* of exposure and onset of response, which includes the time interval necessary to induce the response;
- *consistency* of an association in various populations, under various circumstances, observed in studies with different designs;
- *strength* of an association, expressed in terms of relative risk estimates. The higher the relative risk, the more unlikely an association based on bias or confounding;

- *dose–response relationship*. Higher exposures should lead to increases in response intensity or frequency (though less simple dose-response relationships may be found);
- *biological plausibility*. Agreement with knowledge on underlying mechanisms.

The latter criterion takes us back to toxicology. Epidemiology may validate the relevance of toxicological data for human health. Conversely, toxicological data, combined with clinical findings, may be necessary to validate associations suggested by explorative epidemiology.

20.3.4 Future

In the coming years, the role of epidemiology in health risk assessment is likely to become more prominent. Several developments will contribute to this.

- The reliability of exposure assessment will increase as a result of the development of appropriate biomarkers for exposure as well as the improvement of exposure monitoring and modeling. An example of exposure assessment using biomarkers is the monitoring of DNA-adducts in white blood cells of humans exposed to genotoxic substances. The amount of adducts found in human DNA samples not only reflects exposure to certain genotoxic substances, but also overloading of certain defense mechanisms, such as DNA-repair.
- Biomedical research provides a growing number of biological variables (biomarkers) that are early (accurate) indicators of the development of diseases before the symptoms appear. Examples of early indicators of diseases are changes in immunoglobulin composition, (liver) enzyme activity and lung function parameters. Spectacular developments in molecular biology provide tools to relate specific mutagenicity of substances to activation of oncogenes in human tumor cells. The use of such biological response indicators in epidemiology may improve the specificity of associations between exposure and response' and may enable early interventions.
- Improvement of disease monitoring will offer possibilities to study the distribution of diseases in time and space more adequately.
- Cooperation between epidemiologists and toxicologists may lead to more sensitive methods to analyze and extrapolate available data on human health risks.

Reference and reading list

Bertazzi, P.A., C. Zocchetti, A.C. Pesatori, S. Guercilena, M. Sanarico, L. Radice, Ten-Year Mortality Study of the Population Involved in the Seveso Incident in 1976, in: *Am. J. Epidemiol.* 129, 6, 1187–1200, 1989.

Buffler, P.A., The evaluation of negative epidemiologic studies: the importance of all available evidence in risk characterization, in: *Regul. Toxicol. Pharmacol.* 9, 34–43, 1989.

Checkoway, H., N.E. Pearce, D.J. Crawford-Brown, *Research methods in occupational epidemiology*. New York/Oxford, Oxford University Press, 1989.

Cohen, B.L., Catalog of risk extended and updated, in: *Health Physics* 61, 317–335, 1991.

Doll, R., Health and the environment in the 1990s. *Am. J. Publ. Health* 82, 933–941, 1992.

Feinstein, A.R., Scientific standards in epidemiologic studies of the menace of daily life, in: *Science* 242, 1257–1263, 1988.

Fries, J.F., L.W. Green, S. Levine, Health promotion and the compression of morbidity, in: *Lancet*, 1989, 481–483.

Hattis, D. and K. Silver, Human interindividual variability: a major source of uncertainty in assessing risk for noncancer health effects. *Risk Analysis* 14, 421–432, 1994.

Hertz, Picciototto, Epidemiology and quantitative risk assessment: a bridge from science to policy. *Am. J. Public Health* 85, 484–491, 1995.

Kaldor, J., N. Day, The use of epidemiological data for the assessment of human cancer risk, in: Hoel, D.G., R.A. Merrill, F.P. Perera, (Ed.), *Risk Quantitation and regulatory policy*. Banbury Report 19, Cold Spring Harbor Laboratorium, 1985.

Krewski, D., M.J. Goddard, D. Murdoch, Statistical considerations in the interpretation of negative carcinogenicity data, in: *Regul. Toxicol. Pharmacol.* 9, 5–22, 1989.

Krewski, D., D. Wigle, D.B. Calyson, G.R. Howe, Role of epidemiology in health risk assessment, in: *Cancer Res.* 120, 1–24, 1990.

Lilienfeld, A.M., D.E. Lilienfeld, *Foundations of Epidemiology*. New York/Oxford, Oxford University Press, 1980.

Ozonoff, D., Conceptions and misconceptions about human health impact analysis. *Environ. Impact Assess. Rev.* 14, 499-415, 1994.

Rinsky, R.A., RE: Benzene and leukemia: a review of the literature and a risk assessment, in: *Am. J. Epidemiol.* 129, 5, 1084–1086, 1989.

Rose, G., Environmental factors and disease: the man made environment, in: *Brit. Med. J.* 294, 963–965, 1987.

Rothman, K.J., *Modern Epidemiology*. Boston/Toronto, Little, Brown and Company, 1986.

Rothman, K.J., Methodologic frontieres in environmental epidemiology. *Environ. Health Perspect.* 101, 19–21, 1993.

Skene, S.A., I.C. Dewhurst, M. Greenberg, Polychlorinated Dibenzo-*p*-dioxins and Polychlorinated Dibenzofurans: The Risks to Human Health. A Review, in: *Hum. Toxicol.* 8, 3, 173–203, 1989.

Taubes, G., Epidemiology faces its limits. *Science* 269, 164–169, 1995.

VeFlorey, C. du, Weak associations in epidemiological research: Some examples and their interpretation, in: *Int. J. Epidemiol.* 17, 950–954, 1988.

World Health Organization (WHO), *Guidelines on studies in environmental epidemiology*. Geneva, Environmental Health Criteria 27, 1983.

chapter twenty-one

Risk assessment, risk evaluation, and risk management

C.J. Henry

21.1 Introduction

Most human activities carry some degree of risk. Many risks are known to a relatively high degree of accuracy, because data have been collected on their historical occurrence. For example, the number of deaths caused by motor vehicle accidents in 1 year is divided by the total number of people at risk (e.g., the entire US population) for an actual individual risk of 1/4500 from dying of such an accident. Based on a 70-year lifetime, people in the US have a 1/65 probability of dying in a car accident over an entire lifetime.

The risks associated with many other activities, including exposure to food-borne microbes or various substances found in or associated with food, cannot be readily assessed or quantified. Considerable historical data exist on the risks of some types of chemical or microbial exposure (e.g., the annual risks of death from intentional overdoses or accidental exposures to drugs, pesticides, and industrial chemicals, or the annual risk of food-borne disease). Such data, however, are usually restricted to those situations in which a single high exposure resulted in an immediately observable form of disease or injury, thus leaving little doubt about the cause. Assessment of the risks of exposures to substances or microbes that do not cause immediately observable forms of injury or disease (or only minor forms such as transient eye or skin irritation) is far more complex.

This type of risk assessment is of great concern to scientists and those involved in regulation and, just as importantly, to the general public. For industries associated with food and food production, as well as any type of chemical, food additive, or drug, the health risk assessment associated with the substance is critically important.

Some confusion exists regarding the terms used in microbial and chemical risk assessment since many are commonly used but have slightly different meanings. The term risk, and terms associated with it, such as safety, form an example of the difficulties that can arise when scientists and technical experts use common words in a context different from everyday language. For example, safety — the probability that harm will not occur under specified conditions — has been described as the converse of risk. This probabilistic statement clearly differs from the common definition associated with safe, which suggests "free from harm or risk." In addition, slightly different terms are used in the US, Europe, Japan, and the global scientific community for similar topics or procedures.

Understanding the hazards associated with food is also complex because food safety and food safety assessment rely on two different scientific disciplines: one concerned with assessing the microbiological safety and the other with assessing the chemical (or toxicological) safety of food. In the first case, the microbiological hazards and risks associated with preparation and storage of foods in all links of the food chain must be controlled and evaluated. In the second case, the toxicological risks from substances present in food must be assessed and evaluated. The terms used to describe these two areas are not rigidly defined. The microbiological hazards may include toxic substances of microbial origin. The term chemical risk assessment generally refers to assessment of synthetic substances to which humans may be exposed. The term toxicological risk assessment emphasizes the potential toxic effects of substances, and is also referred to as biological risk assessment, which emphasizes the biological effects of substances. In this chapter the term chemical risk assessment will be used.

21.2 Elements of hazard and risk

Risk is the probability of injury, disease, or death under specific circumstances. A hazard is a set of circumstances that may cause adverse effects, and the likelihood that a hazard will cause such effects is the risk associated with it. Risk may be expressed in quantitative terms, taking values from zero (certainty that harm will not occur) to one (certainty that it will). In many cases, risk can only be described qualitatively, as, for example, high, low, or minimal.

For example, a speeding car constitutes a hazard. The faster the car is driven, the more cars or people in the vicinity of the speeding car, the higher the risk that someone will be injured or killed. All hazards do not pose the same risks. The circumstances must be such that there is a likelihood of injury, harm, or death.

21.3 Microbial risks associated with food

Among the classes of hazards associated with food, microbial and viral contamination of the food supply are among the most important. Table 21.1 provides a list of microorganisms associated with food contamination and illness, and their principal food sources. For the most part, the sources of contamination are well-known microorganisms found in poultry, eggs, dairy products, and meat, and associated with improper handling. If these microbiological hazards are not controlled, this may result in a shorter shelf life of products, spoilage, and food-borne illnesses. Although the yearly number of reported illnesses

Table 21.1 Microorganisms and their principal food sources associated
with food contamination and illness

Microorganism	Food source
Salmonella	Raw meat and poultry, raw milk, eggs
Clostridium perfringens	Meats, poultry, dried foods, herbs, spices, vegetables
Staphylococcus aureus	Cold foods (handled during preparation), dairy products, especially if prepared from raw milk
Bacillus cereus and other Bacillus spp	Cereals, dried foods, dairy products, meat and meat products, herbs, spices
Escherichia coli	Many raw foods
Vibrio parahaemolyticus	Raw and cooked fish, shellfish, and other seafoods
Yersinia enterocolitica	Raw meat and poultry, meat products, milk and milk products, vegetables
Campylobacter jejune	Raw poultry, meat, raw or inadequately heat treated milk, untreated water
Listeria monocytogenes	Meat, poultry, dairy products, vegetables. shellfish
Viruses	Raw shellfish, cold foods prepared by infected food handlers

Source: Roberts, 1990. With permission.

related to the consumption of contaminated food in the US is only a few thousand (because of their generally transient and innocuous character), estimates of the actual total number of cases vary between 20 and 40 million.

Traditional approaches to food safety, hygiene, protection, and sanitation have not made a significant impact in reducing reported food-borne diseases, even in developed countries. Inspection has been the major process in microbiological food safety programs. Inspection programs, however, have serious limitations, including the practice of observing only part of an operation and overlooking critical factors. Vague laws and the lack of giving priority to compliance also limit the effectiveness of food safety programs. Other approaches that have major limitations include inadequate or faulty microbiological testing and examination of workers or their tissue, urine, or blood specimens. A different approach — the Hazard Analysis and Critical Control Point (HACCP) system has been developed to attempt to make a significant impact on food-borne disease.

HACCP consists of a series of interrelated actions that should be taken to ensure the safety of all processed and prepared foods at critical points during production, storage, transport, processing, marketing, preparation, and service (see Table 21.2). The uses and applications of this system are discussed in *Microorganisms in Food* published in 1988 by the

Table 21.2 Hazard analysis and critical control point (HACCP) system

Determine hazards and assess their severity and risks
Identify critical control points
Institute control measures and establish criteria to ensure control
Monitor critical control points
Take action whenever monitoring results indicate criteria are not met
Verify that the system is functioning as planned

Source: Bryan, 1992. With permission.

International Commission on Microbiological Specifications for Food as well as in other international publications.

Although the public perceives that the toxicological risks from manufactured or synthetic chemicals (food additives, pesticides, etc.) are greater than microbiological risks, the scientific evidence does not support this perception.

In a nationwide survey in the US in 1990 entitled, "The Environment: Public Attitudes and Individual Behavior," a random sample of people were questioned about such environmental concerns as water pollution from industrial waste products, radiation from X-rays and microwave ovens, and pesticide residues in food eaten by humans. More than half of those surveyed indicated concern over pesticide residues in food. Of the food safety issues of concern to scientists, microbial contamination of the food supply has been identified as the most important food safety issue to affect public health in industrialized countries. This issue does not receive the public attention it deserves.

Estimation of the risk associated with microbiological contamination of foods suggests that the risk of morbidity, that is, the number of people who become ill, is in the order of 1 in 100 in a given year, while the risk of mortality, that is, the number of people who die directly or indirectly as a result of exposure to food-borne pathogens, is approximately 1 in 100,000 in a given year. Such risks should be compared with those associated with pesticide residues in food, which are in the order of 10^{-6} to 10^{-8}, and risks associated with naturally occurring toxic substances in foods, particularly carcinogens, which may be in the order of 10^{-3} or 10^{-4}.

21.4 Elements of chemical risk assessment

Risk assessment is the tool used to evaluate the safety of food and food additives. As noted by the Joint FAO/WHO (Food and Agricultural Organization of the World Health Organization) Expert Committee on Food Additives (JECFA) safety evaluation of food additives is a two-stage process. In the first stage relevant data are collected, including results of studies on experimental animals and, where possible, human observations, including epidemiology studies. In the second stage data are assessed to determine whether a substance is acceptable for its intended use as a food additive. This scientific process determines the possible adverse effects in humans resulting from exposure to a substance.

In 1983, the US National Academy of Science (NAS) issued a report entitled *Risk Assessment in the Federal Government: Managing the Process*. The report delineated research, risk assessment, and risk management, but emphasized the separation between the scientific exercise of risk assessment, and the policy exercise of risk management (Figure 21.1). The NAS report gave a formalized structure to the risk assessment process. It described four elements: *hazard identification* (see Section 21.4.1), *dose–response assessment* (see Section 21.4.2), *exposure assessment* (see Section 21.4.3), and *risk characterization* (see Section 21.4.4), with recommendations and examples given for the types of scientific information needed for each element.

Risk assessment was defined as the process of assessing the possible adverse health effects in humans resulting from exposure to substances or other potential hazards. This definition allows ordering of the data, identifying data gaps and uncertainties, assigning priorities, and determining research needs. Based on the information in the risk assessment, a regulatory agency can then develop regulatory options, evaluate the public health, economic, social, and political consequences of the regulatory options, and implement agency decisions and actions. These decisions and actions are the core of the risk management process. The four elements associated with risk assessment (see Figure 21.1) are briefly described in the following sections. The types of information used in health risk assessment are summarized in Table 21.3.

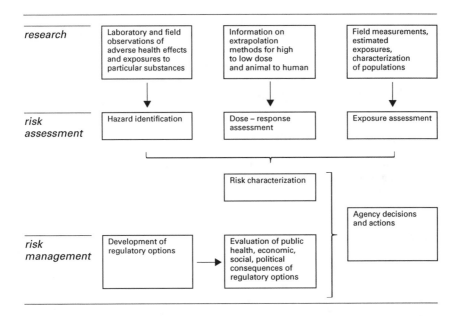

Figure 21.1 Elements of risk assessment and risk management. Reprinted with permission from Risk Assessment in the Federal Government. Copyright 1983, National Academy of Sciences. Courtesy of the National Academy Press, Washington, D.C.

21.4.1 Hazard identification

In the first step of hazard identification, data are gathered and evaluated on the types of health injury or disease that might be caused by a substance and on the conditions of exposure under which injury or disease is caused. It tackles the question: does the agent cause the adverse effect? Information on adverse effects can be found in a variety of studies, including animal toxicology or bioassay studies, *in vitro* studies, structure–activity relationships, epidemiology, human clinical studies, and human volunteer studies (Table 21.4).

Hazard identification may also involve characterization of the disposition of a substance within the body and the interactions it undergoes with the body and with organs, cells, or even cell components. Such data may be valuable in answering the ultimate question of whether the types of toxic effects known to be induced by a substance under experimental conditions or in one population group (children, elderly persons, etc.) are also likely to be induced in humans as a whole. Hazard identification can be considered as a qualitative risk assessment, i.e., it determines whether and to what degree it is scientifically correct to infer that toxic effects observed in one setting will also occur in other settings. For example, are substances found to be carcinogenic or teratogenic in experimental animals likely to have the same effects in humans?

Of the over 300 substances, industrial processes, and complex mixtures ranked by the International Agency for Research on Cancer (IARC), approximately 39 chemicals and chemical processes have been categorized as carcinogenic to humans, while for another 25 or so there is limited evidence of carcinogenicity to humans. All known human carcinogens ranked by IARC have proved to be capable of inducing cancers in some (but not all) species of experimental animals, with the exception of arsenic. It has been suggested that arsenic may not have been adequately tested as yet. Thus, it is prudent to use the results of cancer bioassays as an important element in hazard identification.

Table 21.3 Information used in health risk assessment

I		Hazard identification
	A	Human data
		Monitoring and surveillance
		Epidemiological studies
		Clinical studies
	B	Animal data
	C	In vitro data
	D	Structure-activity relationships
II		Hazard characterization (dose–response assessment)
	A	Human studies
		Epidemiological studies
		Clinical studies
	B	Animal studies
		Minimal effects determination
		Dose–response modeling
		Special issues, including interspecies conversion and high-to-low-dose extrapolation
	C	Pharmacokinetic studies based on physiology
III		Exposure characterization
	A	Demographic information
	B	Ecological analyses
	C	Monitoring and surveillance systems
		Animals
		Humans
	D	Biological monitoring of high-risk individuals
	E	Disposition and transport modeling (mathematical)
	F	Integrated exposure assessments
		Over time
		Over hazard (synergism)
IV		Risk determination
	A	Mathematical
		Unit and population risk estimates
		Threshold determination (e.g., safety factor approach, NOAEL)
	B	Formal decision analysis
	C	Inter-risk comparisons
	D	Qualitative panel reviews
	E	Quantitative informal scientific advice
	F	Risk-benefit analysis

Source: US Department of Health and Human Services, 1986.

21.4.2 Dose–response assessment

Dose–response assessment answers the question: what is the relationship between the dose of a substance and the incidence of adverse effects of it in animals and subsequently in humans? It requires describing the quantitative relationship between the amount of exposure to a substance and the extent of toxic injury or disease. Data are obtained from animal studies or from studies in exposed human populations. The latter are preferred, but not always available. There may be a different dose–response relationship for each endpoint of a substance if it induces different toxic effects (e.g., cancer, birth defects) or when the conditions of exposure are different (e.g., single compared with repeated exposures) (Table 21.5).

The biologically effective dose or the concentration of a toxic substance at the target organ can be determined from pharmacokinetic or toxicokinetic studies. Such studies can

Table 21.4 Elements to consider in hazard identification

Animal bioassay data
What are the most common data available?
Assume that results from animal experiments are applicable to humans.
Epidemiological data
What is the association between exposure to a substance and disease?
Risk is often low, number of people exposed is small, latency period is long, and exposures are mixed and multiple.
Structure–activity relationships
What chemicals are known to cause adverse health effects?
What substances are structurally related and/or have similar mechanisms of toxicity?

Table 21.5 Dose–response assessment

Define the relationship between dose and response.
In general, as the dose of many toxicants increases, toxicity increases; however, the manner in which toxicity increases varies.
It is customary to extrapolate from the high doses administered to animals to low doses experienced by humans.
The validity of these extrapolations must be considered, and the statistical and biological uncertainties defined.

be carried out using simulations of rates of absorption, distribution, biotransformation, accumulation, and excretion of administered agents. These physiologically based pharmacokinetic (PBPK) models use actual blood flow rates to organs and biochemical reaction rates for major physiological systems to estimate the delivered dose to a target tissue and to predict effects in humans qualitatively or even quantitatively from data on experimental animals (see Part 3, Chapter 18). Such predictions assume that the target tissues in different species have the same types of responses. Such models potentially reduce the need for relying on assumptions and uncertainty associated with the purely statistical models of dose–response relationships.

Dose–response evaluation generally requires two extrapolations: one for species differences in body size, lifespan, and basal metabolic rate, and one for differences in doses between animal experiments (high doses) and human studies (lower doses to which humans are likely to be exposed). The dose–response assessment should describe and justify the methods of extrapolation used to predict incidence and should characterize the statistical and biological uncertainties in these methods.

21.4.3 *Human exposure assessment*

Exposure assessment answers the question: what exposures are currently experienced (actual exposure of people) or anticipated under different conditions (potential exposure)? It requires determining the amount or concentration of a substance to which humans are exposed, the nature and size of the population exposed to the substance, and the duration of exposure (Table 21.6). Exposure assessment has been defined as determining what the actual contact is likely to be with a chemical or physical agent. The magnitude of exposure is the amount or concentration of the agent available at the human exchange boundaries (skin, lungs, gut) during a specified time. The evaluation could concern past or current exposures or exposures anticipated in the future. The exposure assessment should describe and justify the methods of measurement as well as characterize the assumptions and uncertainties associated with the exposed population.

Table 21.6 Exposure assessment

What is the concentration or quantity of chemical or substance to which humans are
   exposed?
Over how long a period of time does exposure occur?
Identify the populations exposed to the chemical or substance.
Use analytical measurements or estimates of concentrations, monitoring data, and
   mathematical models to estimate exposures.

Exposure is the critical connection between potentially harmful factors (substances, trace elements, microbes) and human health effects. It is usually difficult to measure, because substances are often present in very low concentrations, move unevenly through several environmental pathways, persist for varying periods of time, and are absorbed by humans in varying amounts depending on individual characteristics such as age, behavior, and nutrition. The types of information used in exposure assessment are listed in Tables 21.6 and 21.7, all of which are aimed at questions such as: how do people become exposed? How can information be obtained on whether they have been exposed? What happens after exposure? What are the implications for public policy or further research?

For substances such as aflatoxin, a toxin produced by strains of the fungus *Aspergillus flavus*, the human populations with potential exposure would be all individuals who consume aflatoxin-contaminated food, such as peanuts and peanut products, rice, cereal, and corn. Because not all individuals in the population of interest will be exposed to identical doses, the assessment should attempt to understand the distribution of the dose in the population.

21.4.4 Risk characterization

Risk characterization generally requires integration of the data and analysis of the first three elements of risk assessment to determine whether and to what extent humans will experience any of the various types of toxicity associated with exposure to a substance (Table 21.7).

Risk characterization deals with the question: what is the estimated incidence and severity of the adverse effect? It is in characterizing the risks that the major assumptions, scientific judgments, and uncertainties should be identified so that the risk estimate can be better understood. In Europe, the term risk estimation is used to characterize risk. Many uncertainties exist, and several approaches have been suggested to improve the characterization and understanding of risks. The US Environmental Protection Agency (US EPA) has considered guidelines for risk characterization. Such guidelines could include, for example, the use of sensitivity analysis on data sets employed in risk extrapolation, expressing variability in risks from a given extrapolation model, statistical levels used to project risks (e.g., median, 95th percentile, range), and ways to evaluate risks quantitatively when the qualitative weight of evidence is low.

21.4.4.1 Limitations and assumptions in risk assessment

Risk assessment is a process that provides a framework for evaluating information and presenting that information in a form useful to decision makers. Risk assessment, however, is limited by:

Table 21.7 Risk characterization

Combine and integrate exposure and dose–response assessment.
Estimate quantitatively some measure of risk.
Identify major assumptions, scientific judgments, and estimates of uncertainties.

- lack of data on substances and adverse health effects;
- uncertainty about the cause of disease;
- uncertainty in extrapolating human risk from animal data.

These limitations have resulted in applying a set of assumptions, or default positions. The assumptions and uncertainties that abound in the risk assessment process have generated much controversy. When there is uncertainty or a lack of data, public health officials tend to use assumptions that will not underestimate risk. Nine of the most generally agreed-upon assumptions in risk assessment have been emphasized, although many more have been identified:

1. In the absence of adequate human data, adverse effects in experimental animals are regarded as indicative of adverse effects in humans.
2. Dose–response models can be extrapolated outside the range of experimental observations to yield estimates or estimated upper bounds on low-dose risk.
3. Observed experimental results can be extrapolated from one species to the other.
4. No threshold doses (i.e., doses below which no adverse effects will occur) exist for carcinogenesis, although threshold levels may apply for other toxicological outcomes.
5. Average doses give a reasonable measure of exposure when dose rates are not constant over time.
6. In the absence of toxicokinetic data, the effective or target dose is assumed to be proportional to the administered dose.
7. The risks from multiple exposure and multiple sources of exposure to the same chemical are usually assumed to be additive.
8. Regardless of the route of exposure (dermal, oral, or inhalatory), 100% absorption across species is assumed in the absence of specific evidence to the contrary.
9. Results associated with a specific route of exposure are potentially relevant for other routes of exposure.

21.4.4.2 Cancer vs. non-cancer risk assessment

The risks from a substance cannot be ascertained with any degree of confidence unless dose–response relationships are quantified. In the US, the regulatory distinction between substances that cause cancer and those that do not has a major impact on the extrapolation methods used to characterize the dose–response curve in the non-observable low-dose range. All carcinogens, whether characterized as genotoxic or non-genotoxic, are considered by US regulatory agencies to pose a risk, no matter how finite, at all doses, while for non-carcinogens a threshold dose is assumed. As will be discussed in the following sections, this distinction results in a different characterization for these two classes of substances. Most European regulatory agencies, by contrast, distinguish between carcinogens characterized as genotoxic and non-genotoxic. For genotoxic carcinogens, it is assumed that there is no threshold. For non-genotoxic carcinogens, the existence of a threshold is assumed, provided the mechanism of carcinogenesis is understood. JECFA has indicated that carcinogens vary in the degree of risk they represent, and the intentional use of a food additive known to be a carcinogen should be considered only under very restricted circumstances.

21.4.4.3 Characterization of non-cancer risks

For noncarcinogens, a threshold dose or level of exposure is assumed below which no effect is observed (Table 21.8). The dose–response evaluation requires estimation of the threshold dose and determination of the no-observed-adverse-effect level (NOAEL) from observations in experimental animals or exposed people. The acceptable daily intake (ADI) (also called the tolerable daily intake, or TDI, see also Chapters 16 and 17) is estimated by dividing the NOAEL by a safety or uncertainty factor. Scientific guidelines and recommendations on the development and use of ADIs have been adopted by the Joint FAO/WHO Food Standards Program (Codex Alimentarius Committee on Food Additives), the FAO Committee on Pesticide Residues, and the WHO Expert Committee on Pesticide Residues. If the maximum daily intake of a non-carcinogenic substance is estimated to be lower than the ADI, then no risk is assumed for almost all members of the general population. Critical to this estimate, however, is the magnitude of the safety or uncertainty factor, which can range from 10 to 10,000 based on the data and on the policy of different regulatory organizations. For example, for non-nutrient substances, the Center for Food Safety and Applied Nutrition at the US Food and Drug Administration (US FDA) uses safety factors of between 100 and 2000, depending on the availability and type of data for analysis. The safety factor accounts for uncertainties concerning interspecies and intraspecies variation.

Where the WHO uses tolerable daily intake instead of accepted daily intake, the US EPA uses reference dose (RfD) to avoid the value judgment implicit in the calculation of an acceptable dose. The no-observed-adverse-effect level (NOAEL) and/or the lowest-observed-adverse-effect level (LOAEL) (Lowest found concentration or amount of a substance, which causes an adverse effect) are determined for each study and type of effect. To determine the RfD, uncertainty factors are applied to the NOAEL (or LOAEL if a NOAEL has not been established).

21.4.4.4 Characterization of cancer risks

From a scientific standpoint, substantial progress has been and is being made in understanding the mechanisms of toxicity, including carcinogenesis, and the causal relationships on which safety assessments are based. It is recognized to an increasing extent that "carcinogen" is difficult to define and that distinctions can be made among carcinogens based on the differing underlying mechanisms. Some substances initiate cancer directly and others are only involved secondarily in carcinogenesis. Thus for some carcinogens, as for non-carcinogens, there may be levels of exposure for which the possibility of harm to humans can be ruled out with reasonable certainty, a threshold dose determined, and for which instead a safety-factor or uncertainty-factor evaluation may be appropriate. A scheme for determining how chemical carcinogens could be identified is presented in Figure 21.2.

In the US, however, under Section 409 of the Food, Drug and Cosmetic Act, the Delaney Clause prohibits the use of food additives found to induce cancer in animals or humans.

Table 21.8 Characterization of risks

Non-carcinogens
> For food additives, the no-observed-adverse-effect level (NOAEL) is divided by a safety or uncertainty factor to estimate an acceptable daily intake (ADI).
> For systemic toxicants, US EPA developed the reference dose (RfD) approach, where the NOAEL is divided by an uncertainty factor and a modifying factor. Generally, the RfD is an estimate of a daily exposure to the human population (including sensitive subgroups) that is likely to be without an appreciable risk of harmful effects during a lifetime.

Genotoxic carcinogens
> Risk is estimated from the cumulative dose and/or the dose–response curve extrapolation.
> Mathematical models are used to extrapolate to low–dose response.
> A range of risks might be produced using different models and assumptions about dose–response curves, relative sensitivities of humans and animals, and for different estimates of human doses.

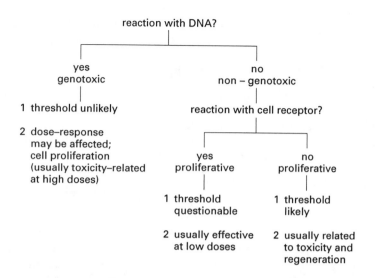

Figure 21.2 Proposed scheme for classification. Source: Cohen and Ellwein, 1990. With permission.

In contrast to the general safety standard for non-carcinogens, which recognizes the impossibility of assessing the complete absence of risk, the Delaney Clause has been interpreted as taking a zero risk approach to substances implicated as carcinogens. It should be stressed that this clause was enacted during a period when relatively few carcinogens had been identified and even fewer were believed to be present or associated with food. The result in the US was the assumption that there was no threshold dose for carcinogens and that oncogenic risk was a function of the cumulative lifetime dose (Table 21.8).

The non-nutritive sweetener saccharin has been shown to induce bladder cancer in rats. It is not metabolically activated when ingested by humans or animals, does not react with DNA, and is not mutagenic in short-term tests. Therefore, it is considered non-genotoxic. The lowest dose for an effect in rats is 2.5% sodium saccharin in the diet, while there is no effect with acid saccharin at 7.5% in the diet. The NOAEL in rats is 1.0% in the diet; there is no effect in the animals if the urine is acidified. At higher doses, increased cell proliferation of the adult rat bladder epithelium is observed. Urinary silicate precipitates and/or microcrystals are critical phenomena in the development of the lesions in rats. There is no evidence of any interaction of saccharin with cell receptors; relatively high doses are required for the effect in the bladder. Thus, as suggested in Figure 21.2, the proliferative response is probably related to toxicity and cellular regeneration, and a threshold dose is likely for this effect.

21.4.4.5 Characterization of risk using mathematical modeling

As most of the information on whether a substance is capable of inducing cancer is obtained from animal studies at high doses, statisticians developed mathematical models to extrapolate from these high-dose level studies to determine the risk at the low doses to which humans would be potentially exposed. This modeling process is used for quantitative risk assessment of chemical carcinogens and involves eight steps (Table 21.9). It has been termed an *empirical risk assessment model* or *default carcinogen risk assessment methodology*. Starting with carcinogenicity in the rodent bioassay, the procedures and calculations are outlined to reach the exposure analysis and risk-benefit analysis needed to determine exposure levels and cancer risks that society can tolerate.

Table 21.9 Empirical risk assessment model or default carcinogenic risk assessment methodology

---

1. Positive response in rodent bioassay
2. Appropriate dose measure; typically mg/kg body weight per day
3. Dose–response function selected for risk to rodents; typically linearized multi-stage model
4. Estimate of the variability of the dose–response function; typically 95% confidence interval
5. Linearized upper 95% bound on risk in the 1 in 10^{-6} region selected to determine quantitative value for risk assessment for rodents
6. Interspecies extrapolation to estimate risk for humans in dose-region of interest
7. Extrapolation from one exposure route to the other
8. Exposure analysis and risk benefit analysis to determine exposure levels and risks society can tolerate

---

The modeling and extrapolation processes employed in quantitative risk assessment are considered by many to be the most important sources of uncertainty in the risk assessment process. A quantitative estimate of the risk from a substance at a particular low-dose level is highly dependent on the mathematical form of the presumed dose–response relationship. Differences between models of at least three to five orders of magnitude are not uncommon. One difficulty with low-dose extrapolation is that some methods fit the data from animal experiments reasonably well, and it is impossible to distinguish their validity on the basis of a good fit. From a mathematical point of view, distinguishing between the models on the basis of their fit with experimental data would require an extremely large experiment which, from a practical point of view, is probably impossible. The different approaches used by the various regulatory agencies for assessing risk are, for example, reflected in the acceptable exposure levels set for 2,3,7,8-tetrachlorodibenzo-*p*-dioxin (TCDD) (Table 21.10). Not all agencies assume that no threshold exists for carcinogenesis. Although TCDD has proved to be extremely toxic to some rodents, its carcinogenic potential to humans has been the subject of considerable scientific controversy. TCDD has also been shown to induce a wide spectrum of adverse effects, not only carcinogenicity, in experimental animals. Use of the NOAEL and a general safety standard as well as a cancer dose-modeling approach yields a 2,000-fold range of allowable exposure levels by various regulatory agencies (Table 21.10).

21.5 Characterization and communication of risks

The importance of the risk characterization step is reflected by the distinction between cancer and non-cancer risks. The interpretation of the concept "one in a million risk of cancer" is often the basis of regulatory decisions in the US. Substances with cancer risks estimated to be greater than one in a million are generally not approved at the federal level.

In 1987 the commissioner of the FDA explained the concept of one in a million risk of cancer when he discussed the cancer risk from residues of methylene chloride residues, a solvent used to decaffeinate coffee: "The risk of one in a million is often misunderstood by the public and the media. It is not an actual risk, i.e., it is not expected that one out of every million people will get cancer if they drink decaffeinated coffee. Rather, it is a *mathematical* risk, based on scientific assumptions used in risk assessment. When the FDA uses the risk level of one in a million, it is confident that the risk to humans is virtually nonexistent."

Thus, how a risk is characterized and by whom can make a significant impact on how the risk assessment of a substance is interpreted and received. These aspects of perception and communication are discussed in Part 3, Chapter 22.

Table 21.10 Acceptable daily intakes of TCDD proposed or
adopted by various regulatory agencies

Agency	Dose-response extrapolation	Allowable intake (fg/kg/day)
US EPA	Linearized multi-stage	6.4
CDC[b]	Linearized multi-stage	28–1,428
OME[c]	Safety factor (100)	10,000
SINH[d]	Safety factor (250)	4,000
FEA[e]	Safety factor (100–1000)	1,000–10,000
FDA[f]	Safety factor (77)	13,000
NYSDH[g]	Safety factor (500)	2,000

[a] US Environmental Protection Agency
[b] US Centers for Disease Control
[c] Ontario Ministry of Environment, Canada
[d] State Institute of National Health, The Netherlands
[e] Federal Environmental Agency, Germany
[f] US Food and Drug Administration
[g] New York State Department of Health
Source: Paustenbach, 1989. With permission.

21.6 Risk Evaluation/Risk Management

Risk assessments have many uses, but a major one is to assist decision makers with the complex choices regarding the options in managing or reducing the potential human health risks associated with a substance or product. *Risk management* is defined in the US as the process of evaluating alternative regulatory actions and selecting among them. It has been characterized as an agency decision-making process that entails consideration of political, social, economic, and engineering information along with risk-related information to develop, analyze, and compare regulatory options and to select the appropriate regulatory response to a potential chronic health hazard. The European Union and WHO use the term *risk evaluation*, defined as a process of analysis that takes into account value systems that cannot be measured in ways described for risk estimation. Risk evaluation relies on social and political judgment, and is aimed at determining the importance of hazards and the risk of harm from the point of view of communities and individuals facing that risk. It is the decision maker or risk manager who must be able to compare risks, risk trade-offs, and risks with the potential benefits of using the material. A series of questions that can be posed regarding risk management are compiled in Table 21.11. These questions were developed in a publication by the US Government Accounting office in Health Risk Analysis. Using experience and judgment, the manager must determine a level of risk that is acceptable.

21.7 Summary

Interspecies extrapolation as well as *high-to-low dose extrapolation* play a major role in estimating health risks associated with exposure to chemicals. The qualitative or quantitative characterization of a risk also has a major impact on the risk assessment. Issues in risk characterization are important in both the assessment and the management processes. Throughout the risk management process, regardless of the agency, decisions affecting risk and safety are made in varying degrees of uncertainty. The *risk assessment/risk management*

Table 21.11 Development of risk management options

Were the variables or factors such as costs and benefits associated with each option specified?
Were the methods and assumptions used in the development of such variables as costs and benefits specified?
Were value judgments for each risk management option specified?
Were uncertainties associated with the development of each risk management option specified?
Did the agency use an analytical approach other than or in addition to worst-case analysis?
Were the risk management options compared with earlier risk management options for similar hazards as a validation check?
Was the development of risk management options independent of the risk assessment work?
Were the risk management options reviewed with respect to practicality?
Was the achievable risk reduction estimated for each option?
Were both population and individual risk indicators examined?
Was the relationship between risk reduction and cost examined for each option?

Source: US Government Accounting Office, 1991.

paradigm is a critical and useful tool, providing a framework in which to _harmonize_ complex and diverse _scientific regulatory issues_ and attempt consistency in them. As _global harmonization of risk issues_ continues to evolve, it will be increasingly important for the scientific community to understand these issues and communicate effectively about them.

21.8 Epilogue

Since the preparation of this chapter, risk analysis has become an increasingly visible and controversial topic in the United States. The discussion has centered on sweeping legislative reform proposed during the 104th Congress (1995-1996) and included suggested reforms in _risk assessment, risk management,_ and _risk communication._ Omnibus risk/_regulatory reform_ legislation, which in some cases would supersede existing law, has been proposed, as well as reforms of specific legislation, including the Safe Drinking Water Act, the Comprehensive Environmental Remediation and Cost Liability Act (Superfund), and the Federal Food, Drug & Cosmetic Act (FFDCA, and specifically the _Delaney Clause_). While none of these measures have been passed into law, the discussions are likely to be on-going for several years, but already have had impacts on agency approaches and policies.

Key issues identified from these complex discussions include _risk-_ and _cost-benefit analysis, risk characterization_ policy, peer review policy, priority setting policy for agencies, relationship between individual statutes and "supermandate" provisions which could supersede individual statutes; and provisions for judicial review of agency actions.

Some selected examples of how these issues translate into legislative language are illustrated by the following, which have appeared somewhat consistently in many of the proposed bills:

- risk assessments based on the most "scientifically objective and unbiased information;"
- risk characterizations to state the "reasonable range of scientific uncertainties" and, in addition use the "best estimate" of risk, and the "plausible upper-bound or conservative estimates in conjunction with plausible lower bound estimate:"

- comparison of risks to establish priorities for allocation of resources for risk reduction activities;
- establishment of peer review programs that "shall not exclude peer reviewers with substantial and relevant expertise" because they represent entities "that may have a potential interest in the outcome" of the peer review, though full disclosure of such interests may be mandated;
- for major rules ($50-100 million), benefits of the rule must "justify" the costs; if the statutory basis for the rule prohibits the cost-benefits balancing, the "least net cost" option among reasonable alternatives must be adopted;
- judicial review of agency actions and compliance with the law;

This intense attention to risk reform by the 104th Congress follows the actions of the 103rd Congress which mandated, under the Federal Crop Insurance Reform and Department of Agriculture Reorganization Act of 1994, the creation of an Office of Risk Assessment and Cost Benefit Analysis. The office has responsibility to assess risks to human health and the environment and prepare cost-benefit analyses for proposed "major" regulations—those having an impact on the economy of $100 million of more.

It has also been suggested that the current debate over *regulatory reform* and risk assessment overlooks the *public health* perspective. Whereas much of what is in the regulatory arena is based on the premise of public health protection, the ongoing arguments over *risk reform* may not translate into public health protection. The Centers for Disease Control, which has major responsibility for *public health protection* and disease prevention, has been virtually absent from the discussions over *risk reform legislation*. To increase understanding and communication about risk analysis among diverse audiences, senior federal agency representatives drafted a set of principles for using risk analysis that could be adopted by agencies on an individual basis. The principles identify terms and briefly describe the elements that comprise risk assessment, risk management, risk communication, and priority setting using risk analysis.

Much of the pending environmental and regulatory reform legislation have requirements that have to be met by state health and environment departments, which lack the tools necessary for performing the proposed enhanced risk assessment and cost-benefit analyses. While states have the capacity for the present framework, some of the pending legislation would require significant and probably costly changes to current state procedures.

During 1995-1996, the *President's Commission on Risk Assessment and Risk Management* held hearings across the United States from a variety of experts and the public. The bipartisan commission was established under the Clean Air Act (CAA) Amendments of 1990, following an earlier period of intense debate over "residual risk," the risk levels remaining after the CAA technology-based standards have been implemented. The commission's report is expected during 1996 and would have jurisdiction and impact on all federal risk assessment and risk management policies, not just those associated with air emissions and contaminants.

Changes in the food safety and inspection practices have been or are being proposed at both the United States Department of Agriculture *(USDA)* and the Food and Drug Administration *(FDA)*. The Department of Agriculture has proposed a "paradigm" shift in the way meat and poultry products are inspected. The USDA's *Food Safety and Inspection Service* is pursuing a strategy that over time will improve the safety of meat and poultry products and significantly reduce the risk of *food-borne illness*. The strategy relies upon the Hazard Analysis Critical Control Point *(HACCP)* inspection system, which focuses on prevention of infectious agent contamination and on mandating microbial testing to detect the presence of pathogens. A similar approach using HACCP has been proposed by the FDA to apply to seafood to increase the safety of the seafood supply.

More recently, the United States Environmental Protection Agency released in the Federal Register new Proposed *Guidelines for Carcinogen Risk Assessment*. These proposed guidelines update the 1986 guidelines and are intended to allow incorporation of the increasing knowledge of the mode of action of cancer formation and the uncertainties associated with not only whether a chemical causes cancer in humans, but how it might do so. Other EPA related activities at various stages of preparation include guidelines on *Ecological Risk Assessment*, Exposure Assessment, Non-cancer Risk Assessment, *Benchmark Dose*, and *Endocrine Disruptors*. While this approach to consideration of *carcinogenicity mechanisms* has received much attention in the United States during the last several years, it is already in use in various forms in Western European countries and in The Netherlands since 1978.

And finally, numerous scholarly studies have been released and published on topics touched on in this chapter, most notably by the National Research Council on carcinogens and *anticarcinogens* in the human *diet* and the means by which cancer risks associated with food intake and the diet are estimated. The diet in the United States contains both naturally occurring and synthetic substances that are known or suspected to affect cancer risk. It has been suggested from such studies that toxic chemicals that occur naturally in foods may pose a greater-though still small-risk of cancer than the residues of synthetic pesticides that people consume in their diets.

Thus, the *risk paradigm* in the United States is under vigorous scrutiny from the scientific community, the regulatory community, decision-makers, and the public. In response to this scrutiny, federal and state agencies with responsibility for protection of the public health and environment, and which use risk analysis as a tool to fulfill their responsibilities, have implemented major changes and reforms. The success of these reforms will require several years to be measured.

Reference and reading list

Amdur, M, J.D. Doull, C.D. Klaassen, *Casarett and Doull's Toxicology: the Basic Science of Poisons*, 4th ed. New York, Pergamon Press, 1991.

Ames, B.N., L.S. Gold, Too many rodent carcinogens: mitogenesis increases mutagenesis, in: *Science* 249 (1990), 970.

Andersen, M.E., H.J. Clewell, M.L. Gargas, *et al.*, Physiologically based pharmacokinetics and the risk assessment process for methylene chloride, in: *Toxicol. Appl. Pharmacol.* 87 (1987), 185.

Barnes, D.G., M.L. Dourson, Reference dose (RfD): description and use in health risk assessments, in: *Regul. Toxicol. Pharmacol.* 8 (1988), 471.

British Medical Association, *Living with Risk: the British Medical Association Guide*. New York, John Wiley & Sons, 1987.

Burke, T.A., N.L.Tran, J.S. Roemer, and C.J. Henry, Regulating Risk: The Science and Politics of Risk. Washington, D.C., ILSI Press, 1993.

Bryan, F.L., *Hazard Analysis Critical Control Point* Evaluations. Geneva, World Health Organization, 1992.

Cohen, M.S., L.B. Ellwein, Cell proliferation in careinogenesis, in: Science 249 (1990), 1007.

Davies, J.C. Comparing Environmental Risks. Washington, D.C., Resources for the Future, 1996.

Emerson, J.E., C. Stone, Toxicity testing and risk assessment, in: *First Asian Conference on Food Safety*. Selangor Darul Ehsan, Malaysian Institute of Food Technology, 1990.

Engler, R., E. Rinde, C. Frick, J. Quest, The weight of evidence among group C carcinogens, in: *Quality Assurance: Good Practice, Regulation, and Law* 1 (1991), 51.

ENVIRON Corporation, *Elements of toxicology and chemical risk assessment*. Washington, D.C., ENVIRON, 1988.

Flamm, G.A., Critical assessment of carcinogenic risk policy, in: *Regul. Toxicol. Pharmacol.* 9 (1989), 216.

Frederick, C.B., A.G.E. Wilson, Comments on incorporating mechanistic data into quantitative risk assessment, in: *Risk Analysis* 11 (1991), 581.

Goldstein, B.D., The problem with the margin of safety: toward the concept of protection, in: *Risk Analysis* 10 (1990), 7.

Graham, J.D. The Role of Epidemiology in Regulatory Risk Assessment. Washington, D.C., Elsevier, 1995.

ILSI Risk Science Institute, Review of research activities to improve risk assessment for carcinogens. Washington, D.C., 1987.

International Commission on Microbiological Specifications for Foods, *Microorganisms in Food: Application of the Hazard Analysis Critical Control Point (HACCP) System to Ensure Microbiological Safety and Quality.* Oxford, Blackwell Scientific Publications, 1988.

Joint WHO/FAHO Expert Committee on Food Additives, Toxicological evaluation of certain veterinary drug residues in food. Geneva, World Health Organization, 1991.

Knowles, M.E., J.R. Bell, J.A. Norman, D.H. Watson. Surveillance of potentially hazardous chemicals in food in the United Kingdom, in: Food Additives and Contaminants 8 (1991), 551.

Kraus, N.N., P. Slovic, Taxonomic analysis of perceived risk: modeling individual and group perceptions within homogeneous hazard domains, in: *Risk Analysis* 8 (1988), 435.

National Research Council, *Risk Assessment in the Federal Government: Managing the Process.* Washington, D.C., National Academy Press, 1983.

National Research Council, *Improving Risk Communication.* Washington, DC, National Academy Press, 1989.

National Research Council, *Frontiers in Assessing Human Exposures to Environmental Toxicants.* Washington, DC, National Academy Press, 1991.

National Research Council, Pesticides in the Diets of Infants and Children. Washington, D.C., National Academy Press, 1993.

National Research Council, Science and Judgment in Risk Assessment. Washington, D.C., National Academy Press, 1994.

National Research Council, Carcinogens and Anticarcinogens in the Human Diet. Washington, D.C., National Academy Press, 1996.

Norris, J.R., Modern Approaches to food safety, in: *Food Chemistry* 33 (1989), 1.

Office of Science and Technology Policy, Chemical carcinogens: review of the science and its associated principles, in: *Federal Register* 50 (1985), 10372.

Olin, S., W. Farland, C. Park, L. Rhomberg, R. Scheuplein, T. Starr, J. Wilson, Low-Dose Extrapolation of Cancer Risks: Issues and Perspectives. Washington, D.C., ILSI Press, 1995.

Paustenbach, D.C., *The Risk Assessment of Environmental and Human Health Hazards: a Textbook of Case Studies.* New York, John Wiley & Sons, 1989.

Park, C.W., Mathematical models in quantitative assessment of carcinogenic risk, in: *Regul. Toxicol. Pharmacol.* 9 (1989), 236.

Reitz, R.H., A.L. Mendrala, R.A. Corley, et al., Estimating the risk of liver cancer associated with human exposures to chloroform using physiologically based pharmacokinetic modeling, in: *Toxicol. Appl. Pharmacol.* 105 (1990), 443.

Risk Policy Report, Federal Commission to Weigh Science Basis of Mandates, in: Risk Policy Report 2 (1995) 12.

Risk Policy Report, Scientists Urge Dioxin Reassessment Fixes, in: Risk Policy Report 2 (1995) 4.

Risk Policy Report, EPA Takes Risk Reform Path, in: Risk Policy Report 2 (1995), 23.

Risk Policy Report, EPA Cancer Guidelines Expand Info Used to Assess Cancer Risks, in: Risk Policy Report 3 (1996), 4.

Roberts, D., Sources of infection: food, in: *Lancet* 336 (1990), 859.

Risk Policy Report, Newly Elected Republicans Gear Up to Push Through Risk Legislation, in: Risk Policy Report 1 (1994), 3.

Rodricks, J.V., T.B. Starr, M.R. Taylor, Evaluating the Safety of Carcinogens in Food—Current Practices and Emerging Developments, in: *Food and Drug Cosmetic Law Journal* 46 (1991), 513.

Rodricks, J., M.R. Taylor, Application of risk assessment of food safety decision making, in: *Regul. Toxicol. Pharmacol.* 3 (1983), 275.

Rodricks, J., *Calculated Risks.* Cambridge, Cambridge University Press, 1992.

Roper Organization, *The Environment: Public Attitudes and Individual Behavior.* Roper Organization, 1990.

Samuels, S.W., R.H. Adamson, Quantitative risk assessment report of the Subcommittee on Environmental Carcinogenesis, National Cancer Advisory Board, in: *Journal of the National Cancer Institute* 74 (1985), 945.

Scheuplein, R., Risk assessment and food safety: scientific regulator's view, in: *Food, Drug and Cosmetic Law Journal* 42 (1987), 237.

Slovic, P., Informing and educating the public about risk, in: *Risk Analysis* 6 (1986), 403.

UK Department of Health Committee on Carcinogenicity of Chemicals in Food, Consumer Products and the Environment, *Guidelines for the Evaluation of Chemicals for Carcinogenicity.* London, HMSO Publications, 1991.

US Department of Health and Human Services, Task Force on Health Risk Assessment, *Determining Risks to Health: Federal Policy and Practice.* Dover, MA, Auburn House, 1986.

US Environmental Protection Agency, Guidelines for carcinogen risk assessment, in: *Fed. Regist.* 51 (1986), 33992.

US Environmental Protection Agency, Proposed guidelines for carcinogen risk assessment. (EPA/ 600/p-32/003C April 1996) Internet at http://www.epa.gov/ORD.

US Environmental Protection Agency, Office of Health and Environmental Assessment, *Exposure Factors Handbook.* Washington, DC, EPA/600/8-89/043.2, 1989.

US Government Accounting Office, *Health Risk Analysis: Three Case Studies*, GAO/PEMD-8714. Washington, D.C., GAO, 1991.

U.S. Office of Management and Budget, Office of Information and Regulatory Affairs, The regulatory program of the United States Government. Washington, D.C., Office of Management and Budget, 1990.

Whittemore, A.S., S.C. Groffer, A. Silvers, Pharmacokinetics in low-dose extrapolation using animal cancer data, in: *Fund. Appl. Toxicol.* 7 (1986), 183.

World Health Organization, *Environmental health criteria: principles for the safety assessment of food additives and contaminants in food.* Geneva, WHO, 1987.

Young, F., Risk assessment: the convergence of science and the law, in: *Regul. Toxicol. Pharmacol.* 7 (1987), 179.

Young, F.E., Weighing food safety risks, in: *FDA Consumer* 23 (1989), 8.

chapter twenty-two

Behavioral change and risk perception

G.J. Kok, P. van Assema, and R.M. Meertens

22.1 Behavioral Change

22.1.1 Model for planned behavioral change

Consumers do not always behave in a way food experts consider healthy or safe. Therefore, interventions have been and will be developed to change dietary behaviors of the population. Based on experiences with other risk behaviors, such as smoking, systematic knowledge about planned behavioral change is now available. This knowledge can be applied to changes in dietary behavior.

Before developing interventions for behavioral change, the health or safety problems concerned should be analyzed. Then the behavior related to this problem, the determinants of that behavior, the possible intervention methods, and the possible implementation strategies can be analyzed (see Figure 22.1).

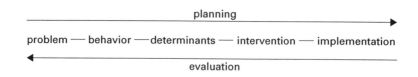

Figure 22.1 Model for planned behavioral change.

The model in Figure 22.1 consists of a planning phase and an evaluation phase, each with a five-step sequence. The *first* step in the *planning phase* is analysis of the problem: what exactly does the problem imply and how serious is it? Health problems of concern might be cardiovascular disease, cancer, or food intoxication. Only if the health problem is serious does the planning process move onto the next step.

The *second* planning step is behavior analysis: to what specific behaviors is the problem related? In the case of cardiovascular disease, the behavior might be consumption of too much food, too much salt, too much fat, etc. With regard to food intoxication, the behavior might be negligence in keeping food fresh, or insufficient heating during preparation. Analysis of behavior is intended to select specific behaviors to which the problem is strongly related.

The first two planning steps are primarily within the field of epidemiology. The Dutch Nutrition Council has published consensus reports on health and safety problems in relation to dietary behavior. These reports include advice on behavior, both general advice and specific advice for various target groups, such as children and the elderly. Assuming that the behaviors which should be changed are known, the planning is continued.

The *third* planning step is analysis of the determinants of behavior: the reasons why people behave as they do. Why do people eat too much fat? Why do people keep food insufficiently fresh? A behavior is often determined by more than one factor and different behaviors often have different determinants. Therefore, these have to be analyzed specifically. In Part 1B (Chapter 7), a model is presented that can be used for analyzing determinants of dietary behavior, in combination with the results of a number of studies on determinants of specific dietary behavior.

The *fourth* planning step is analysis of intervention possibilities and the development of an intervention. Which intervention is likely to be most effective depends largely on the specific behavior and its determinants, and on the characteristics of the target group. Different determinants often need different interventions. In Section 22.1.3 the four types of intervention will be discussed: health education, provision, pricing, and regulation. Nutritional education as part of health promotion will be emphasized in particular.

The *fifth* and last step of the planning phase is implementation of the intervention. This step includes an analysis of the political situation and the policies of the different factors involved in the intervention. Especially important for dietary behavioral change is the achievement of cooperation between retailers, caterers, (local) governmental authorities, etc.

The sequence of the five planning steps can be illustrated by the following example. A national government is worried about the incidence of cancer in its country. Cancer is a serious problem that is related to a number of unhealthy behaviors such as smoking' and eating too much fat. High fat intake is mainly attributed to the amount of high-fat meat in the daily eating pattern. A (hypothetical) determinant of high-fat meat consumption is the consumer's perception that there are no alternatives to high-fat meat. Therefore, the intervention should present alternatives, i.e., low-fat food products. These could, for example, be displayed at the place of purchase: the supermarket, where the consumer can taste them. The next step is then to increase the assortment of low-fat food products.

The *evaluation phase* consists of the same five steps, but in reverse order. Now, the first step concerns implementation; in the example above: was the cooperation between the educators and the supermarkets successful? The second step is to find out whether the intervention has been carried out as planned: have people seen and tasted the alternatives? In the third step, the determinants are queried: are the consumers now better informed about possible low-fat alternatives? The fourth step concerns behavioral change: do people eat less high-fat meat? In the last step, the result of the intervention is the focal point: has the health problem, in this case cancer incidence, decreased in size? This question can only be answered in the long term because of the latency period of cancer. Consequently, on its own, reduction of the initial health problem is often not a realistic evaluation criterion. In such cases, a change in problem-related behavior is the best criterion for success of the intervention. The evaluation phase has been introduced here after the planning phase. In practice, however, evaluation starts right from the first step of the planning phase with data collection. To interpret evaluation results, it is important to compare pre-intervention measures with post-intervention measures. For large-scale interventions where there is no control group, e.g., a public campaign, longitudinal pre-intervention measures are the only basis for concluding whether or not an intervention has been successful.

Alcohol abuse is a behavior (step 2). The physical problem may be liver damage. Alcohol abuse may be caused by social pressure (in bars or at parties), or by the fact that alcohol is seen as a way to cope with problems such as unemployment or loneliness. An intervention may be developed that teaches people to resist social pressure. This intervention can then be used by organizations working with people who have an alcohol problem.

22.1.2 *Importance of planned behavioral change*

Planned behavioral change is an intensive and time consuming activity. This is one of the reasons why practitioners have their doubts about the need for such an elaborate approach. In order to investigate this problem, a meta-analysis of a large number of studies on the effectiveness of different educational methods has been carried out (Mullen et al., 1985). The first finding was that there appeared to be no difference in effectiveness between the various methods; every method can be effective depending on the behavior involved, the determinants, the target group, etc. The second finding was that a positive result of interventions only depends on *the quality of the planning process*, i.e., conscientious answering of the planning questions. As a result, the success of the intervention also depends on the strength of the relationships between the various planning variables: problem–behavior, behavior determinants, etc. If the answer to one planning question is not clear, e.g., there is no consensus on the specific behavior that might be related to the problem, the subsequent questions can not be answered conscientiously. Unfortunately, this situation is not uncommon. Governments and experts are often forced into action, including health education, against a certain health threat at a time when the epidemiological data are not yet conclusive. In a pragmatic approach, the responsible authority will develop interventions based on the best estimation of the answers to the planning questions. At the same time, research projects will be started to fill the gaps in the knowledge that is necessary to answer those questions conclusively.

Besides planning, evaluation is important. Without evaluation, the effects of the intervention remain unknown. During the planning all kinds of decisions have to be made. Planned evaluation can examine the decisions and adjust them, if necessary. Evaluation is also needed for the identification of unexpected and undesirable side effects of the intervention.

As stated above, planned behavioral change is a time consuming and costly affair. In the long term, however, it is effective in preventing wrong decisions, and wasted money and effort because of them. Wrong decisions are often made if the planning steps are not conscientiously taken. A number of pitfalls which are quite common and often result in wrong decisions are listed below. The pitfalls are related to the various steps.

Pitfall 1: Development of an intervention for a problem that does not really exist. An example is a campaign against the use of alcohol, and especially the decision of the Dutch government to pay particular attention to the use of alcohol by pregnant women. Researchers were requested to analyze the planning steps. First, they found that the data collected showed that the assumed relationship between behavior (the mother's alcohol drinking) and the health problem (for the child) was not confirmed for small amounts of alcohol. Actual problems with the child only occurred if the mother drank more than one glass a day. Secondly, it was found that only about 3% of the population of pregnant women drank more than one glass a day. Pregnant women appeared to be very well-informed about the risks due to alcohol and almost all complied with the advice on drinking. These results made clear that the campaign would be a waste of money, since the problem did not really exist. Moreover, the campaign might even have had undesirable side effects, such as feelings of guilt in mothers who found out about their pregnancy rather late.

Pitfall 2: Development of an intervention intended to change a behavior for which there is no consensus about the relation to the problem. This can be illustrated by the discussion on cholesterol testing. Only recently has some consensus been reached about the relationship between cholesterol plasma levels and specific cardiovascular health problems. In addition, currently the relationship between diet and cholesterol levels is being questioned. As a result, interventions intended to lower the cholesterol content of diets may have limited effects on health. Changing behaviors which are not unequivocally related to a problem will not be helpful in reducing the problem.

Pitfall 3: Development of a behavioral change intervention, based on a misconception of the determinants of the behavior in question. This is one of the most common mistakes made in health education as well as in nutritional education. The main reason for this probably is a lack of money and time. Notably, health risk, though a cause for concern for nutrition experts, toxicologists, and health educators (e.g., cancer, intoxication), is almost never a major determinant for consumers in their choice of behavior. Taste, convenience, costs, etc., are the determining factors while health and safety are seldom of overriding importance. Just informing the public about health risks and preventive behavior hardly ever results in behavioral change. (If it did, the habit of smoking, for example, would be eradicated by now).

Pitfall 4: Development of a wrong intervention, e.g., an intervention that is directed at the wrong target group, or a message that is too complex. This can be illustrated by a school project on (un)healthy food aimed at children and adolescents. The target groups are not really involved in decisions on their own dietary behavior. The important decisions are made by their parents, for example, the choice of vegetables in the main meal.

Pitfall 5: Development of an intervention that pays insufficient attention to its implementation. An example is the development of a potentially effective program to be used by local health services, without an adequate system of distribution or without even the means for disseminating the knowledge in the program.

Pitfall 6: Unjustified satisfaction with the effects of an intervention. Without serious evaluation, people use marginal criteria to determine the success of the intervention, such as the quantity of educational materials that have been handed out or the number of participants in the educational meetings. Important as these successes may be, they do not give any information about the desired behavioral change or the reduction of the problem, which are the only real criteria for success.

Falling into these pitfalls may seem unnecessary but is in fact quite common. Mistakes like these can be prevented by careful planning and evaluation. Using the planning model is no guarantee for success, but the (few) successful programs known, followed the planning model. As discussed above, the analysis of problems and behaviors is within the field of epidemiology. Information on determinants of behavior can be found in Part 1B; this chapter continues with a further study of the development, implementation, and evaluation of interventions that are directed at changing dietary behavior to prevent adverse health effects following food intake.

22.1.3 Methods of behavioral change: policy

Four methods of interventions aimed at behavioral change can be distinguished: health education, provision (e.g., assortment), pricing, and regulation. This distinction is only an abstraction, others are also possible. *Health education* is based on voluntary behavioral change. People receive information, help, or training, but eventually they themselves decide about their behavior. *Regulation* is on the other end of the continuum and involves enforcement, with sanctions and control if necessary. *Provision* and *pricing* are interventions in between the two extremes of enforcement and voluntary change. Often they are governmental decisions, but the individual's choice is still voluntary. Pricing is based on the psycho-economic principle that people buy what is cheaper. If healthy food were less expensive than unhealthy food, pricing might well be the decisive behavior determinant for many people. Figure 22.2 shows the so-called "food health policy matrix": the different intervention types vs. the actors in the nutritional field.

					Food chain					
	producers	processors	manufac- turers	interme- diaries	distributors	caterers	consumers	government	pressure groups	treatment consultants
education										
facilities										
pricing										
provision										
regulation										

Figure 22.2 Food health policy matrix.

The choice of interventions in dietary behavior takes up an essential place in governmental policy. With regard to nutrition, however, policy is not always consistent, not even in the case of a crucial subject like smoking. An exception is the Finnish antismoking policy, visualized in Figure 22.3.

A change policy based on only one type of intervention will seldom be effective. Integrated approaches are needed. Two examples can illustrate the feasibility of an integrated nutrition policy.

Norway is one of the few countries with an integrated nutrition policy. Since 1976 this country has worked to achieve four goals:

1. promotion of a healthy diet (reduction of fat intake, increase in vegetables and fiber);
2. stimulation of the country's own food production: fulfilling 52% of its need in 1990, compared to 39% in 1976;
3. stimulation of agricultural developments in thinly populated areas, taking environmental protection requirements into consideration;
4. contribution to world food safety and stimulation of food production in developing countries.

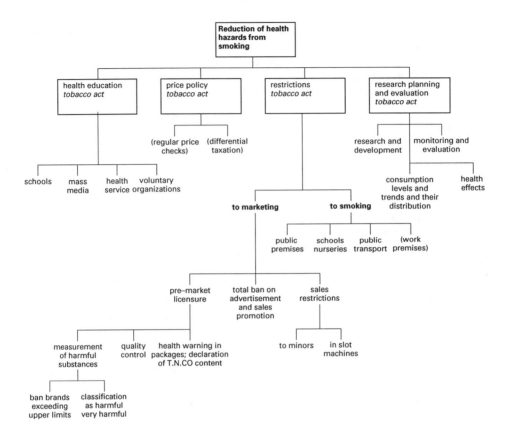

Figure 22.3 Objectives and means of the Finnish antismoking policy. (Source: Leppo and Vertio, 1986. With permission.)

A necessary condition for this integrated policy was that it should have no negative macroeconomic implications. The government financed the necessary investments out of the oil revenues. Due to the short period of time since the implementation of the policy, no improvement in public health has been proven. But from an economic point of view, the Norwegian project has already proved to be a success. The investments appear to be cost-effective and the economic situation of the farmers has improved.

What are the characteristics of the Norwegian nutrition policy? First, it is an integrated policy. Knowledge in the fields of agriculture, economics, nutrition, and health care are integrated. This means that a number of interventions are combined: regulation, long-term planning, health education, financial and economic stimulation, and continuous evaluation of the effects of the interventions concerned. The implementation of the interventions is the task of an intersectoral management team, of all the actors involved, called the Norwegian Interdepartmental Nutritional Council. Why does Norway succeed in the implementation of an integrated nutrition policy, while other countries do not? On the one hand the Norwegian situation was characterized by long-term lobbying activities by individuals as well as organizations, combined with the fact that the risks from negative macroeconomic effects were low. On the other hand, all actors agreed on the integrated policy, including the government itself.

The principle of an intersectoral, i.e., integrated policy being more effective than only one type of intervention, is also applicable at the local level. Integrated local nutritional interventions have been developed and implemented in several countries, using the so-called *community-based approach*. Community-based programs are directed at all members of a certain community, in the sense of an existing social network, such as a neighborhood, village/city, work site, school or organization. Important characteristics of the community-based approach are:

1. it makes use of the existing means of communication of the social network;
2. it stimulates participation of the target groups themselves;
3. it uses intersectoral programs, i.e., including health education, provision, pricing, and regulations, that are carried out by existing organizations, such as the local government, the distributors, and health organizations;
4. it involves multi-media educational interventions including interpersonal communication;
5. it includes training paraprofessionals or volunteers from within the target group as educators;
6. it considers dietary behavior as part of people's lifestyle.

Evaluations have confirmed that community-based nutrition interventions are more promising than the more classical approaches with only one type of intervention, e.g., health education. Further, community-based programs are particularly successful in reaching "difficult" target groups, such as groups with a low socio-economic status and immigrants, who have a lower state of health, partly caused by an unhealthy lifestyle. This is very important, because in general, interventions have been most successful for people that need it the least, such as groups with a higher income and higher education. In interventions effective in reaching consumers with lower education and lower income, the above six characteristics of community-based programs are easily recognized. It may be clear that at this local level too, coalition formation and intersectoral cooperation are essential conditions for success with regard to characteristic 3.

As stated above, local interventions are only partly educational. Programs intended to affect the consumers' environment have been reported to be effective in changing dietary behavior. These programs have some characteristics in common with those based on the community-based approach since they:

1. stimulate changes in the availability of products: low fat, low calorie, low salt, more fiber; healthy menus in restaurants; training of servers (intersectoral cooperation, multi-media approach);
2. provide nutrition information at the moment of choice: in supermarkets, in restaurants, including work sites (existing means of communication of the social network);
3. cooperate with distributors by giving information, training (intersectoral cooperation, multi-media approach);
4. carry out interventions at the work site: weight loss programs during work time, availability of testing (existing means of communication of the social network, intersectoral cooperation);
5. bring changes in health care organizations: training of dieticians and doctors (intersectoral cooperation, multi-media approach).

An integrated approach is more effective. Behavior is often determined by several factors and some of these, for instance price or inavailability of products, cannot be altered by health education.

22.1.4 *Education as a method of dietary behavioral change*

As mentioned above, nutritional education should always be considered as part of an intersectoral health promotion policy. An approach that is solely based on educational activities will hardly ever be effective. As part of an intersectoral program, however, in combination with the other types of intervention, planned health education can indeed be effective in changing people's behavior.

The development of theories concerning behavioral change by communication has resulted in a matrix with the stages of the change process vs. the communication variables (see Figure 22.4).

	message	*receiver*	*channel*	*source*
attention				
comprehension				
attitude change				
change in social influence				
change in self–efficacy				
maintenance of behavioral change				

Figure 22.4   Matrix of behavioral change by education (Adapted from McGuire, 1985.)

On the left of the matrix the stages that lead to a continuous change in behavior are listed:

- attention for and comprehension of the message;
- alteration of the determinants of behavior, such as attitudes, social influence and self-efficacy (see Part 1B), in combination leading to a change in behavior;
- maintenance of the behavioral change.

Above the matrix the communication variables can be found: message (form and content), receiver (target group), channel (medium), and source (an expert, organization, authority, etc.). The questions that need to be answered in the development of communication interventions can be formulated as: what is communicated *(message)*, to whom *(receiver)*, how *(channel)*, and by whom *(source)*? The boxes of the matrix represent large numbers of decisions which have to be made in order to develop an effective educational intervention. The first decision concerns the desired behavioral change (message) in a particular target group (receiver). The second decision is directed at the determinants of behavior, i.e., what message is needed to affect attitude, social influence, and self-efficacy (see Part 1B). Subsequent decisions concern the communication of the various messages to target groups, choosing appropriate channels. A decision is necessary on how to attract attention, as well as on how to get people to understand the message.

A good deal of knowledge is available concerning the various boxes in the matrix. In order to make optimal decisions, the existing empirical data and theoretical insights should be carefully weighed. It should be noted, however, that decisions in one box may conflict with decisions in another. A source that attracts the most attention may not necessarily be a credible source for attitude change. A mass medium that reaches many people may not be the best medium to improve skills (e.g., the ability to prepare low-fat meals). These dilemmas have also been studied. In the next paragraphs, a number of conclusions from research in the various boxes are summarized. The reader who wants to try and develop a new intervention should always bear in mind that there is no such thing as "one intervention method is always more effective than the other." The "rules" that are presented here should always be weighed carefully against the specific characteristics of a new problem. The following remarks are relevant here as regards the effects of the four communication variables.

Message effects. In health education, messages' explicit conclusions are often more effective than implicit conclusions. The more a message is repeated, the more effectively it is received. After three times, however, repetition may no longer be effective, or may even lead to opposite effects. New and valid arguments may change someone's attitude, but information about the opinions and behaviors of others may also contribute to changes in attitude and perceptions of social standards. The discrepancy between the position of the source regarding the issue at hand, and the position of the receiver should not be too large, especially in the case of high involvement of the receiver in the issue under consideration. Improvement of self-efficacy by giving specific instructions is important, particularly in the case of fear-arousing messages (e.g., about the relationship between diet and cancer). An individual's efforts and perseverance in attaining a behavioral change are stimulated if a challenging but realistic goal is stated in the message and care is taken to give feedback information about the effects of the change, as has been shown in weight-loss programs.

Receiver effects. Characteristics of the receiver may affect the results of health education. Changes in attitudes and behavior are curvilinearly related to age. Up to nine, the higher the age the more change. For older people the reverse has been found. Public commitment of the receivers to a certain behavior may cause them to resist change. However, public commitment to the new (changed) behavior can lead to greater maintenance of the behavioral change. Low self-efficacy can be the cause of lapses or relapses (people falling back into their former unhealthy behaviors). It may even keep a person from trying to change. Improving self-efficacy by relapse prevention techniques can result in behavioral change and maintenance of that change. In relapse prevention, four steps are distinguished:

1. convince the person that the reasons for former failure are not stable, but change-able;
2. identify high-risk situations and find adequate responses to high-risk situations;
3. put those responses into practice until they become automatic;
4. in case of (re)lapse, use the lapse as a learning experience and start again with 3, 2, or 1.

Relapse prevention techniques have proved to be effective in changing compulsive and habitual behavior, such as (non)compliance with diet advice.

Channel effects. As far as the effects of communication channels are concerned, it is important to pay attention to mass media on the one hand and interindividual communication on the other. Mass media can be very effective in reaching many people, and in attracting attention to the message. However, interpersonal communication is essential in improving self-efficacy, resistance to social pressure, skills, and maintenance of behavioral change. The simplest conclusion is of course to combine the strengths of both channels in a multi-media intervention. Recently, a number of technological innovations have made it possible to use mass media in combination with interindividual communication, e.g., interactive computer programs and computerized individualized feedback.

Source effects. The source of an educational message is not always the educator himself or herself. It can also be the organization represented by the educator, e.g., a national cancer foundation. Important sources are opinion leaders from people's own environment: medical doctors and paraprofessionals such as dieticians, but also colleagues and friends. The effectiveness of sources is positively related to their competence, their integrity, their attractiveness (including similarity), and power. Powerful and attractive sources may lead to compliance, i.e., do what the source wants you to do. However, this change will only continue while the source is present. On the other hand, competent and trustworthy sources will lead to acceptance, i.e., an internalized change of attitudes and behavior that will be maintained by itself.

The requirements for the source to motivate people to a certain change are:

- explicit conclusions,
- some repetition, but not too much,
- new and valid arguments,
- information on the opinion and behaviors of others,
- information that is not too discrepant from the receiver's position, and
- specific instructions.

And his/her own characteristics are preferably: attractiveness (similarity), competence, integrity, and power.

The various stages of the change process need completely different interventions. Educational programs should have the possibility of differentiated interventions for people at different stages. In the case of people not aware of a possible problem with their diet, the accent is on attention, comprehension, and attitude change. For people who are aware of problems but do not know how to change their behavior, interventions should be directed at attitude change, social influence, and in particular, at self-efficacy and skills (including resistance to social pressures). For people who try to change their behavior, the intervention is focused on maintenance of that behavioral change, and methods such as feedback and relapse prevention are used.

Approaching people who are not aware of a problem with interventions directed at training skills is as useless as approaching relapsers with information about the nutritional problem: it may even be counterproductive.

Community-based programs (see Section 22.1.3) are a specific application of the matrix approach. The six characteristics of these programs, particularly the use of the social network and the participation of para-professionals from the target group, improve the chances of achieving attention and comprehension. Messages will be better adapted to the psychological, social, and structural situation of the receivers, resulting in more changes in determinants and behavior, and in maintenance of behavioral changes.

Even professional educators tend to underestimate the differences between the target group and themselves regarding receiver characteristics. This explains also the effectiveness of paraprofessionals from within the target group itself. The source–receiver dissimilarity can be approached by careful *pretesting* of all materials that are used in the intervention. Pretesting consists of a critical review of the educational materials by the following groups:

- experts in the field of the subject concerned;
- experts in communication and health education;
- a subpopulation of the target group itself, preferably a random sample.

Pretesting by the latter group is extremely important in order to prevent costly mistakes.

The effectiveness of nutrition education programs has not been studied much. In addition, the range of behaviors involved is rather wide. This, combined with the low quality of the research designs, makes it difficult to draw well-considered conclusions.

Programs aimed at students were found to impart knowledge, but not to lead to behavioral change. As far as high-risk groups such as patients with diabetes, cancer, kidney problems, high blood pressure, high cholesterol level, and obesity are concerned, maintenance of behavioral change (diet compliance) is the major problem. Mass media nutrition education programs do not yield positive results. Successful nutrition education programs are characterized by interpersonal contact, social support, self-control, and feedback. Adaptation of the intervention to the individual's attitudes and skills is effective. Interventions aimed at groups (e.g., patient groups) and those based on relapse prevention techniques are promising. For example, in the Netherlands, a large number of programs are available, but subsequent evaluation is often lacking. The "Way of Life" television campaign, intended to combine health education with entertainment, appeared to attract people's attention, and to heighten their awareness of the importance of a healthy lifestyle (including nutrition) for the prevention of cardiovascular diseases and cancer. Group interventions as well as individual counselling have been proven to be effective in weight-loss programs.

In conclusion:

1. effective nutritional education interventions are possible if they are carefully planned and evaluated. They should be based on the results of an analysis of the determinants of the behavior involved;
2. education interventions alone are not sufficient to change an individual's behavior. They should be integrated in an intersectoral approach, not only at national level but also at local level (community).

22.2 Risk perception, risk acceptance, and risk communication

22.2.1 Introduction

The way experts perceive toxicological risks due to food intake often differs considerably from the way the public perceives them. For example, people worry about the health consequences of the presence of antioxidants and colorants in food, and food irradiation, while experts consider these risks involved to be minimal. Experts warn the public of bacterial infections and intoxications, and of consuming too much fat, while the public generally does not worry about these issues (see Part 3, Chapter 16).

A few explanations of the differences in risk perception between experts and lay persons will only be touched upon in this section, as they are so plausible that they do not need much discussion. First, in practice, the opinions of experts may sometimes not reach the public. In this case individuals who want to eat "healthy" foods have no other choice than to base their risk judgments on hear-say or on their own intuition. Secondly, even when the message of the expert does reach the public, they may prefer to rely on their own intuition, because the expert's message is distrusted. Lack of confidence in experts may have several reasons. For example, experts may have lost their credibility because they have made mistakes before. Disagreement among the experts themselves may also lead people to conclude that experts are not to be believed and that they "don't know either." However, the gap between experts and the public also appears to exist when the public receives the expert's message and considers the experts trustworthy. A third explanation of differences in risk perception between lay persons and experts concerns the way the former perceive probabilities and risks.

This section will be mainly devoted to the last explanation, as it is less trivial than the others, and therefore warrants more discussion. The consequences of the ways in which people normally assess risks for the practice of risk communication are also included.

In Section 22.2.2 expert risk assessment methods are summarized (preceding chapters have discussed this in more detail). It will be stressed that experts normally cannot be 100% sure of their risk assessments, as they themselves realize quite well. This uncertainty in the expert's judgment is relevant here, because messages which contain an element of uncertainty are difficult to communicate adequately to the public. Furthermore, the ways people deal with probabilities and risks are discussed. This will be largely based on risk perception research that does not directly concern food and food safety. Hardly any studies have been reported in this area. In the last subsection, recommendations are given on how to communicate toxicological risks arising from food intake, based on information in the preceding subsections on the way experts assess risks and the public perceives them.

22.2.2 Expert risk assessment

In general, experts base toxicological risk assessment on data from animal as well as epidemiological studies. They are well aware that there may be differences in physiology between species and interindividual differences in sensitivity to a toxin. For that reason, scientists apply safety factors in their recommendations (see Chapters 17 and 18).

A certain degree of caution in risk communication, however, remains warranted because of some other limitations of risk assessment methods. First, combined actions (synergism or interaction) may occur (see Part 2, Chapter 13). For example, a high-fat diet may promote azoxymethane-induced intestinal cancer. Secondly, long-term effects, which are sometimes difficult to determine in research, may occur. For example, two-generation studies on the possible mutagenicity of saccharin have revealed effects which had not been observed in one-generation studies. Thirdly, in some cases, the safety factors which used to account for inter– and intraspecies variations, may not be appropriate.

Although the above aspects of expert risk assessment cannot be denied, one should realize that guaranteed zero risk levels cannot be given. The risks lay persons run without any further thought in everyday life are usually considerably higher than those from food intake, which experts judge relatively safe. The important point here is that experts often cannot give the 100% certainty the public wants. People do not like uncertainty, however small the risk involved is. They want the experts to tell them which food is absolutely safe, and which food is absolutely hazardous. An American senator once stated: "We need one-armed scientists, who cannot answer questions with: On the one hand the evidence is so, but on the other hand ..."

The next subsection deals with the lay person's perception of probabilities. Some illustrations will be given of the preference people have for certainty.

22.2.3 Probability perception

Scientists use special data sets and advanced statistical computer programs to assess the risk from toxins. Of course, the public does not have such data and tools at hand, and therefore relies upon much simpler methods. For instance, people may apply the so-called availability heuristic, i.e., memories which have some relevance to the problem are used as starting points to estimate the risk involved.

The way in which such a heuristic works is illustrated by the following example. Imagine a man trying to find out whether food colorings are toxic. A quick memory search results in two examples of people who always eat incredibly pink cakes and puddings. One example concerns a child, the other a woman of 85. Since the child looks really pale and the woman has stomach complaints, the man is likely to conclude that colorings are toxic.

This shows how the way lay persons handle data differs essentially from that of epidemiologists. First, the lay person relies heavily on small biased data sets. Salient, vivid, or emotional memories are likely to come to mind first. Secondly, the man only considers one class of data, viz. data on individuals who have a bad state of health and ingest a considerable amount of colorings. He does not take into account any data on individuals who did not ingest large amounts of colorings, or who are healthy. Few people would understand that one would need (at least) four categories of data to establish the association in question: (1) data on healthy people who consume colorings; (2) data on unhealthy people who consume colorings; (3) data on healthy people who do not consume colorings; (4) data on unhealthy people who do not consume colorings. The speed with which people establish associations should not be considered "dumb" without further thought. Heuristic processing is quite adequate for situations where evidence is strong and there is little time to think. Since sophisticated statistical processes only begin to yield results after one has managed to survive in relatively simple situations where quick decisions are needed, heuristic processing appears to be valuable, all things considered.

Of course, the lay person's way of establishing associations would not pose any problem to risk communicators if only people would easily change their minds on the basis of information given by the experts on the risks they assessed. However, psychological research shows that heuristics even affect risk judgments when objective statistical evidence is at hand. This is intuitively plausible. Many people defend their smoking and drinking behavior by referring to the old man they once knew who smoked two packets and drank a bottle of liquor a day, and lived healthily until the age of 96.

Although the evidence in this respect is scarce, it seems plausible that lay persons' associations play an important role in risk and probability perception. Because some types of chemicals are dangerous, all chemicals (even vitamin C if it is designated as ascorbic acid) are perceived to be toxic. Foods of natural origin, however, are considered to be healthy. A complicating factor is that, once an idea has been formed about the riskiness of a substance, the idea is often resistant to change. People are either selective in the information they are looking for, or they interpret the facts in a way that supports the already existing ideas. For example, the accident at Three Mile Island was interpreted by nuclear energy supporters as evidence that nuclear energy is safe (there were no large negative consequences), and by nuclear energy opponents as evidence that it is dangerous (there had been an accident).

22.2.4 *Certainty and uncertainty*

Another characteristic of people's ideas on probability is that all probabilities are perceived to have much in common. A probability of 30% is not viewed different from a probability of 50%, as one would expect from a rational point of view. The difference between certainty (100% probability) and 80% probability, however, is perceived in a completely different way.

In an experiment on this issue, a group of subjects were asked to choose between two options, A and B, of which option A meant 20% chance to win $4000, and option B, 25% chance to win $3000. From a rational point of view, subjects should choose option A, because the expected benefits of option A are more substantial ($1/5$ of $4000 = $800) than those of option B ($1/4$ of $3000 is $750). Indeed, the results show that 65% of the subjects chose option A. However, another group also had to choose between two options, A and B. Now, option A meant 80% chance of winning $4000, while for option B the chance to win $3000 was 100%. Using the same approach as before one would again predict a preference for option A ($8/10$ of $4000 = $3200, a larger sum than $3000). In this case — and that will be intuitively plausible to most people — 80% of the subjects preferred option B; a certain benefit is preferred to an uncertain one. In the case of losses, empirical evidence was obtained for the behavior to be exactly the reverse. In general, to be sure of losing a certain amount of money is less preferred than the risk of loosing a larger amount. The preference most people have for certainty was also shown in a study in which two groups of subjects were asked whether they would decide to become vaccinated in a certain situation. One group of subjects was informed that 20% of the population was expected to be infected with the disease in question, and that the vaccination would protect half of the group against the disease. The second group was told that the disease has two variants, each of which would infect 10% of the population. The vaccination would provide protection against one type of the disease, but not against the other. In both cases, vaccination reduces the chance of infection from 20 to 10%. However, in the second case the subjects more often chose to be vaccinated than in the first case. In the second case, vaccination gives 100% certainty of protection against one type of the disease. Therefore, vaccination seemed more useful than in the first (uncertain) case.

22.2.5 *Perception of probabilities and risk communication*

What are the consequences of the lay person's comprehension of probabilities for risk communication in food safety practice? First, risk communicators themselves should realize that people cannot be expected to deal accurately with the probability aspects of risk messages. Secondly, risk communicators should realize that the public feels that all probabilities have much in common, but that people make a clear distinction between chance and certainty. Thus, although experts claim that the chance of a certain type of additive causing cancer is really small (for example: 0.0000016%), people may still perceive this probability as greater than one would expect. By formulating the message slightly differently, the facts may sometimes be conveyed better. People may comprehend more easily that a danger is actually small, when the message says that the particular type of additive increases the risk of dying from cancer from 24.7% to 24.7000016%.

In literature, more examples can be found of the way presentation influences probability perceptions and decision making. For example, a study on patients revealed that a therapy was chosen more often when the percentage of patients that survived (e.g., 90%) was given, than when the percentage of patients that died (10%) was mentioned.

22.2.6 *Safety scale*

Several scientists have proposed the introduction of so-called safety scales as a solution for the relatively poor comprehension people have of the concept of probability. A safety scale, as proposed by Urquhart and Heilman (1984) shows how safe or unsafe a certain activity is on a 1 to 8 scale: the higher the score, the safer the activity. For example, smoking cigarettes (a packet a day) scores 2.3, flying is 5.9. Numbers are assigned to activities by logarithmic transformation of statistical data. When, for example, a certain activity takes the life of 1 person a year per 100 people who carry out the activity, the score on the safety scale is 2. Urquhart, however, is of the opinion that it is not necessary for people to know how the score on the scale is calculated. He argues that people also have an idea of what an earthquake of five on the Richter scale means without knowing the background to measurement and calculation. Although such a scale might give the public more insight into the magnitude of risk in some cases, not all risk communication problems seem to be solved by it. First, people need to be informed about risks which are relatively unknown and of which a detailed estimate in terms of human lives lost per year is not possible. Secondly, comparing risks which differ considerably in nature does not make anything clearer. For example, flying by plane and ingestion of food additives are so different that a direct comparison does not make any sense. Normally, one survives the intake of food containing additives (possible health effects only show themselves after some time), but one either survives a flight or not. Thirdly, there are often more reasons than only the deaths an activity causes for calling the activity risky. The activity, for example, may also be considered risky because it may lead to illness or handicaps. Related to this is the fact that people do not define risk as the probability that an activity leads to death. This point will be considered in more detail in the next subsection.

22.2.7 Risk perception

Risk, as defined by the lay person, seems to be a concept that has many qualitative dimensions. By using multivariate techniques like factor analysis, two fundamental dimensions have been shown in several studies. The first dimension is called "threat," the second "observability." A third dimension that sometimes shows itself is "the number of people that may be affected." The first two dimensions and their characteristics are summarized in Figure 22.5. The figure shows that activities are perceived to be less risky if they can be carried out voluntarily. It also shows that activities are perceived as less risky if the activity is more or less familiar, and its consequences are manageable and non-global. Further, risks are considered to be more acceptable if many advantages are attached to the activity.

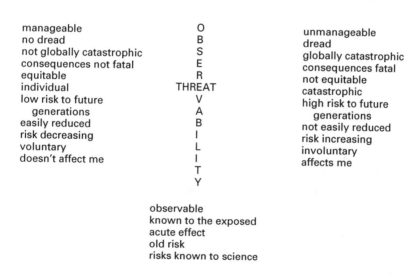

Figure 22.5 Dimensions underlying risk perceptions (results of a factor analysis) and the characteristics of the dimensions. (Source: Slovic et al., 1982. With permission.)

The recognition that the lay person's risk perceptions are related to the qualitative aspects of risky activities leads to several recommendations for risk communication. First, whenever possible, people should be left to choose whether they want to take the risk, however small that risk may be in the expert's eye. People should be given the information they need to decide whether they want to eat food to which colorings have been added, or food that has been produced by new biotechnological techniques. Adequate labeling of food products is one necessary step to enable consumers to choose freely. Furthermore, the advantages of the risky activity should be mentioned, especially the advantages it has for the consumer. Also, the measures that have been taken to manage the risk and its possible negative consequences should be stressed. Further, not only the new aspects of the risk should be mentioned, but also the "old" ones.

Some authors have suggested that "catastrophic potential" is the most important determinant of the lay person's risk perception. The more severe and the more globally effective the theoretically possible negative consequences are, the stronger the risk is perceived. At first sight this seems a fairly simplistic and crude way of evaluating risks since several important aspects of risk, such as the probability of negative consequences and the advantages of the risky activities, are not taken into account. However, this way of risk evaluation may be understood better from a "survival of the species" point of view. In a (micro-)evolutionary sense, it makes quite a difference whether an activity costs one life per year during a million years, or whether it costs the lives of a million people in one catastrophe once in a million years. In the eyes of general public, it makes a difference whether in a worst-case scenario the whole world population may die, or just one individual. So, the perception of risk in terms of catastrophic potential may have some value after all.

The research referred to above is merely descriptive, and largely empirical. A theoretical basis may be found partly in the stress theory and in research. The stress theory states that stress originates from something (an event, an activity, a substance) that is perceived in a primary appraisal process as a threat ("a risk"). However, only after an inventory has been made of the available coping mechanisms (secondary appraisal), and these coping mechanisms do not seem adequate to handle the threat, an individual will experience stress. Thus, the stress theory seems to underline the importance of the manageability of the threat or risk and its possible negative consequences.

In health education, self-efficacy, a concept related to manageability, has proven its value in predicting people's behavior. Self-efficacy is the judgment the person has of his or her own efficacy to display a certain behavior, i.e., to manage and eliminate barriers and difficulties. To impart self-efficacy and a sense of manageability, risk communication should give detailed and clear information on the way in which risks and their negative consequences can be managed.

Of course, in some cases it is not possible to give detailed information. Many severe threats cannot be removed easily. Some authors have suggested that the so-called reactance and learned helplessness theories should be integrated, to gain insight into situations where a threat occurs without clear possibilities of managing it. In such situations, individuals will sense a loss of control, and will first show reactance. They will try to regain control in several ways. However, when they do not succeed in doing so, feelings of helplessness take over. They do not take actions to reduce the threat anymore, and stay apathetic even when the threat has become manageable again.

When such a situation occurs, there may be a tendency to conclude that the public is not worried about the threat, because only a small group shows actual concern. However, the majority may consist of people who take the threat seriously, but do not see any solution and hence have decided (either consciously or unconsciously) to repress feelings of threat-related concern. For example, some people may "decide" not to worry about *Salmonella* infections of food, because they do not know what to do to enhance food safety. What should be done in such a situation is to tell them how they can reduce the risks effectively: explain in detail how the food should be stored, prepared, and cooked. Once they feel capable of managing the risk, people will be more inclined to admit that the risk really exists.

A last interesting characteristic of the lay persons' risk perception is that in general individuals think they are less susceptible to negative events such as disease, divorce, and robbery than the average person (of course, this is logically implausible). One explanation for this so-called unrealistic optimism is that people only believe what they like to believe. Another explanation is that individuals only think of the steps they themselves already take to reduce the risk, and do not realize that others also take measures to reduce it. Anyhow, it may be useful to stress that everybody runs risks and that one should take all available measures to reduce them.

22.2.8 Recommendations

After the above discussions of experts' risk assessments and the lay persons' perception of risks and probabilities, this Section provides a number of recommendations on communicating toxicological risks due to food intake. The recommendations must necessarily be tentative, for they are formulated on the basis of the limited information available. This information is not only obtained from studies on toxicological risks due to food intake, but also from studies on technological risks and gambling studies. Thus, some reserve is in order here.

The first recommendation urges experts to indicate clearly (as they generally do already) that their risk assessments may be incomplete. Risk assessment is always based on the present state of knowledge and this may be incomplete.

Secondly, the public should have freedom of choice. There are two reasons for this: first, there may be uncertainty about the conclusions of risk management; secondly, people will find risks more acceptable when they can choose voluntarily. However, the public should of course be informed of the expert's judgments of the risks involved.

As a third recommendation, the experts should stress the advantages of the risky activity, especially the advantages to the public.

Fourthly, the measures already taken to manage the risk should be emphasized, especially the measures that reduce the possibility of negative consequences.

The fifth recommendation stresses the point that particular attention must be paid to the way risks and probabilities are presented. General guidelines for how risks should be presented cannot be given. A comparison of risks may be helpful, but only if the risk one wants to communicate does not differ too much from the risk with which it is compared. Further, one should be especially careful in communicating very small and very large probabilities (because of the psychologically big difference between certainty and probability), and give a framework of reference when possible (e.g., the chance of dying from cancer anyhow). It may also be useful to check a message for emotions releasing concepts. Mentioning death, for example, seems to augment the perceived threat considerably. The recommendations given above are of particular importance when the first aim in risk communication is reassurance of the public (e.g., you can safely eat food with licensed additives) and not behavioral change. However, sometimes the public underestimates the risk involved (e.g., bacterial contamination of food), while the risk communicator would like people to take risk-reducing steps. In this case, clear advice should be given on the steps one should take to reduce the risk. Clear advice may enhance the individual's self-efficacy, and consequently he or she will be more strongly inclined to act according to the communicator's advice. Furthermore, because of the "unrealistic optimism," it should be emphasized that everybody may be exposed to the negative consequences of the risk, and that the advice to reduce the risk should be followed completely.

Finally, communicators should realize that behavioral change requires more than an adequate perception of the risks involved. Generally, factors such as self-efficacy, attitude and social influence are regarded as more important.

Reference and reading list

Bandura, A., *Social foundations of thought and action: a social cognitive theory*. Englewood Cliffs, NJ, Prentice Hall, 1986.

Berkowitz, (Ed.), *Adv. in Exp. Soc. Psychol.* 8, 178–236.

Bull, A.W., B.K. Soullier, P.S. Wilson, M. Tan Hayden, N.D. Nigro, Promotion of azoxymethane-induced intestinal cancer by high-fat diet in rats, in: *Cancer Res.* 39, 4956–4959, 1979.

Contento, I., The effectiveness of nutrition education and implications for nutrition education policy, programs and research: A review of research. *J. Nutr. Educ.* 27, 277–418, 1995.

Den Boer, D-J., G. Kok, H.J. Hospers, F.M. Gerards and V.J. Strecher, Health education strategies for attributional retraining and self-efficacy improvement, in: *Health Educ. Res.* 6, 239–248, 1991.

Ellwein, L.B., S.M. Cohen, The health risks of saccharin revisited, in: *Crit. Rev. Toxicol.* 20, 311–326, 1990.

Glanz, K. and R.M. Mullis, Environmental interventions to promote healthy eating: a review of models, programs and evidence, in: *Health Educ. Quarterly* 15, 395–415, 1988.

Green, L.W. and M.W. Kreuter, *Health Promotion Planning: an Educational and Environmental Approach.* Palo Alto, Cal., Mayfield, 1991.

Hendrickx, L, C. Vlek, H. Oppewal, *Relative Importance of Scenario Information and Frequency Information in the Judgment of Risk. Acta Psychol.* (in press).

Janz, N.K., M.H. Becker, The health belief model: a decade later, in: *Health Educ. Quarterly* 11, 1–47, 1984.

Jonker, D., R.A. Woutersen, P.J. van Bladeren, H.P. Til, V.J. Feron, 4-week oral toxicity study of a combination of eight chemicals in rats: comparison with the toxicity of individual compounds, in: *Food Chem. Toxicol.* 28, 623–631, 1990.

Kahneman, D., A. Tversky, Prospect theory: an analysis of decision under risk, in: *Econometrica* 47, 263–91, 1979.

Kar, A. van de, R.M. Meertens, G. Kok and A. Knottnerus, Determinants of consulting the general practitioner and patients worry: An experimental and an observational study compared, in: *Determinants of consulting the general practitioner.* Thesis, University of Limburg, 1992.

Kok, G., H. Schaalma, H. de Vries, G. Parcel and Th. Paulussen, Social psychology and health education, in: W. Stroebe and M. Hewstone, Eds., *Eur. Rev. Soc. Pscyhol.* 7. Chichester, Wiley, 1996.

Lazarus, R.S., *Psychological Stress and the Coping Process.* New York, McGraw-Hill, 1966.

Leppo, K. and H. Vertio, Smoking control in Finland; a case study in policy formulation and implementation, in: *Health Promotion —An International Journal* 1, 5–16, 1986.

Liedekerken, P.C., R. Jonkers, W. de Haes, G. Kok and J.H. Saan, *The Effectiveness of Health Education.* Assen, Van Gorcum, 1990.

Locke, E.A. and G.P. Latham, *A Theory of Goal Setting and Task Performance.* Englewood Cliffs, N.J., Prentice Hall, 1990.

Marlatt, G.A. and J.R. Gordon, *Relapse Prevention; Maintenance Strategies in the Treatment of Addictive Behaviors.* New York, Guilfort, 1985.

McGuire, W.J., Attitudes and attitude change, in: A. Lindsay and E. Aronson, Eds. *The Handbook of Social Psychology, Volume 2*, 233–346. New York, Random House, 1985.

McNeil, B.J., S.G. Pauker, H.C. Sox Jr., A. Tversky, On the elicitation of preferences for alternative therapies, in: *New Engl. J. Med.* 306, 1259–62, 1982.

Milio, N., Promoting health through structural change: analysis of the origins and implementation of Norways farm-food-nutrition policy, in: *Soc. Sci. Med.* 15, 721–734, 1981.

Milio, N., *Nutrition Policy for Food-Rich Countries: a Strategic Analysis.* Baltimore/London, John Hopkins University Press, 1988.

Mullen, P.D., L.W. Green and G. Persinger, Clinical trials of patient education for chronic conditions: a comparative meta-analysis of intervention types, in: *Prev. Med.* 14, 753–781, 1985.

Nisbett, R.E., L. Ross, *Human Inference: Strategies and Shortcomings of Social Judgment.* Englewood Cliffs, NJ, Prentice Hall, 1980.

Pascal, G., *Risk Assessment: Governmental Aspects.* Paper presented at the 1991 EUROTOX Congress, Maastricht, The Netherlands, 1–4 September 1991.

Paterson, R.J. and R.W.J. Neufeld, The stress response and parameters of stressful situations, in: R.W.J. Neufeld, *Advances in the Investigation of Psychological Stress.* New York, John Wiley & Sons, 1989.

Petty, R.E. and J.T. Cacioppo, *Communication and Persuasion; Central and Peripheral Routes to Attitude Change.* New York, Springer, 1986.

Puska, P., Community-based prevention of cardiovascular disease: the North Karelia project, in: J.D. Matarazzo et al. Eds., *Behavior Health: a Handbook of Health Enhancement and Disease Prevention,* 1140–1147. Silver Spring, Wiley, 1984.

Rogers, R.W., C.R. Mewborn, Fear appeals and attitude change: effects of a threats noxiousness, probability of occurrence and the efficacy of coping response, in: *Journal of Personality and Social Psychology* 34, 54–61, 1976.

Siero, S., G.J. Kok, J. Pruyn, Effects of public education about breast cancer and breast self examination, in: *Soc. Sci. Med.* 18, 881–888, 1984.

Slovic, P., B. Fischhoff, S. Lichtenstein, Facts and fears: understanding perceived risk, in: R. Schwing, W.A. Alberts Jr., (Eds)., *Societal Risk Assessment: How Safe is Safe Enough?* New York, Plenum, 1980.

Slovic, P., B. Fischhoff, S. Lichtenstein, Facts vs. fears: understanding perceived risk, in: D. Kahneman, P. Slovic, A. Tversky, (Eds.), *Judgments Under Uncertainty: Heuristics and Biases.* Cambridge, Cambridge University Press, 1982a.

Slovic, P., B. Fischhoff, S. Lichtenstein, Response mode, framing and information processing effects in risk assessment, in: R. Hogarth, (Ed.), *New Directions for Methodology of Social and Behavioral Science: Question Framing and Response Consistency.* San Francisco, CA, Jossey Bass, 1982b.

Svenson, O., Are we all less risky and more skillful than our fellow drivers?, in: *Acta Psychol.* 47, 143–8, 1981.

Tversky, A., D. Kahneman, The framing of decisions and the psychology of choice, in: *Science* 211, 1453–8, 1981.

Urquhart, J. and K. Heilman, *Risk* watch. New York, Facts on file, 1984.

Verhagen, H., *Toxicology of the food additives BHA and BHT.* Thesis, University of Limburg, 1989.

Vlek, C., P. Stallen, Judging risks and benefits in the small and in the large, in: *Organizational Behavior and Human Performance* 28, 235–71, 1981.

Weinstein, N.D., E. Lachendro, Egocentrism as a source of unrealistic optimism, in: *Personality and Social Psychology Bulletin* 1982, 195–200.

Weinstein, N.D., Why it won't happen to me: perspectives of risk factors and susceptibility, in: *Health Psychol.* 3, 431–57, 1984.

Wortman, C.B., J.W. Brehm, Response to uncontrollable outcomes: an integration of reactance theory and the learned helplessness model, in: L. Berkowitz, (Ed.), *Advances in Experimental Social Psychol.* 8, 236, 1975.

chapter twenty-three

Food safety policy

M.J. van Stigt Thans

23.1 Introduction

Policy making involves making choices and implementing them. Food safety policy is no exception to this general principle. This chapter describes briefly the process of government policy making, before the specific elements, tools, and instruments of food safety policy are discussed.

Food safety is an issue for which the government in almost any country is likely to adopt a certain degree of formal responsibility. The reasons are obvious. Food safety is related to public health, agriculture, food manufacture, and trade and, consequently, has substantial impact on the actual and potential (economic) strength of the country. The health and well-being of the population, and in particular of high-risk groups, are closely related to the availability and safety of food and drinking water. Moreover, the food and agricultural production system suffers from, and at the same time is responsible for, the presence of residues and contaminants (e.g., pesticides and nitrate) in the environment that may interfere with food safety.

Consequently, government measures taken to protect food safety clearly have an effect on:

- public health
- national economy
- international trade in food and agriculture raw materials and consumer products
- income situation of farmers and others involved in the food production chain
- consumer food prices
- quality of the environment

Nowadays, the consumer is usually only aware of issues related to food safety, particularly in developed countries. Both accidents and policy measures in this area receive wide coverage through mass media. Public attention is easily focused on events such as an outbreak of salmonellosis, abuse of hormones in cattle breeding, TCDD contamination of milk, and the detection of pesticide residues in drinking water. Also, related policy measures like new standards for nitrate levels in vegetables, or proposed budget cuts in the food inspection services are likely to attract attention, particularly if consumer legal advisors rouse interest in such issues.

To ensure the safety of the food on the consumer's plate, the government's policy will have to call on the responsibility of every link in the food chain, including the consumer. The reason for this is that food may become unsafe or even unsuitable for consumption due to improper handling at almost any step on the way from raw material to consumer.

Food safety policy does not only involve the participation of several government agencies. It is influenced by a number of non-governmental organizations, experts, and lobbyists. Their specific role in and impact on the decision-making process varies, like in any other area of government policy where scientific evidence and politics have to merge. Government agencies most closely related to food safety policy are the departments who are responsible for:

- public health
- agriculture
- social welfare and employment
- environmental protection
- economic affairs

Of course, there are also international aspects in food safety policy. The international trade in raw materials and food products is only one of them. The European Union has taken authority over food safety issues such as legislation on additives and contaminants in its member states. Further, the priorities for food safety issues in developing countries with food shortages and drinking water supply problems may well be quite different from those in developed countries.

Obviously, food safety policy in a particular country involves many societal aspects, and concerns issues that may have local impact only, or raise questions with (inter)national, if not global dimensions. This chapter is aimed at stimulating the awareness of the fact that food safety policy is not so much a matter of independent scientific achievement and sophisticated mathematics, but rather a reflection of (political) acceptability of government interventions. Therefore, the subject is discussed in a descriptive way. Practical examples will be given that illustrate the man-made character of food safety policy. It is also stressed that food safety involves a compromise, colored by culture, era, and actors, rather than the application of models or systems.

23.2 Food safety policy making: science meets politics

23.2.1 Introduction

Chernobyl, USSR, April 26th 1986. An explosion in a nuclear plant brings a considerable quantity of radioactive material in the atmosphere, spreading across a number of European countries which decide to take protective measures. In the Netherlands, dairy cattle grazing is temporarily prohibited to prevent contamination of milk through ingestion of radioactive fallout-contaminated grass. The spinach harvest is considered to pose a health hazard and is destroyed according to a government statement.

In the USSR, however, during a period of several days, if not weeks after the event, the government holds the view that Western countries deliberately suggest the "accident" to be a disaster, in order to discredit the Soviet Union. The government "food safety policy" is to make it clear to the Soviet population that there is no reason to worry. In fact, the central government message to the population is that it would display an anti-Soviet attitude, if it asks critical questions about government action after the Chernobyl accident.

Only some time after the event, protective measures are taken to reduce health risks due to consumption of contaminated agricultural products. However, priority appears to be given to reducing the damage to the Soviet ideology. Until ten days after the event, Moscow persists in stating that the radioactivity level in Kiev is only marginally above average. At the same time, however, local radio is recommending the inhabitants to close their windows and to avoid contamination of fruits and vegetables.

The government of France maintains complete silence for a week after the Chernobyl accident. Then it states that there was no hazard whatsoever because of the large distance. However, it does not mention the fact that radioactivity had been detected in the atmosphere during the days before, but had been removed largely as a result of a change in wind direction. In its May 12th edition, the respected newspaper Le Monde considers this selectivity of information to the public symptomatic for the French government's attitude towards nuclear energy, and should be looked at in the light of the fact that 65% of the electricity in France is obtained from nuclear plants.

23.2.2 *Identification of safety risks due to food intake*

Generally, most people are of the opinion or believe that the food they ingest will not be harmful to their health. Still, from time to time the media reports cases that give rise to concern about food safety. Pesticide residues in food or drinking water, antibiotics in animal feed, hormones in meat, nitrate in green vegetables, new techniques in food processing such as application of biotechnology, and contamination are some examples of causes for concern.

Consumers, when asked to indicate major health risks associated with food consumption, usually put food additives and contaminants first. They are prone to consider high fat consumption and other undesirable eating habits, or inadequate hygienic conditions when handling food in the kitchen, of lower importance.

The consumer's perception of risk is not always associated with reality (see also Chapters 16 and 22). As a result, food safety policy is questioned, unless scientific evidence is not only accounted for, but also the way in which risks are perceived by the consumer. This means that food safety policy should include solving technical problems as well as health education and risk communication.

Food safety is aimed at the prevention and/or reduction of toxicological risks due to food intake. Exposure to food depends on the choice of the total diet, and the composition of individual diet components. Food safety measures should be based on available data and/or relevant research. Information may originate from sources such as:

- epidemiological evidence, health statistics, data on the incidence of diseases resulting from contamination of food with bacteria etc., contamination of drinking water (see Figure 23.1)
- monitoring of the levels of relevant food components and (potential) food contaminants in the environment (water, soil, air, plants, animals) which may involve hazards
- monitoring of food consumption patterns, eating habits, and other behavioral factors relevant to the safety of food handling practices

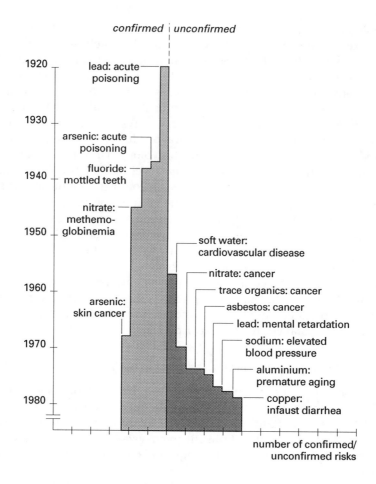

Figure 23.1 Epidemiological evidence of health risks caused by drinking water components. Results presented in the order of publication year of pioneering work. The number of unconfirmed risks increases much more rapidly than the number of confirmed risks. Source: Grimvall and Ejvengård, 1986.

- specific surveys and investigations intended to confirm or reject hypotheses on food-borne health risk factors
- notification of problems associated with food-borne health risks, crossing the border from an adjacent country
- specific experience in other countries.

An adequate and coherent system for monitoring and surveillance of the critical factors and parameters, an (inter)national network of expert contacts, and in particular political backing for funding and operation of the relevant government services are of vital importance for a timely identification of (potential) food safety problems.

In this way, risk identification can be a useful tool to optimize the efficacy and preventive potential of food safety policy. Prevention of risks from food intake rather than reduction is the key to a fruitful food safety policy. It is practically unfeasible to examine all, say 15,000, food products on the shelves of an ordinary supermarket for health hazards every day. Protection of the consumer can be achieved more effectively and at lower costs by analyzing and checking the critical points in the food supply system.

Health risks from food intake against which policy measures are issued, can be very diverse, ranging from an acute threat to the majority of the population to a potential risk which poses health problems to a limited number of consumers only in the long term. Illustrative examples of the various types of risks from food intake are

- fraudulent use of veterinary drugs in cattle breeding that may lead to contamination of meat or dairy products with substances that have inadequately been screened and tested for potential toxicity
- deficient education and training of the personnel working in food catering services may result in inaccurate hygienic practices in the handling and preparation of food, giving rise to an increase in food infections or food poisonings due to pathogenic microorganisms
- uninspected dumping of industrial and household waste in agricultural areas may cause contamination of food chains and farming products with harmful chemicals and pathogenic bacteria
- introduction of new varieties of vegetables, cereals, and potatoes may unintentionally result in exposure of consumers to harmful concentrations of substances, originating from wild-type varieties in breeding
- improper use of pesticides such as DDT in developing countries supplying raw materials for animal feed may result in contaminated meat and dairy products in countries importing such raw materials
- availability of new techniques for food packaging and preservation may make it appear that as far as prolongation of the shelf life of perishable food products is concerned, "nothing seems impossible." This may reduce the alertness of the consumer with respect to the risks associated with food infection and spoilage
- scaling-up of food production processes may involve hazards, such as the formation of nitrosamines in beer as a result of the direct contact with traces of nitrogen compounds in the hot vapors from the brewery boiler-house during the large-scale drying of malted barley
- (particularly in affluent societies,) consumers depend more and more on diets based on products that are readily available almost year-round. This requires the application of an immense number of interventions, processes, techniques, processing aids, and food additives, all of which may, alone or in combination, have effects on the safety of such foodstuffs. For example, the nitrate contents of greenhouse-cultivated vegetables may be twice or three times as high as that of summer-harvested crops, and even exceed the safety limits
- food shortages, famine or poverty may lead to the consumption of inferior, partly spoiled, or otherwise harmful products that normally would have been considered unsuitable for human consumption
- unfamiliarity with the hazards of food handling, storage, and preservation at home may result in inadequate use of techniques, especially heating. One result may be the development of the dangerous *Clostridium botulinum* bacteria
- the increasing popularity of microwave ovens may also increase the risk of *Salmonella* infections on heating raw chicken. Further, the use of microwave ovens stimulates the demand for "cold-chain" products which are preserved by techniques that may facilitate infections by psychrophylic pathogenic bacteria such as *Listeria monocytogenes*
- the increasing popularity of farm shopping (buying fruits, vegetables, dairy products, etc. directly from the farm) may involve by-passing the usual control procedures on the way between farm and shop

- food intolerance is not always noticed by the national health services, and the consumers (or parents of children) involved are not aware of this possibility
- changes in government administration and/or priorities, leading to a cutback of government spending for the food inspection services, may make it difficult to meet the minimum requirements that are of vital importance to problem analysis and food safety.

These and other examples of food hazards may be identified and quantified by the government officials responsible for food safety policy. It should be realized, however, that most countries do not have a food and nutrition policy, including food safety aspects. At present, only seven European countries have such a policy, namely Denmark, Finland, Iceland, Malta, the Netherlands, Norway, and Sweden. This means that in many countries, food hazards are not dealt with in an adequate way, i.e., on an ad-hoc basis rather than according to a coherent monitoring and surveillance system. Once a food hazard has been identified, the next step (of the government) should be action to reduce the hazard to an acceptable level.

23.2.3 Objectives

As far as food and nutrition policy is concerned, the government usually sets objectives in rather general terms. For example, the official document on "Food and Nutrition Policy in the Netherlands" includes a statement on the objectives of the government policy which may be summarized as follows. The government's food and nutrition policy intends to

- provide an adequate, safe supply of food, reasonably priced according to quality, from which the consumer can select a healthy, palatable diet at a price he can afford, and
- promote sound eating habits, including information and education concerning a healthy diet, and encourage industry to watch the safety and nutritional aspects of product policy, labeling, and advertising practices.

Further reading shows that the objective concerning the responsibility for food safety is described in terms of 'maintaining a balance between various interests at local, national and international level.'

As mentioned earlier, this illustrates that policy-making largely consists of weighing procedures. Scientific results and dose–response relationships may carry as much weight as political acceptability and estimated feasibility. The lack of unequivocal objectives may damage the credibility of the responsible authorities. A quantitative evaluation of the success or failure of a food safety policy is rarely possible. Its objectives are usually insufficiently specified. Some examples of specified objectives are

- a planned number of inspections by food inspection services
- determination of levels at which exposure to hazardous food components is believed to be safe. Standards may be set on the basis of these levels
- availability of budgets for research projects, specific surveillance activities, and educational programs
- reduction of the use of chemicals in food production
- setting of (amendments to) safety standards of guidelines

In democratic countries, food safety policy objectives often reflect the goals the government considers to be achievable without being accused of either disregard or exaggaration by the actors involved. For example, setting a maximum tolerable TCDD level for cow's milk to be used for direct consumption of dairy products for export, will be closely watched and commented upon by interested parties such as:

- The farmers organizations. A strict standard may result in the banning of dairy cattle from extensive areas in the immediate vicinity of (certain) chemical plants, refineries, or waste incinerators.

- The dairy industry and milk marketing board. The image of the purity of milk is at stake, and sales may decrease if safety standards are too slack according to the public opinion.
- The export trade. Markets may be lost due to competition if the importing country is not satisfied with the safety standard for a particular product.
- The management of waste incinerators and other installations, and their insurance companies. Farmers may submit an insurance claim if strict standards are appplied.
- Consumers organizations/legal advizers. Their point of departure is usually "no risk at all is acceptable."

This example of TCDD illustrates once more that as far as specification of objectives and setting standards is concerned, not only experts are involved in policy making. Quantitative objectives of food safety, such as standards for the acceptable level of potentially harmful substances in food, may be defined on the basis of the following ways of thinking.

According to one line of thought, substance X should not be present at all in a particular food. Starting from here, it may be stated that certain substances, for example pesticide residues, should not pass the water purification processes, and should not be found in drinking water. Every amount detectable in water then exceeds the zero tolerance limit and calls for action. Media reports on such cases are known to confuse consumers readily. Although legally unacceptable, the substance is considered to pose no toxicological risks. Another disadvantage of this approach is that detection limits of yesterday may be altered tomorrow, leading to a reassessment of the zero-tolerance level.

According to a second approach, the maximum tolerable level of substance X in a particular food should be based on calculated (or at least estimated) risk, taking into account the contribution of the food component involved to the total exposure of potential high-risk groups to substance X. This approach will result in relatively more standards for the allowable acceptable levels of harmful substances in food.

In several countries, both approaches are followed in food safety policy. The World Health Organization has developed the campaign "Health for all by the year 2000" which was launched in Europe in 1984. It included no less than 38 objectives for the improvement of health. One of these was concerned with food safety. It was stated that member states should reduce the risk of food poisoning and infection considerably, and take measures to protect consumers against harmful additives and contaminants. However, the WHO did not specify the objectives for the individual countries. It considered "the existence of a national system for food safety inspection and evaluation" to be a sufficient indicator for a country to achieve the food safety objective of the health improvement campaign.

23.2.4 Possibilities for government intervention

Policy, and so also food policy, needs instruments that can be deployed (by the government) to achieve the objectives set. Fundamentally, government interventions may be developed and used either to stimulate or to discourage certain behaviors of specific target groups that are playing a role in the way from raw material to consumer. Government authorities can enforce or prevent certain actions by legislation, economic measures, and communication.

In view of health protection, food should not be contaminated with harmful substances or (micro)organisms to such an extent that normal use poses health risks. Government interventions aimed at health protection through food safety policy measures may include all types of activities from the sponsoring of food safety research projects or educational programs, to straightforward legislation and enforcement, for example:

- setting standards and tolerances for the production, distribution, and informative labeling of foods
- regulating screening procedures required for novel foods or processes
- incorporating basic knowledge of safe food handling and preparation in educational programs
- standardizing education and training of experts involved in food production and preparation

Also economic measures may be applied to promote certain food products or to stimulate the use of certain food production techniques.

A third possibility to intervene in actions is communication. Dissemination of information about health risks in relation to food safety occurs in almost any country where food policy includes food safety. Intervention by communication can take many forms and may vary from official government-made messages distributed or broadcasted through government-controlled channels to government-sponsored educational activities and mass-media campaigns.

It should be noted that the effectiveness of policy measures based on just one approach is usually poor; integration of approaches should be the goal. For example, great efforts may be made to prevent infection of chicken by pathogenic bacteria such as *Salmonella* during production. However, the results may be disappointing if consumers and food catering personnel are insufficiently acquainted with the hygiene requirements for handling and preparation of the meat.

23.2.5 Policy makers must make choices

As remarked in the preceding sections, food safety policy making is largely based on scientific arguments and facts, and the former in particular should be stressed. Further, the costs of health risks (see Figure 23.2) should be set against those of potential interventions. Cost-benefit calculations may in some cases provide extra information for decision makers. For example, an assessment in 1986 of the net benefits of the Canadian Meat Hygiene Program 1970–1984 in terms of human health showed its effectiveness saving money through reducing the health problems (Table 23.1). However, political backing and attainability of positive results are only of decisive importance.

23.3 Food legislation

The "art" of adulteration of foods or making them look better than they actually are, has a long history. For example, milk was often diluted with water, and inferior fats were colored yellow to make them look like butter. Further, society suffers considerable financial losses as a result of food contamination, rodent and insect plagues, or spoilage. To control excessive adulteration and cheating, and to reduce losses of food due to inadequate processing, local and regional authorities have taken various measures. Food acts belong

Table 23.1 Human Health Benefits in Financial Terms
by Canada's Meat Hygiene Program, 1970–1984

	$ Mil.
Tuberculosis	2302
Brucellosis	30
Beef tapeworm	39
Trichinosis	6
Residues/cancer	117
General food poisoning	722
Botulism	16
Roral benefits	3232

Source: Intercambio Limited, 1986. With permission.

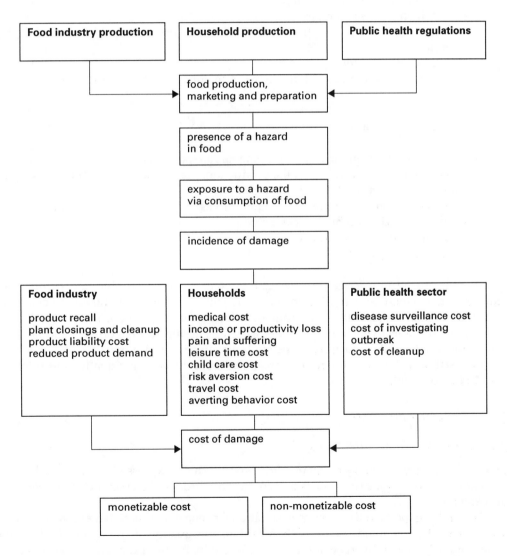

Figure 23.2   Costs from exposure to food-borne disease. Source: Krystynak, 1988.

to the eldest types of acts. Modern food legislation started with the onset of urbanization, i.e., not before the beginning of this century. Nowadays, most countries have food legislation, basically regulating production, handling, marketing, and control of food. Protection against food hazards, fair competition between food manufacturers and distributors, and reliability of information on food are the fundamental points of consideration for each country. In essence, food legislation is usually a translation of these points.

Food acts can be distinguished into two types:

A. Basic acts which include:

- the objective of the act
- definitions of basic concepts
- the scope of the act
- possibilities of implementation
- control facilities and procedures
- enforcement and sanctions
- procedures for the registration of food additives, processing aids, and packaging materials
- standardization of food products
- procedures for the preparation and amendment of regulations for implementation of the act

B. Specific acts including:

- quality standards and safety requirements for specific (groups of) food products
- adequate hygienic practices in the production, preparation, processing, packaging, transport, storage, and distribution of foods
- safety conditions for the use of food additives, pesticides, and techniques such as irradiation and biotechnological techniques
- informative labeling of foods
- procedures for food control

In order to introduce acts and regulations which are practicable and enforceable, experts from the food industry should cooperate with scientists, government officials, and consumer organizations in the development of food safety legislation. To execute the acts and regulations effectively, a network of inspectors, administrative officers and analysts with adequate laboratory facilities is needed. An effective and impartial enforcement of food legislation will contribute to the consumers' confidence in the quality and safety of his food assortment.

23.4 Food safety policy in practice

23.4.1 Intercountry differences

Food safety policy may have many national aspects. This is expressed in issues such as the regulation on hazardous substances present in food through the assessment of tolerances and standards.

Generally, experts in the various international agencies and national authorities agree with each other on main points with regard to the quality of the available scientific information. National divergencies may be reflected in the interpretation of research data. This is where science meets with national politics, and where conflicting interests between countries may result in different views on food safety issues in discussions in international settings such as the Commision of the European Community, or in FAO/WHO expert committees. An illustrative example are the conclusions on a particular food contaminant reached by groups of experts reviewing the collected scientific data. The conclusions may differ from one country to another. The relative importance of different aspects of such an evaluation may well be affected by national circumstances. There may be two reasons for this. First, the actual or potential extent of exposure to a particular food-borne hazard strongly depends on the relative importance of the particular food involved to different target groups, e.g., the general population and a specific high-risk subpopulation. This is determined by consumption patterns and eating habits, which are well-known to be liable to differ at national, or even regional and local levels. Secondly, food safety policy measures which are implemented effectively in one country may well be of limited use in another country, due to differences in national circumstances.

The above may be illustrated by a number of examples.

- If fish is the most important protein-providing food in a developing country, the presence of a particular contaminant in fish is likely to be judged more linearly, bearing the relevance to the national diet in mind. In a country where the same contaminant would affect a different food which does not play a key role in the food supply, the judgment will probably be different.
- Countries like Greece that produce, as well as utilize olive oil, would judge olive oil adulteration quite differently from what dairy-oriented countries like Denmark, which have little interest in olive oil, would do.
- If the export of cocoa beans forms the backbone of the national economy of a country, contamination of cocoa with aflatoxin may obviously be more relevant than in a country that does not produce the beans and where cocoa consumption is usually moderate. The interest of the exporting country in international discussions may lead to relatively liberal aflatoxin standards which do not interfere considerably with that country's sale of its national product on the world market. Other countries, however, may be of the opinion that the standards should be more strictly aimed at reduction of risk irrespective of the economic consequences for the producing countries.
- The highly persistent pesticide DDT was already banned in quite a few countries more that 25 years ago. It was believed to pose an unacceptable health risk, as it accumulates in human and animal adipose tissue. Nevertheless, DDT is still widely used in malaria-sensitive regions where cost-benefit comparisons turn out in favor of DDT.

These national differences are not only reflected in the political processes of standard setting and assessment of tolerances or standards but also in other elements of the national food safety policy.

- The distribution of leaflets, circulars, and other written material on food hygiene and safe food-handling practices for the information and education of the public or the personnel working in food supply, catering establishments, etc., may be a very effective tool of government food safety policy in Western countries, but will fail in many developing countries where illiteracy may amount to 90% of the population.
- A sophisticated food safety inspection service may be organized effectively in countries where the food production and distribution system is dominated by a small number of relatively large food manufacturers and distributors. However, such a service may be less succesful in other countries where the food production

and distribution system is operating mainly at the level of small-scale farmers who provide rural villages with their produce.

– Marking of the shelf-life date on the labels of perishable food products may be useful in countries where prepackaged foods make a considerable contribution to the food range, but obviously will fail to reduce the risk of food poisoning and food infection in countries where the majority of the food products are sold in its original form or after preparation on the spot.

23.4.2 *European Union*

Whithin the European Union, the general principle of free trade between member states implies that each member country has to accept the import of food products produced legally in one of the other member states. This means that the food acts in all member states have been declared valid for those products. However, there are a few exceptions that permit a member to close its borders to a particular food product. Such a barrier to free trade is only acceptable if the product:

– poses a health hazard to the consumer
– would seriously mislead the consumer

The Commission of the European Union harmonized food acts through issuing EU directives for certain aspects of food legislation. Implementation of these directives in the national food acts of the member states is mandatory. Some important food safety issues for which the EU Commission has assumed authority are

– positive lists for a number of food additives, such as food colorings, preservatives, and antioxidants
– biotechnology and the production of "novel" foods
– contaminants
– food irradiation
– services for inspection and control of food quality and safety

From January 1993, Europe is an open market of 12 member states with together about 350 million consumers. In itself, the European food safety policy is not very different from that in some member countries. However, the EU principle has now been accepted that first responsible for the inspection and control of the quality and safety of food is the member state that has produced a particular food or has imported it from a non-EU country. The quality and effectiveness of the food inspection service is clearly not at the same level in the various European countries. This imposes a heavy burden on the process of harmonization for procedures and methods used in the food inspection agencies. If standardization and good cooperation between the inspection services in the European countries were to fail, this could have adverse effects on the protection of consumers against food products of inferior quality. This could even pose health risks, if less responsible traders find a suitable channel to avoid thorough quality and safety checking procedures. A Dutch initiative has resulted in the establishment of an EU-working party of food inspectors to deal with this potential problem.

Reference and reading list

Anon, Government, consumer groups, industry discuss future of food policy at conference. *World Food Regulat. Rev.* 3, 18–20, 1994.

Dawson, R., Food laws; why do we need them?, in: *Food Nutr.* 9, 38–40, 1983.

Ehiri, J.E., G.P. Morris, Food safety control strategies: a critical review of traditional approaches. *Int. J. Environ. Health Res.* 4, 254–263, 1994.

Gardner, S., Food safety: an overview of international regulatory programs. *Eur. Food Law Rev.* 6, 123–149, 1995.

Grimvall, A. and R. Ejvegård, The dynamics of scientific uncertainty and its implications for the use of conservative procedures in risk analysis, in: *Fourth International Conference on Environmental Mutagens*, Liss, A.R. New York, 1986.

Guidelines on the assessment of novel foods and processes, Department of Health report on Health and Social Subjects 38. London, HMSO, 1991.

Jacob, M., Legislation, in: B.C. Hobbs and D. Roberts, Eds., *Food poisoning and food hygiene*. London, Edward Arnold, 1993.

Krystynak, R., Regulating food safety and the consumer interest, in: *Food market commentary* 10, 16–22, 1988.

Ministry of Welfare, Health and Cultural Affairs, *Food and Nutrition Policy in the Netherlands*, 1984.

Labuza, T.P., W. Baisier, The role of the federal government in food safety. *CRC Crit. Rev. Food Sci. Nutr.* 31, 165–176, 1992.

Rosenberg, I.H., Could food labels be dangerous to health?, in: *Nutr. Reviews* 50, 298–299, 1992.

Index